Course 400

Fiber Optic Installation and Testing

Student Manual

Author
Dr. Stephen C. Paulov

Co-Authors
Ronay L. Wolaver
Nathan E. Moore

Editors
Felicity Reed-Sumner, Sr. Editor
Jim Jackson, Associate Editor

Desktop Publisher
Kirsten Pierard

McGraw-Hill
New York   Chicago   San Francisco   Lisbon   London   Madrid
Mexico City   Milan   New Delhi   San Juan   Seoul
Singapore   Sydney   Toronto

McGraw-Hill

A Division of The **McGraw·Hill** *Companies*

1 2 3 4 5 6 7 8 9 0 1PBT/1PBT 0 6 5 4 3 2 1

ISBN 0-07-139128-2

Printed and bound by Phoenix Color/Book Technology.

| **Performance Objective** | Given an interactive discussion, hardware, tools and equipment, and associated materials, the trainee will perform the following tasks to 100% mastery. |

- Understand the fundamentals of fiber optic technology, including history, advantages, disadvantages and applications.

- Perform the process of removing the jacket off fiber optic cable and placing the cable in the cabinet connected to other hardware.

- Demonstrate how to make a variety of connectors.

- Fuse two fibers together using either a mechanical splice or a fusion splice.

- Examine the function of the coupler, including how the coupler distributes optical signals.

- Examine the hardware used in specific areas to house the cable and terminating points of the fiber cable.

- Understand building construction from interior walls to outside brick work, and learn pertinent building codes such as National Electrical Code including designations OFNP, OFCP, etc.

- Describe the thought process of a designer when considering both fiber and copper issues.

- Understand the basics of local area network concepts as they relate to fiber optic cabling.

- Demonstrate how to find and correct system problems using accepted testing practices.

Instructional Flow

This course is presented in 24 hours. The course will provide participants with opportunities to demonstrate their skills immediately upon returning to work. Students receive feedback and improvement and suggestions during the classroom. The sections are taught in consecutive order, each section building upon that which was learned in the previous section.

Instructional Method

Interactive Discussion
Hands-on Demonstration

Instructional Media

Instructional Manual
Overhead Projection
Cabling Materials, Tools and Equipment

Duration

3 Days

Business Need

Fiber optics is the fastest growing field in communications. All communications network including telecommunications, CATV, LANs, and security are using more fiber every day. Fiber optics has become the transmission medium of choice for most communications. Its high speed and long distance capability make it the most cost-effective communications medium. While higher performance and lower cost components are in continual development, it has become critical to train competent personnel to design, install and maintain state-of-the-art fiber optics networks. Fiber isn't hard to learn, but the unique aspects of fiber optics can make on-the-job training expensive, since little mistakes cost big money.

Course Goal

The goal of this course is to provide the participants with an understanding of the attributes and operating characteristics of fiber optics network infrastructures and be able to place, terminate and test within recommended standard parameters. In addition, participants will have a working knowledge of all related standards, how they apply in a commercial environment, and which references or articles are applicable. The outcome measure of the course is to train and certify participants for 100% mastery of the knowledge and skills needed for a job position as a telecommunications fiber optics technician.

Course Overview

This course is the fourth course in the Cabling Business Institute's curriculum for telecommunications cabling professionals. This course is taught within the instructional guidelines of EIA/TIA, IEEE, ANSI, ETA, and NFPA (NEC). Participants in the course will receive a certificate of completion and can apply for accreditation through the Association of Cabling Professionals and Cabling Business Institute. This course is approved for twenty-one (21) BICSI RCDD continuing education units. In addition, the program will support further certifications through the Texas A & M University System's Telecommunications Training Program.

Course 400

Informative References:

Glossary and Acronyms
Standards

Section 1.0
Overview of Fiber Optics

Objective

Understand the fundamentals of fiber optic technology, including history, advantages, disadvantages and applications.

Outline

- *History of Fiber Optics*
- *Evolution of Fiber Optics*
- *Is Fiber Right for Everyone?*
- *Fiber Applications*
- *Parts of a Fiber Optic System*

Learning Activity

Assessment: Test Module 1
Lab Exercise: None

History of Fiber Optics

The recorded history of light as a communications medium dates back to ancient times, approximately 1184 B.C., when the Greeks used signal fires on a chain of islands to announce their victory over the Trojans to Queen Clytemestra some 900 km (559.23 miles) distant.

John Tyndall, an Irish physicist and scientist living in England is best known for experiments with light and light scattering. The Tyndall Effect states that particles of strong light that are normally invisible are easily discernable when viewed from the side. These particles become visible because they reflect some of their incident light. In 1880, Tyndall demonstrated the principal of guiding light through internal reflection using water. (See Figure 1.1.)

In 1880, Alexander Graham Bell invented a device called the photophone. This device used reflected light and vibrating mirrors to transmit a voice signal to a receiver in line of sight some distance away. (See Figure 1.2.)

Today, fiber optic networks and their applications are integrated into all aspects of today's communications and telecommunications, both commercial and public. Fiber networks are used in long-distance services, cable television and is the backbone of the information highway.

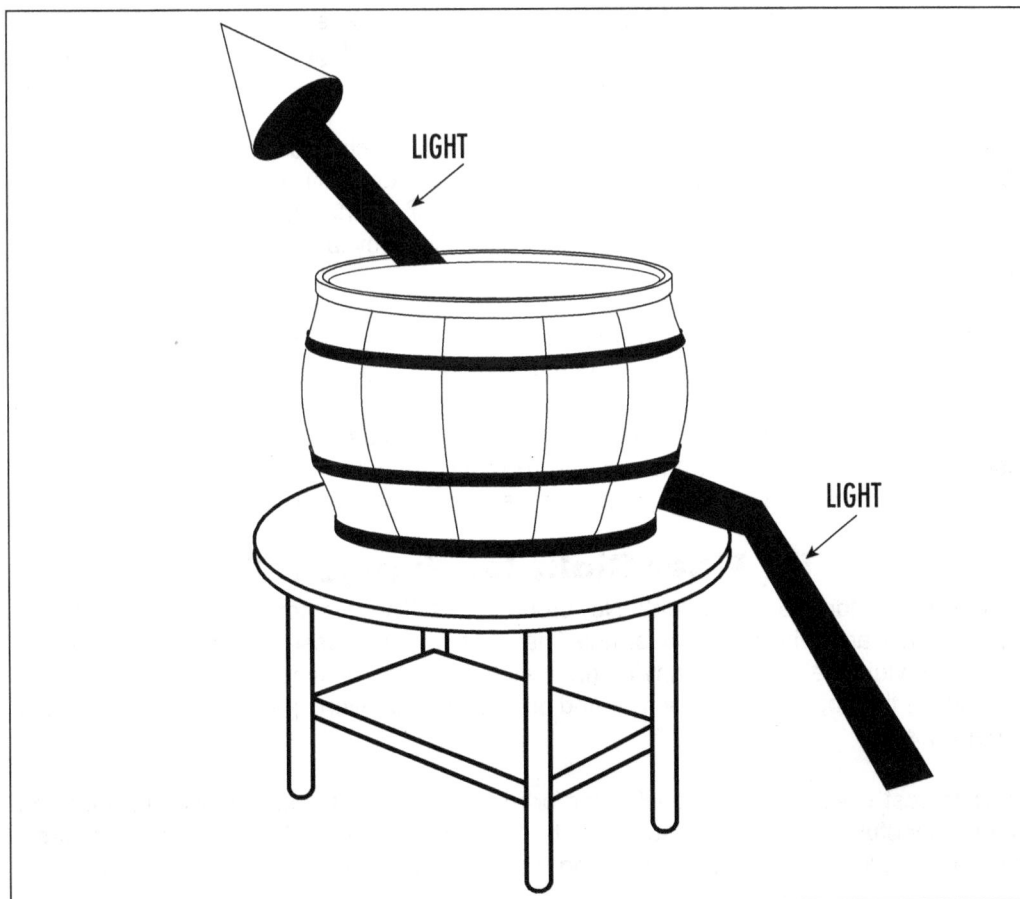

Figure 1.1

LIGHT

LIGHT

Figure 1.2

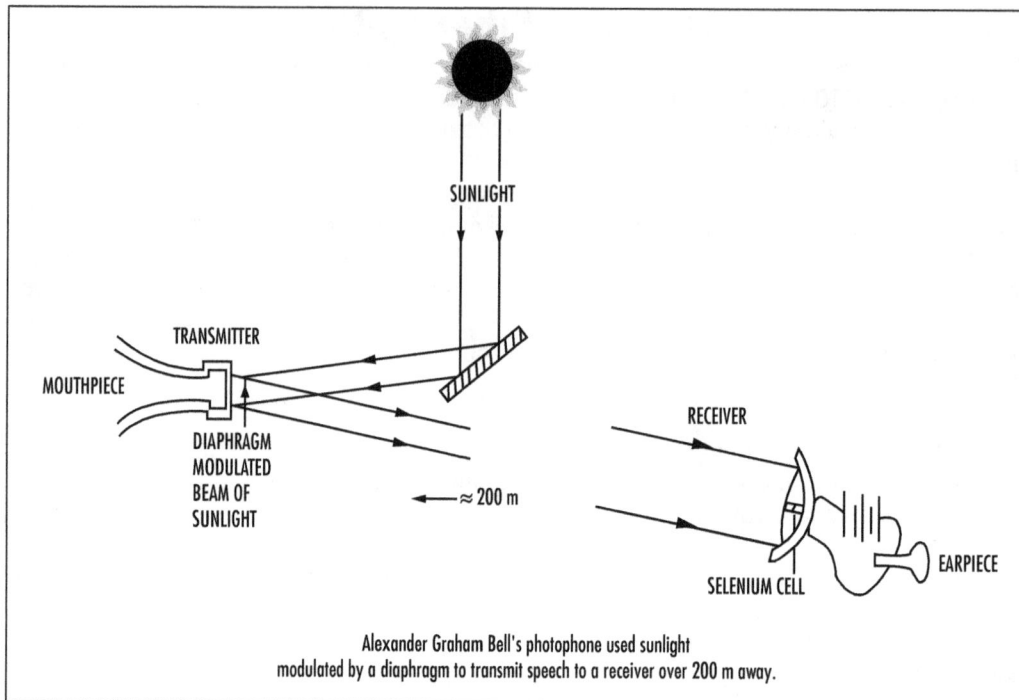

Alexander Graham Bell's photophone used sunlight
modulated by a diaphragm to transmit speech to a receiver over 200 m away.

Evolution of Fiber Optics

The term fiber optics was first used in the 1950s with the development of the flexible fiber-scope, which is still widely used in the field of medicine. These types of fibers are also used in industrial applications to inspect inaccessible areas, such as the inside of jet air-craft engines to search for signs of stress or metal fatigue. Light loss in these early fibers was extreme, but the length of the fibers was so short, this was not of particular concern.

In the 1960s, the laser became the topic of much research and the possibility of optical communications became a topic of great interest. Extremely wide bandwidth was not ignored by the telecommunications industry. However, interference by rain, fog, dust and other obstructions showed that line of sight transmission by laser through the atmosphere was simply not practical.

Finally, in 1970, Corning Glass Works announced the development of optical fiber with less than 20 dB/km loss. At last, telecommunications by fiber optics was to become a reality. By the mid- to late 70s lasers, LEDs, photodetectors, connectors, splices and other components were developed for use with these low loss fibers, and working systems were tested and made available to the market. Shortly thereafter, long haul lines were being installed up and down both coasts, and east and west across the United States.

Is Fiber Right for Everyone?

Is fiber perfect for every situation? No, probably not. There is not a set of standards that dictate when and where you must use fiber, but a quality assessment of a company's needs will yield that result. Many companies burden themselves financially by installing a complete fiber optic network only to find out they do not even use a fraction of the fiber's capabilities.

Fiber is best used in large-capacity and/or in long-haul systems. Companies that only have a handful of users would not really benefit from the use of fiber unless there was a problem with line noise, such as is found in many industrial applications. However, new high-end graphics, CAD/CAM and engineering programs create a need for fault-free, high-speed communications that thrives on fiber.

It is still, for the most part, less expensive to hook up a system with copper than with fiber. Although the cost of the fiber optic cable is comparable to the cost of high-grade copper cable, the electronics that must be used to couple into the fiber optic systems are still cost prohibitive. Thus, until the prices come down (and they are headed that way), a lot of people who might benefit from fiber probably will not take advantage of it.

The telephone industry has obviously accepted fiber for its long-haul and intermediate trunking. The military market, one of the fiber pioneers, is expanding its fiber usage. Most large data communications systems utilize fiber in the backbone and in some cases, even to the desktop.

Advantages and Disadvantages

There are definite disadvantages in using copper cabling: the amount of cabling required in large applications, the crosstalk, susceptibility to electromagnetic inference (EMI) and radio frequency interference (RFI) weight of the copper cable, ease of handling, etc. To provide 22,000 voice channels, it would require 900 pairs of copper versus 12 pairs of fiber. (See Figure 1.3.)

Fiber Advantages	Fiber Disadvantages
Wide bandwidth	Cost of tooling
Electromagnetic immunity	Cost of test equipment
Lightweight	Complex terminations
Small size	Cost of electronic equipment
Safety	Terminations somewhat costly
Security
RFI immunity

Fiber overcomes EMI and RFI because it is made of a dielectric material and not susceptible to any type of electromagnetic, radiation or other interference. Thus, users will not suffer a hum on their telephone lines or glitches appearing on their data systems. Fiber optic cable is smaller, easier to install and carries more information. Most users do not realize any communications system that can run on copper can run on fiber. Since most of the fiber optic cables used in today's LANs have no ground wires or metallic sheaths, optical networks are basically free of the ground current problems that can plague copper- based systems.

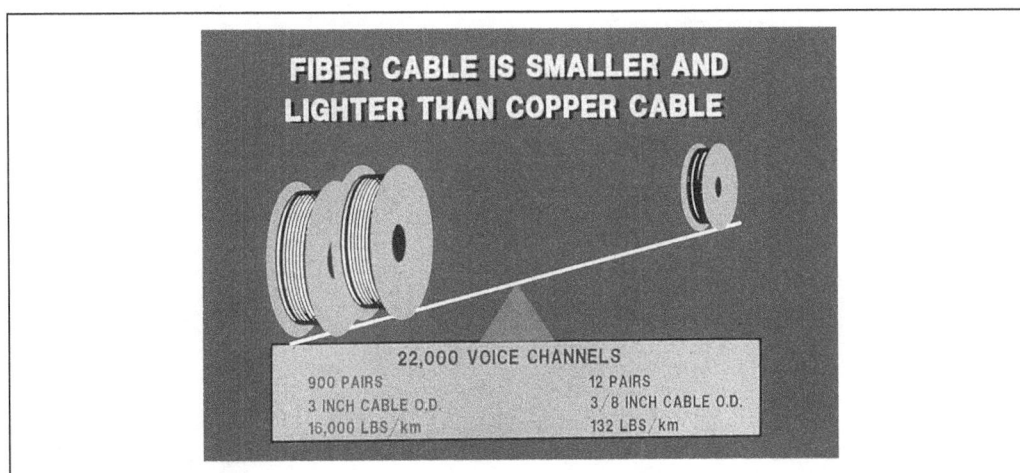

FIBER CABLE IS SMALLER AND LIGHTER THAN COPPER CABLE

22,000 VOICE CHANNELS
900 PAIRS — 12 PAIRS
3 INCH CABLE O.D. — 3/8 INCH CABLE O.D.
16,000 LBS/km — 132 LBS/km

Figure 1.3

In some applications, the physical size and considerable weight of copper cable can be problematic. For example, a coaxial cable weighs about 80 lbs. per 1,000 feet, while an identical length of fiber optic cable weighs only about 9 lbs. and may carry more data. (See Figure 1.4.) Fiber optic cable is intrinsically safe. Breaking the fiber neither produces sparks that could ignite flammable material, nor poses an electrical shock hazard.

Installation is made somewhat easier because fiber optic cable is smaller and lighter than comparative copper cable lengths. However, make no mistake, fiber optic cables and all high-grade copper cables require more installation skills than older types of cabling ever did.

Fiber Applications

Dark or Dormant:

Many large users such as IBM, Prudential, military, local, state and federal government agencies have installed a large amount of fiber paralleling their copper-based system. Some of these will sit dormant or dark (installed and connectorized but not used) until technological changes force the price of interfacing equipment to come down to a respectable level. In this way, they are prepared for the future today.

Point-to-Point Applications:

Many of today's fiber optic communication systems are point-to-point applications. These consist of two nodes (or communications devices) communicating directly and exclusively with each other. Point-to-point applications usually require a fiber pair (transmit and receive) to operate, except in cases such as simultaneous simplex fiber two-way transmission, simplex video and instrumentation applications. Point-to-point applications exploit fiber optic bandwidth and low attenuation, allowing a signal to be sent longer distances at faster speeds than available on coaxial or twisted pair cables. The illustration in Figure 1.5 consists of a simple point-to-point link from one building to another building in a campus environment.

A fiber optic communication system is designed in form division: hardware, cable, terminations and the optic electronics.

Starting at (A) in the illustration (See Figure 1.5.), the fiber starts from the terminal (A) to the optoelectonics and cross-connect panel (B) via the riser cable (C) and is spliced as it enters the building (D). Then cabling to another building, the cable is spliced (E) and the same function occurs in the other building.

Figure 1.4

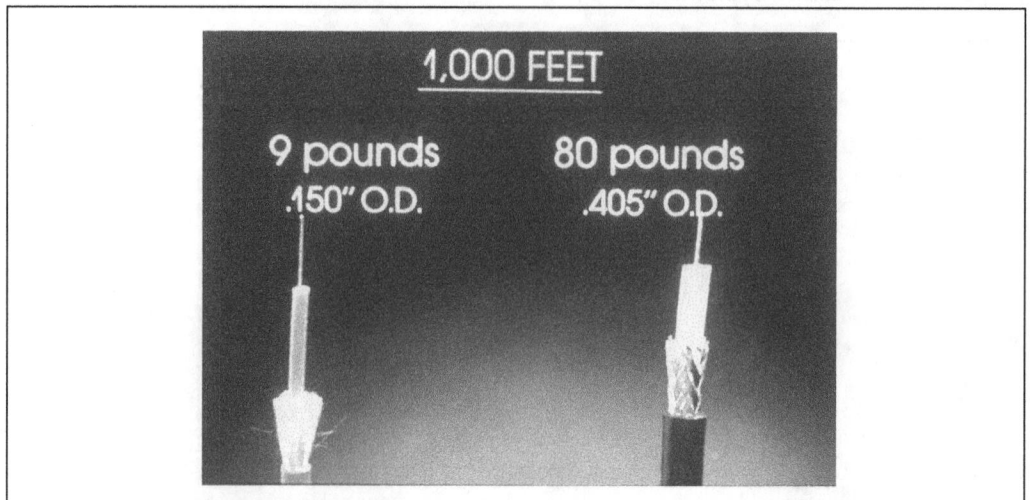

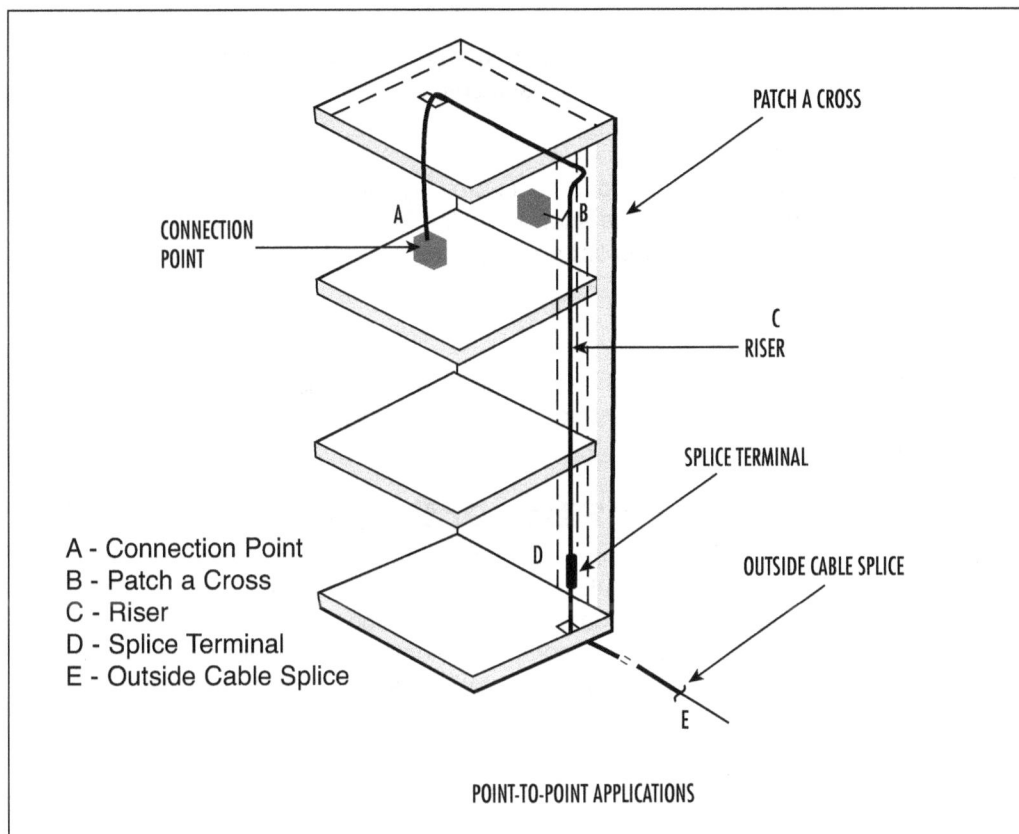

Figure 1.5

PATCH A CROSS

CONNECTION
POINT

A

B

C
RISER

SPLICE TERMINAL

A - Connection Point
B - Patch a Cross
C - Riser
D - Splice Terminal
E - Outside Cable Splice

D

OUTSIDE CABLE SPLICE

E

POINT-TO-POINT APPLICATIONS

Other Point-to-Point Applications

Channel Extender
A fiber optic channel extender allows a mainframe computer to communicate directly with a controller or other peripherals at much greater distances than possible with copper. It allows for such things as a remotely located high-speed channel attached printer, something not possible before fiber optics.

Voice/Data Multiplexer
A fiber optic multiplexer combines several communications channels into a single bit stream for transmission over the fiber. Channels are then separated at the receiving end. Common applications include T-1 communications, large PBX internodal links, and RS-232 and 3270 transmission.

Modem
A fiber optic modem converts electronic signals to light pulse signals. Fiber modems have a multitude of purposes, from extending distances between computers to the linking of peripheral devices. (PC printer sharing is a common use.)

Video
Fiber optic links are often used for video systems. Broadcast quality video transmission over fiber is not susceptible to the inherent problems encountered with coaxial cables but may require special connectorization.

Security Video
Fiber optic point-to-point links are often used in outdoor environments to minimize lightning damage, interference and to extend distances.

LAN Applications
The installation and use of fiber optic LANs is growing rapidly. In a fiber optic LAN, two or more nodes communicate with one another via the fiber optic cable. The nodes can be logically arranged in a ring, star, bus or tree configuration as they would be with any other communications medium.

Fiber-to-the Desktop
In certain high-end, high-speed and high-capacity applications, fiber is required to the desktop. These applications may include CAD/CAM or specialized graphics and engineering programs.

Parts of a Fiber Optic System

Optical fiber is the medium in which communication signals are transmitted from one location to another in the form of light guided through thin fibers of glass or plastic. These signals are digital pulses or continuously modulated analog streams of light representing information. These can be voice, data, computer, video or any other type of information. Fiber optic communication could be compared to Morse transmission as shown in Figure 1.6.

Figure 1.6

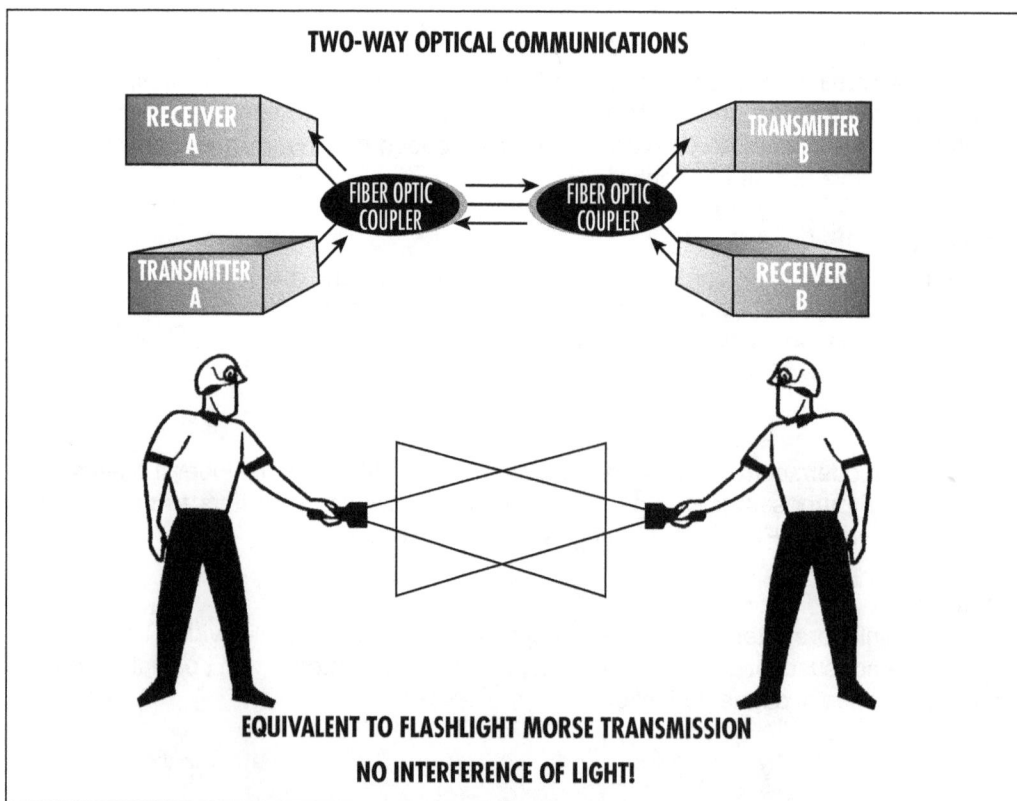

TWO-WAY OPTICAL COMMUNICATIONS

RECEIVER A · TRANSMITTER B · FIBER OPTIC COUPLER · FIBER OPTIC COUPLER · TRANSMITTER A · RECEIVER B

EQUIVALENT TO FLASHLIGHT MORSE TRANSMISSION

NO INTERFERENCE OF LIGHT!

Four Parts of a Fiber Optic System

There are four groups that make up a fiber optic system as defined below:
- The fiber, which acts as a conduit, allowing the light to travel.
- The cable protects the fiber.
- The connecting group provides alignment of both connectors and sleeves.
- Electro-optical devices are required to convert the electrical signals to optical signals and the optical back to electrical signals. (See Figure 1.7.)

Figure 1.7

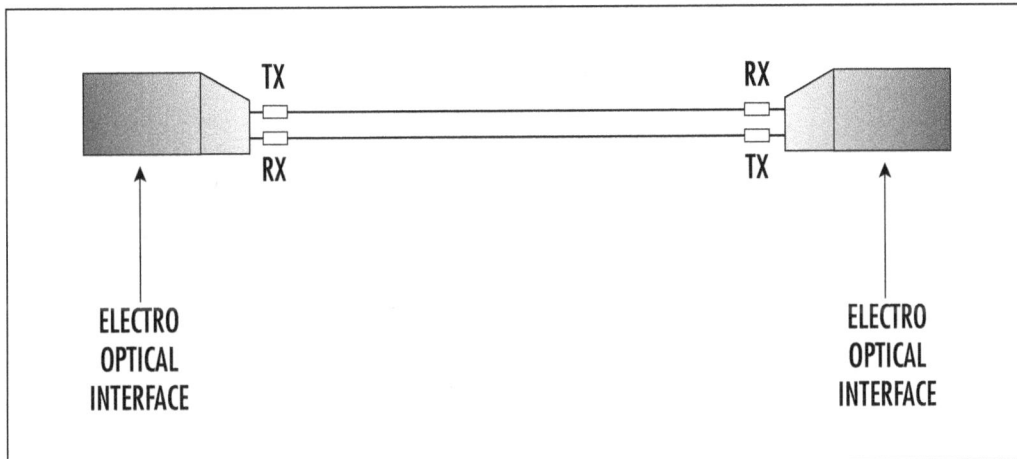

TX   RX

RX   TX

ELECTRO OPTICAL INTERFACE

ELECTRO OPTICAL INTERFACE

TEST MODULE 1

1. Fiber optics was first used in the 1950s with the development of the flexible fiber-scope, which is still widely used in the field of medicine.

   True   False

2. Fiber is best used in small-capacity and/or in long-haul systems.

   True   False

3. Point-to-point applications exploit fiber optic bandwidth and low attenuation, allowing a signal to be sent longer distances at faster speeds than available on coaxial or twisted pair cables.

   True   False

4. Optical fiber is the medium in which communication signals are transmitted from one location to another in the form of light guided through thin fibers of steel or plastic.

   True   False

5. The four groups that make up a fiber optic system are fiber, cable, connecting group and electro-optical devices.

   True   False

Section 2.0
Fiber Manufacture

Objective

Explain how fiber optic cable is manufactured and understand the fundamentals of light propagation, beginning with the speed of light, modes of fibers and losses of fibers.

Outline

- *Fiber Optic Technology*
- *How Fiber Works*
- *Fundamentals of Light Propagation*
- *Optical Fiber Configurations*
- *Comparison of the Types of Optical Fibers*
- *Losses in Fibers*
- *Light Emitting Diodes*
- *Numerical Aperture*

Learning Activity

Assessment: Test Module 2
Lab Exercise: None

FIBER OPTIC TECHNOLOGY

Fiber Manufacturing Process Methods

There are three methods presently used to manufacture moderate to low loss waveguide fibers; modified chemical vapor deposition (MCVD) or inside vapor deposition (IVD), outside vapor deposition (OVD) and vapor axial deposition (VAD).

Modified Chemical Vapor Deposition (MCVD)

This is one of the methods currently being used to manufacture fiber. A hollow glass preform, approximately three feet long and one inch in diameter, is placed in a horizontal or vertical lathe and spun rapidly. A computer-controlled mixture of gases is passed up and down the inside of the tube. On the outside of the tube, a heat source (oxygen/hydrogen torch) passes back and forth along the length of the tube. (See Figure 2.1.)

Each pass of the heat sources fuses a small amount of the precipitated gas mixture to the surface of the tube. Most of the gas is vaporized silicon dioxide (glass), but there are carefully controlled amounts of impurities (dopants) which cause changes in the index of refraction of the glass. As the torch moves and the preform spins, a layer of glass is formed inside the hollow preform. The dopant (mixture of gases) can be changed for each layer so the index may be varied across the diameter.

After sufficient layers are built up, the tube is collapsed into a solid glass rod called a preform. It is now a scale model of the desired fiber, but much shorter and thicker.

The preform is then taken to the drawing tower, where it is pulled into a length of fiber up to 10 km long.

Outside Vapor Deposition (OVD)

This method utilizes a glass target rod, which is placed in a chamber and spun rapidly on a lathe. A computer-controlled mixture of gases is then passed between the target rod and the heat source. (See Figure 2.2.)

On each pass of the heat source, a small amount of the gas reacts and fuses to the outer surface of the rod. After enough layers are built up, the target rod is removed and the remaining soot preform is collapsed into a solid rod.

The preform is then taken to the drawing tower to be heated and pulled into fiber.

Figure 2.1

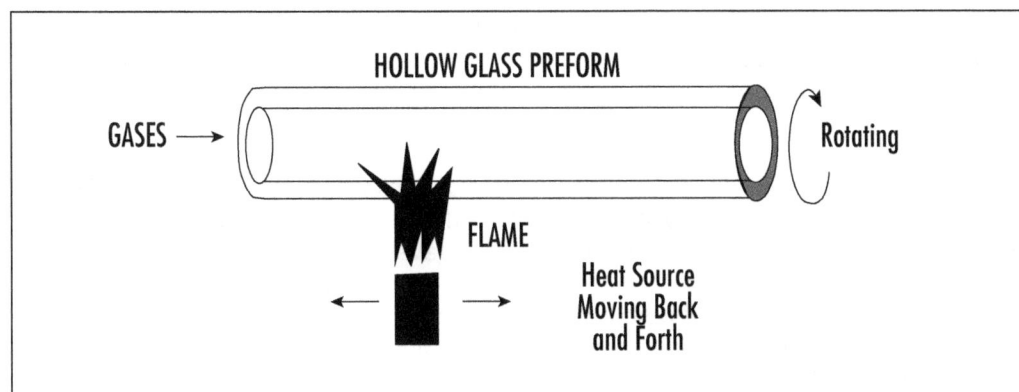

HOLLOW GLASS PREFORM

GASES →

Rotating

FLAME

Heat Source
Moving Back
and Forth

Figure 2.2

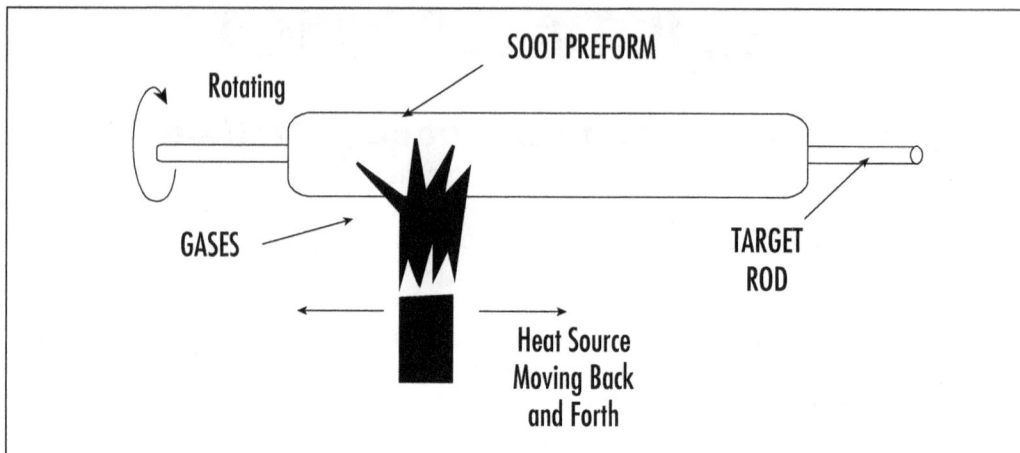

Vapor Axial Deposition (VAD)
This process utilizes a very short glass target rod, which is suspended by one end. A computer-controlled mixture of gases is applied between the end of the rod and the heat source.

The heat source is slowly backed off as the preform lengthens due to the soot buildup caused by gases reacting to the heat and fusing to the end of the rod. (See Figure 2.3.) After sufficient length is formed, the target rod is removed from the end, leaving the soot preform.

The preform is then taken to the drawing tower to be heated and pulled into the required fiber length.

Coating the Fiber for Protection
After the fiber is pulled, a protective coating is applied very quickly after the formation of the hair-thin fiber. (See Figure 2.4.) The coating is necessary to provide mechanical protection and prevent the ingress of water into any fiber surface cracks. The coating typically is made up of two parts: a soft inner coating and a harder outer coating. The coating overall thickness varies between 62.5 and 187.5 micron, depending on fiber applications.

These coatings are typically strippable by mechanical means and must be removed before fibers can be spliced or connectorized.

How Fiber Works
The general consensus seems to be that optical fibers are hollow glass tubes with a mirror-like finish on the inside. While they are not hollow, this idea is close to being correct. One of the problems with describing how optical fibers really work is that you have to get the individual away from the idea that the center (core) of the fiber is less dense (more like hollow) than the glass surrounding it (the cladding). That idea is exactly backward. The center or core of the fiber is more dense than the surrounding glass or cladding. (See Figure 2.5.)

To put this into a context that anyone can understand, picture a dirt road in the middle of winter. The shoulders of the road are frozen solid with two or more inches of solid ice. Due to the slight crown in the road, the ice in the travel lanes is thinner and has broken apart. There are six inches or so of mud and slush on the road. As you drive down the road, your car slides toward the shoulder and the right tire hits the ice on the shoulder. The tires on the ice tend to slide toward the side of the road, while the tires in the mud also have little traction.

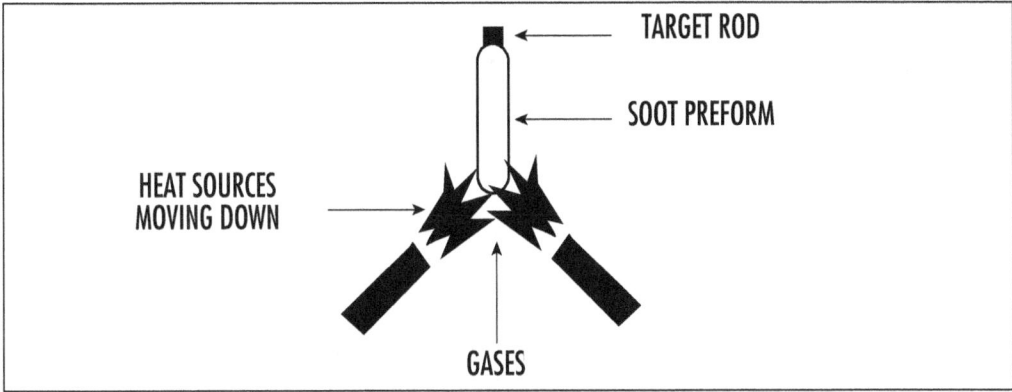

Figure 2.3

TARGET ROD

SOOT PREFORM

HEAT SOURCES
MOVING DOWN

GASES

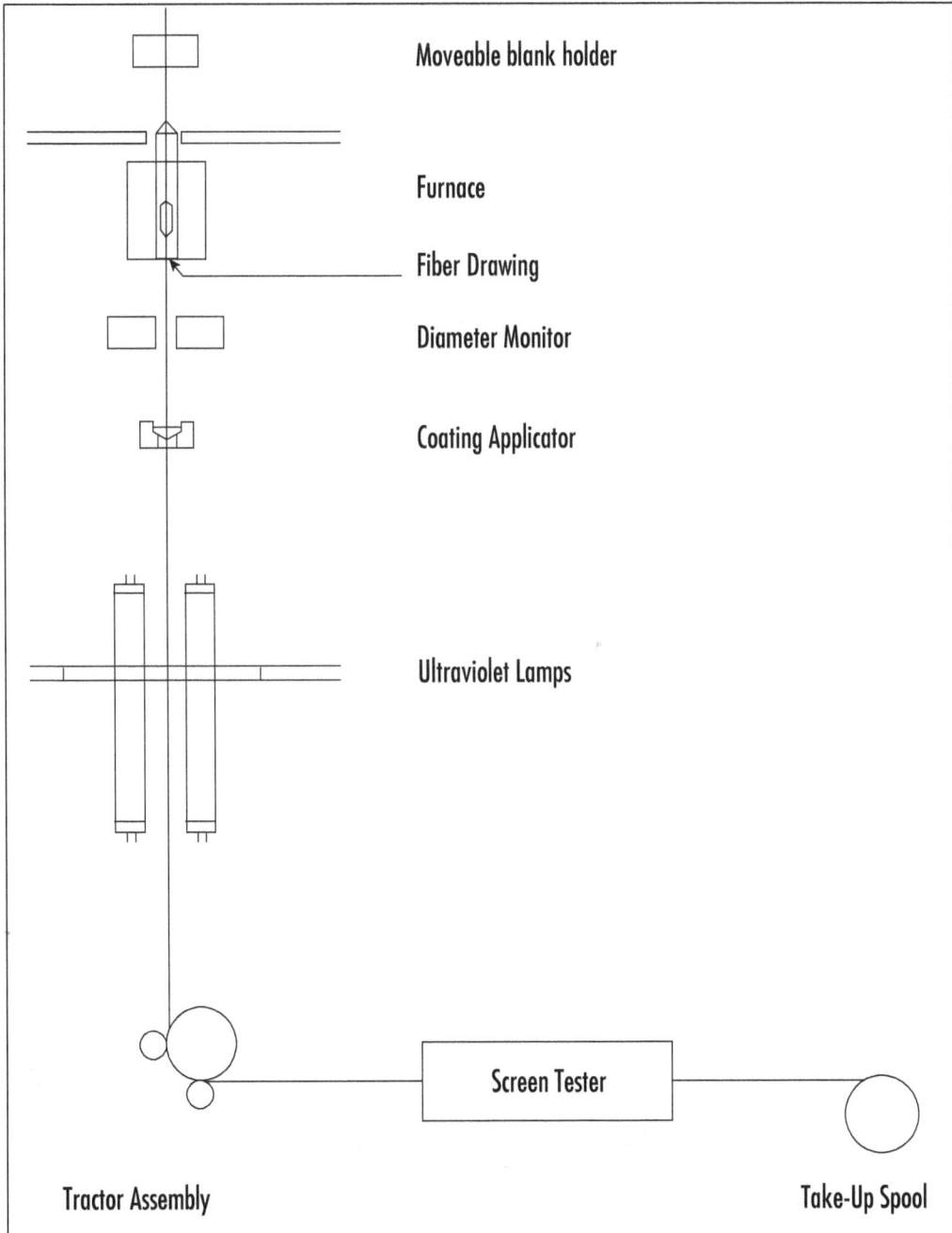

Figure 2.4

Moveable blank holder

Furnace

Fiber Drawing

Diameter Monitor

Coating Applicator

Ultraviolet Lamps

Screen Tester

Tractor Assembly

Take-Up Spool

Figure 2.5

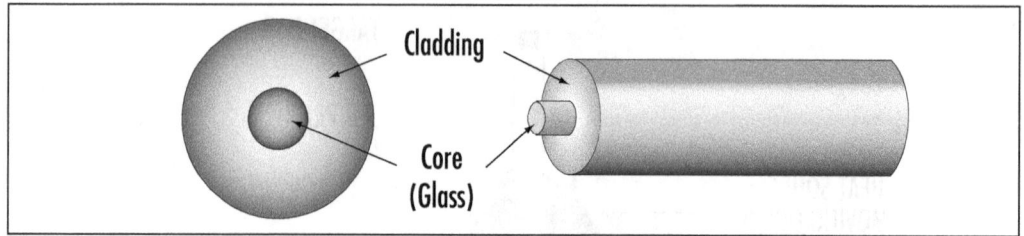

So, as the right tires slide on the ice, the vehicle turns back toward the center of the road, and the car is now headed for the left shoulder. Then, the left tires hit the ice on the left shoulder and, of course, the right tires are in the mud and are having a hard time keeping up with the left tires that are sliding on the ice. The car is once again turned back toward the center of the road. (See Figure 2.6.)

There you go zigzagging back and forth down the road. If you think of the center of the road as the more dense (thicker, if you will) core of a fiber and the shoulders of the road as the less dense (slicker) cladding of the fiber, you will have a very good, though crude, understanding of how optical fibers work.

When we address mechanical and fusion splicing of optical fibers, it will be necessary for you to have a basic understanding of the makeup of the fiber and of the cable or cordage (pigtails/jumpers).

Bare Fiber

Bare fiber is not actually bare but consists of the glass fiber and a primary or protective coating. (See Figure 2.7.) The fiber consists of the glass core, ranging typically from 4 microns in diameter to 100 microns or more. The core is surrounded by the cladding, which usually brings the outside diameter of the fiber to 125 microns, although larger diameters are not uncommon. For protection and added strength, the fiber is coated with any of several types of plastic. This coating is often referred to as the protective coating, but more correctly, it should be called the primary coating. Probably the single most common mistake in fiber optic technology is referring to the primary coating as the cladding. One tool company refers to a tool, which was designed to remove the primary coating, as capable of removing the tight buffer and the cladding in one easy operation. Even if it were physically possible to strip the cladding glass away from the core glass, it would render the fiber useless. Take time to understand the difference between cladding and primary coating.

Core: the inner light-carrying section.

Cladding: the center layer, which serves to confine the light to the core.

Buffer: the outer layer, which serves as a shock absorber to protect the core and cladding from damage.

Total Internal Reflection

Light injected into the core that strikes the core-to-cladding interface at an angle greater than the critical angle will be reflected back into the core. (See Figure 2.8.) Since angles of incidence and reflection are equal, the light ray continues to zigzag down the length of the fiber. The light is trapped within the core. Light that strikes the interface at less than the critical angle passes into the cladding and is lost.

Figure 2.6

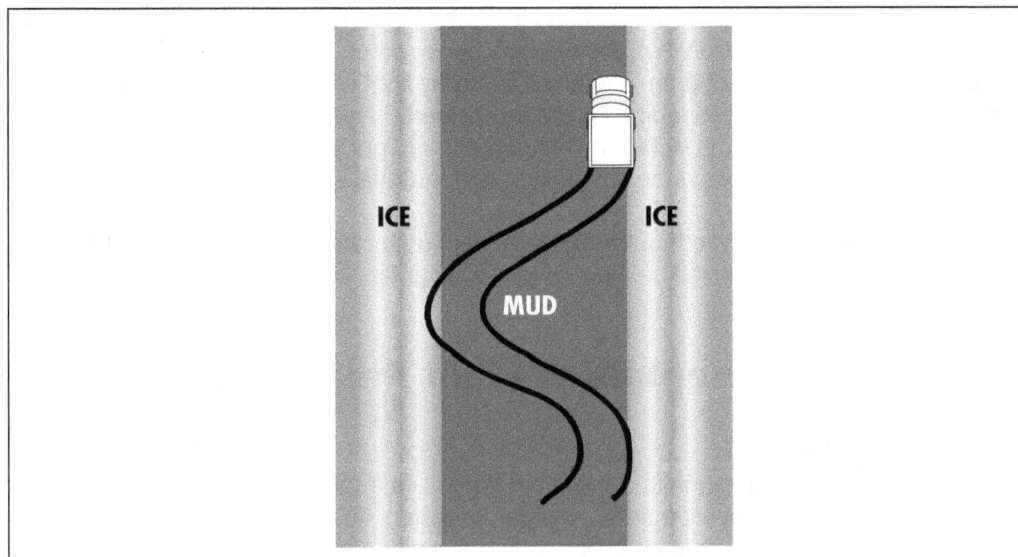

Figure 2.7

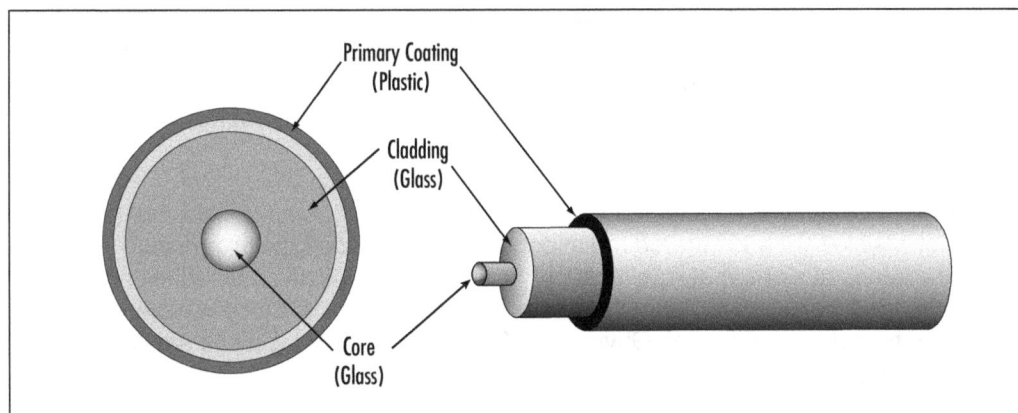

Figure 2.8

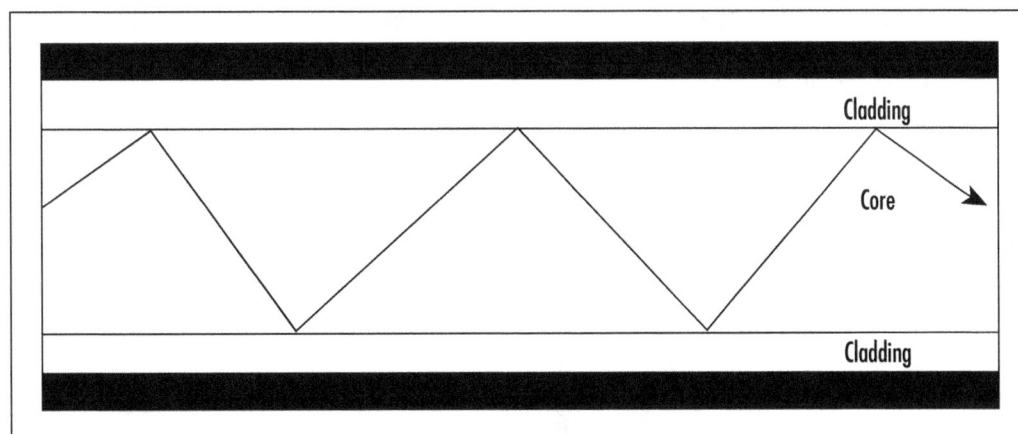

Rays of light do not travel randomly. They are channeled into modes, which are possible paths for a light ray traveling down the fiber. A fiber can support as few as one mode and as many as tens of thousands of modes. While we are normally not interested in modes per se, the number of modes in a fiber is significant because it helps determine the fiber's bandwidth. More modes typically mean lower bandwidth. The reason is dispersion.

As a pulse of light travels through the fiber, it spreads out in time. While there are several reasons for such dispersion, two are of principal concern. The first is modal dispersion, which is caused by different path lengths followed by light rays as they bounce down the fiber. Some rays follow a more direct route than others. The second type of dispersion is material dispersion, which is different wavelengths of light traveling at different speeds. By limiting the number of wavelengths of light, the material dispersion is also limited.

Dispersion limits the bandwidth of the fiber. At high data rates, dispersion will allow pulses to overlap so the receiver can no longer distinguish where one pulse begins and another ends.

Fundamentals of Light Propagation

Wavelength of Light

Since light is an electronmagnetic wave, what then is its frequency or wavelength? In the electromagnetic frequency spectrum, optical radiation lies between microwaves and e-rays. (See Figure 2.9.) It is common to use wavelength rather than frequency when dealing with high-frequency electromagnetic waves such as light. In fiber optics, wavelength is important because it can be measured directly. The wavelength of light is typically measured in nanometers(nm), micrometers (um) and angstroms (A), instead of feet or inches. Nanometers is the unit most often used in fiber optics. One nanometer is 10e - 9m, and 1 angstrom is 10c - 10 m. Therefore, 1 nanometer is equal to 10 angstroms.

Visible light includes all wavelengths between 10 nm and 1 mm. Within this range are ultraviolet, visible light and infrared radiation. Visible light is defined as radiation that stimulates the sense of sight, that is, affects our optic nerve. It includes all radiation from 390 to 770 nm, from violet to red. Light itself does not have color, but these wavelengths stimulate color receptors in the eye. Obviously, the visible spectrum is just a small fraction of the electromagnetic spectrum.

A typical wavelength in fiber optics is 850 nm, which may be expressed also as 0.850 um or 8500A. In the electromagnetic spectrum, this radiation is designated infrared, although it is referred to as light because it can be controlled and measured by instruments similar to those used for visible light.

Bandwidth

In theory, the greater the carrier frequency, the larger the available transmission bandwidth. This also means an increase in the information-carrying capacity of the particular communications system. (Information-carrying capacity is directly related to the bandwidth or frequency of the modulated carrier.) This theory does not hold for light

Figure 2.9

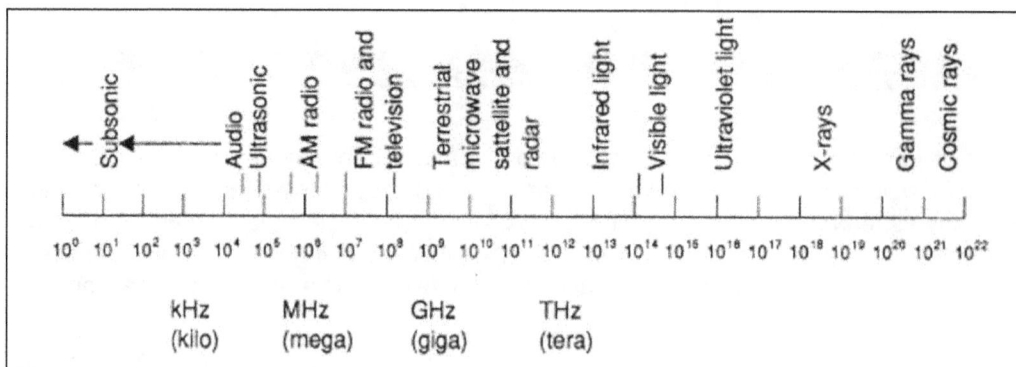

transmission because there are certain wavelength regions that propagate light better. For example, the 1,300 and 1,550 nm wavelengths propagate light better than the 850 nm wavelength and, therefore, have greater transmission capabilities.

Speed of Light

Electromagnetic energy, such as light, travels at approximately 300 million meters per second (186,000 miles per second) in free space. For propagation in free space and in the atmosphere, the speed of light is the same for all wavelengths, however, in other materials, such as water and glass, different wavelengths travel at different speeds. Regardless of the wavelength, when light travels through such materials, its speed is noticeably reduced. Because of this, a light ray moving from air into a solid or a liquid will change direction at the surface of the new medium that is known as refraction. In optics, a medium is any substance that transmits light.

In materials more dense than free space, the velocity of electromagnetic waves is reduced. When the velocity of an electromagnetic wave is reduced as it passes from one medium to another medium of a denser material, the light ray is refracted toward the normal. The normal is simply a line drawn perpendicular to the interface at the point where the incident ray strikes the interface. When a light ray enters a less dense material, the ray bends away from the normal.

Refraction

A light ray is refracted as it passes from a material of a given density into a material of a different density. The light ray is not bent as previously believed, but rather, it changes direction at the interface. The refractive index of a material is wavelength dependent.

Figure 2.10 shows how sunlight, which contains all light frequencies, is affected as it passes through a material denser than free space. Refraction occurs at both air/glass interfaces.

Since colors of light are different frequencies and have different wavelengths, they are refracted differently. For visible light, the violet wavelengths are refracted the most, and the red wavelengths are refracted the least. The color separation of which light in this manner is called prismatic refraction. This phenomenon explains rainbows; water droplets in the atmosphere acting like small prisms that split the white sunlight into the various wavelengths, creating a visible spectrum of color.

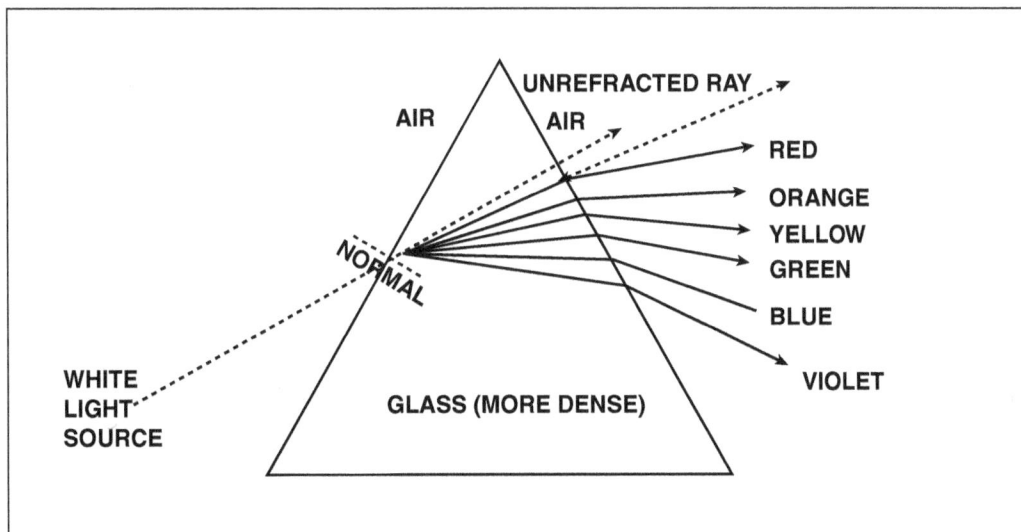

Figure 2.10

Refractive Index

The amount of refraction that occurs at the interface of two materials of different densities is quite predictable and depends on the refractive index of the two materials. The refractive index is simply the ratio of the velocity of propagation of light in free space to the velocity of propagation of light in a given material. Mathematically, the refractive index is speed of light in free space (vacuum) divided by the speed of light in the material.

Refractive index is also a function of frequency, but the variation in most applications is insignificant.

Critical Angle

The critical angle is defined as the minimum angle of incidence at which a light ray may strike the interface of two media and result in an angle of refraction of 90 degrees or greater. (This definition pertains to a situation only when the light ray is traveling from a denser medium into a less dense medium.) If the angle of refraction is 90 degrees or greater, the light ray is not allowed to penetrate the less dense material. Total reflection takes place at the interface, and the angle of reflection is equal to the angle of incidence.

Figure 2.11 shows the source end of a fiber cable. When light rays enter the fiber, they strike the air/glass interface at normal A. The refractive index of air is 1 and the refractive index of the glass core is 1.5. The light entering at the air/glass interface propagates from a less dense medium into a denser medium. The light rays will refract toward the normal. This causes the light rays to change direction and propagate diagonally down the core at an angle that is different than the external angle of incidence at the air/glass interface. In order for a ray of light to propagate down the cable, it must strike the internal core/cladding interface at an angle that is greater than the critical angle.

The ability to couple light into or out of a fiber is represented in its numerical aperture (NA). Numerical aperture is a figure of merit that is used to measure the light-gathering or light-collecting ability of an optical fiber. A high NA fiber will accept light at a greater angle from the axis.

Light Propagation in an Optical Fiber

Light can be propagated down an optical fiber cable by either reflection or refraction. How the light is propagated depends on the mode of propagation and the index profile of the fiber.

In fiber optics terminology, the word mode simply means path. If there is only one path for light to travel down the cable, it is called singlemode. If there is more than one path, it is called multimode. Figure 2.12 shows single and multimode propagation of light in an optical fiber.

Figure 2.11

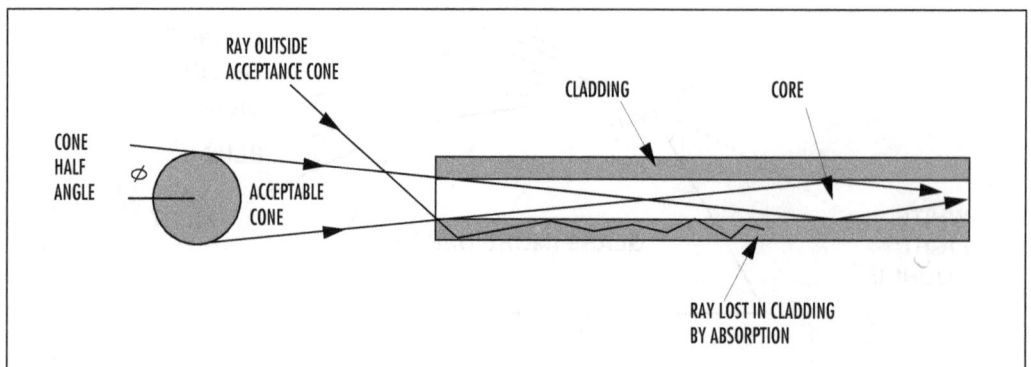

Figure 2.12

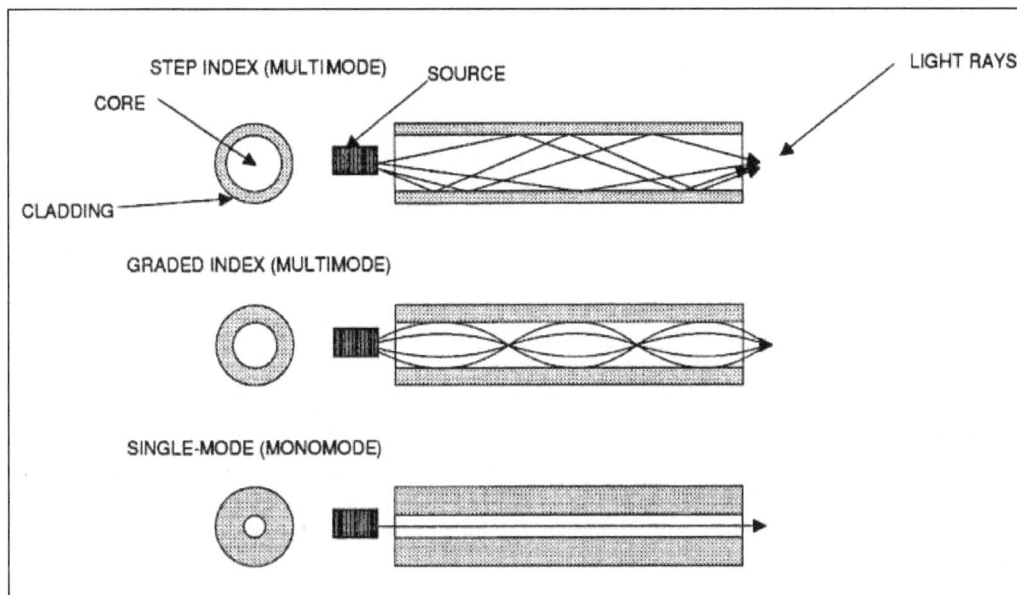

There are two basic types of index profiles: step and graded. A step-index fiber has a cladding with a uniform refractive index less than that of the central core. In a step-index fiber, there is an abrupt change in the refractive index at the core/cladding interface. In a graded-index fiber, there is no cladding, and the refractive index of the core is non-uniform. It is highest at the center and decreases gradually toward the outer edge.

Optical Fiber Configurations

There are three types of optical fiber configurations.
 • Singlemode step-index
 • Multimode step-index
 • Multimode graded-index

Singlemode Step-Index Fiber

A singlemode step-index fiber (Figure 2.13) has a central core that is sufficiently small so there is only one path that light may take as it propagates down the cable. In the simplest form of singlemode step-index fiber, the outside cladding is simply air. The refractive index of the glass core is approximately 1.5, and the refractive index of the air cladding is 1. The large difference in the refractive indexes results in a small critical angle at the glass/air interface. The fiber will accept light from a wide aperture. This makes it relatively easy to couple light from a source into the cable. This type of fiber is typically very weak and of limited practical use.

A more practical type of singlemode step-index fiber is one that has a cladding other than air. The refractive index of the cladding is slightly less than that of the central core and is uniform throughout the cladding. This type of cable is physically stronger than the air-clad fiber, but the critical angle is also much higher. This results in a small acceptance angle and a narrow source-to-fiber aperture, making it much more difficult to couple light into the fiber from a light source. With both types of singlemode step-index fibers, light is propagated down the fiber through reflection. Light rays that enter the fiber propagate straight down the core or, perhaps, are reflected once. All light rays follow approximately the same path down the cable.

Figure 2.13

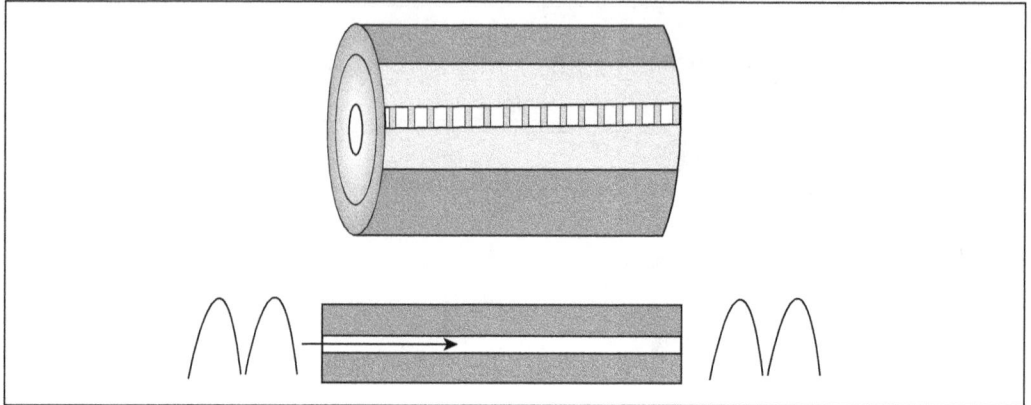

Multimode Step-Index Fiber

Multimode step-index fiber is similar to the singlemode configuration except the center core is much larger. This type of fiber has a larger light-to-fiber aperture and allows more light to enter the cable. The light rays that strike the core/cladding interface at an angle greater than the critical angle are propagated down the core in a zigzag fashion, continuously reflecting off the interface boundary. Light rays that strike the core/cladding interface at an angle less than the critical angle enter the cladding and are lost. There are many paths that a light ray may follow as it propagates down the fiber. As a result, all light rays do not follow the same path and do not take the same amount of time to travel the length of the fiber. (See Figure 2.14.)

Multimode Graded-Index Fiber

A multimode graded-index fiber is characterized by a central core with a refractive index that is not uniform, but it is maximum at the center and decreases gradually toward the outer edge. Light is propagated down this type of fiber through refraction. Light enters the fiber at many different angles. As the light rays propagate down the fiber, the light rays that travel in the outermost area of the fiber travel a greater distance than the rays traveling near the center. The refractive index decreases with distance from the center and velocity is inversely proportional to the refractive index. This allows light rays traveling farthest from the center to propagate at a higher velocity. This allows the lightwaves to take approximately the same amount of time to travel the length of the fiber.

Comparison of the Types of Optical Fibers

Singlemode Step-Index Fiber

Advantages
1. There is minimum dispersion. Because all rays propagating down the fiber take approximately the same path, they take approximately the same amount of time to travel down the cable. A pulse of light entering the cable can be reproduced at the receiving end very accurately.
2. Because of the high accuracy in reproducing transmitted pulses at the receiving end, large bandwidths and higher information transmission rates are possible with singlemode step-index fibers than with other types of fibers.

Disadvantages
1. Because the central core is very small, it is difficult to couple light into and out of this type of fiber. The source-to-fiber aperture is the smallest of all the fiber types.
2. Because of the small central core, a highly directive light source such as a laser is required to couple light into a singlemode step-index fiber.

Figure 2.14

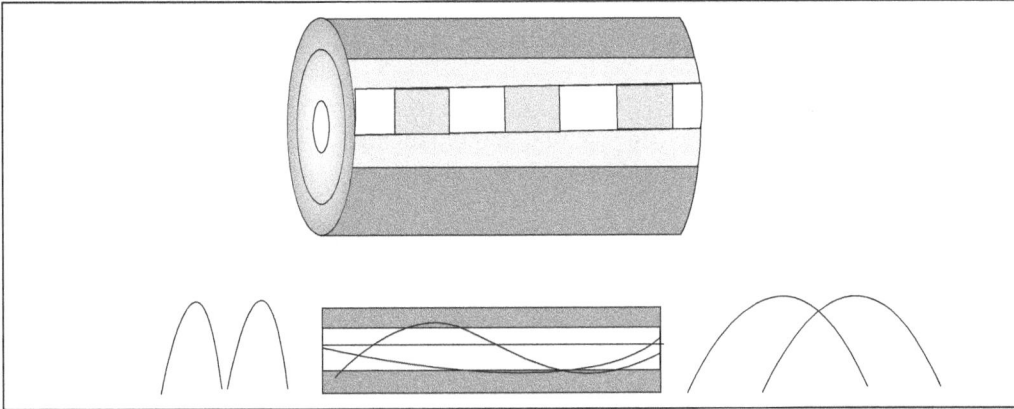

Multimode Graded-Index Fiber

Multimode graded-index fibers are easier to couple light into and out of than singlemode step-index fibers but more difficult than multimode step-index fibers. Distortion due to multiple propagation paths is greater than in singlemode step-index fibers but less than in multimode step-index fibers. Graded-index fibers are easier to manufacture than single-mode step-index fibers but more difficult than multimode step-index fibers. The multimode graded-index fiber is considered an intermediate fiber compared to the other types.

Optical Fibers

Fiber Types

There are three varieties of optical fibers available today. All three varieties are constructed of glass, plastic or a combination of glass and plastic. The three varieties are:
1. Plastic core and cladding
2. Glass core with plastic cladding (often called plastic-clad silica (PCS) fiber)
3. Glass core and glass cladding (often called silica-clad silica (SCS) fiber)

Plastic fibers have several advantages over glass fibers. First, plastic fibers are more flexible and more rugged than glass. They are easy to install, can better withstand stress, are less expensive and weigh approximately 60 percent less than glass. The main disadvantage of plastic fibers is their high attenuation characteristic. They do not propagate light as efficiently as glass. Plastic fibers are limited to relatively short runs, such as within a single building or a building complex.

Fibers with glass cores exhibit low attenuation characteristics, however, PCS fibers are slightly better than SCS fibers. Also, PCS fibers are less affected by radiation. SCS fibers have the best propagation characteristics and are easier to terminate than PCS fibers. Unfortunately, SCS cables are the least rugged, and they are more susceptible to increases in attenuation when exposed to radiation.

The selection of a fiber for a given application is a function of specific system requirements.

Fiber Construction

There are many different cable designs available today. (See Figure 2.15.) Depending on the configuration, the cable may include a core, a cladding, a protective tube, buffers, strength members and one or more protective jackets.

With the loose tube construction, each fiber is contained in a protective tube. Inside the protective tube, a polyurethane compound encapsulates the fiber and prevents moisture.

In a multiple-strand configuration, to increase the tensile strength, a steel central member and a layer of Mylar tape wrap are included in the package. A ribbon configuration is frequently seen in telephone systems using fiber optics.

The type of cable construction used depends on the performance requirements of the systems and both the economic and environmental constraints.

Figure 2.15

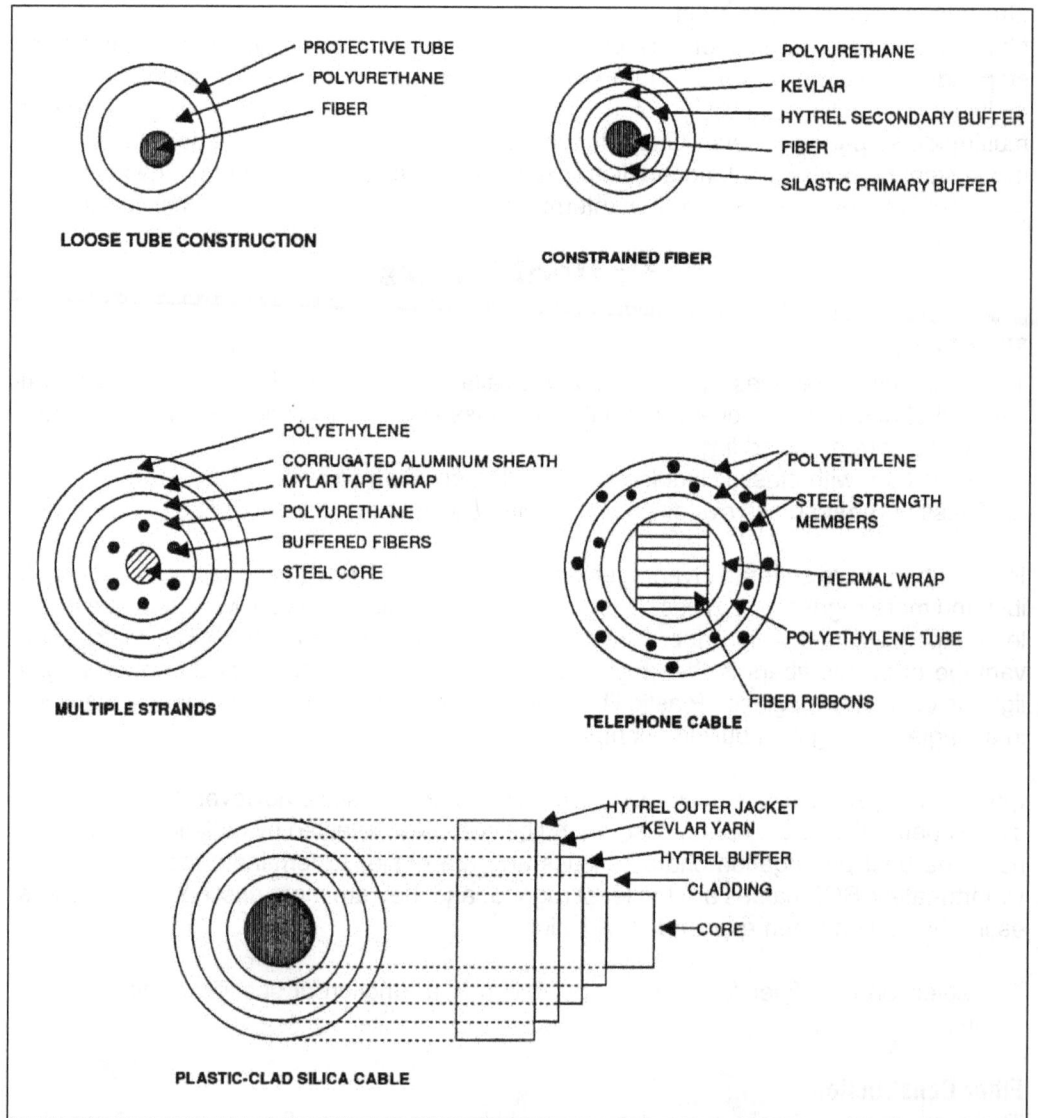

LOOSE TUBE CONSTRUCTION

- PROTECTIVE TUBE
- POLYURETHANE
- FIBER

CONSTRAINED FIBER

- POLYURETHANE
- KEVLAR
- HYTREL SECONDARY BUFFER
- FIBER
- SILASTIC PRIMARY BUFFER

MULTIPLE STRANDS

- POLYETHYLENE
- CORRUGATED ALUMINUM SHEATH
- MYLAR TAPE WRAP
- POLYURETHANE
- BUFFERED FIBERS
- STEEL CORE

TELEPHONE CABLE

- POLYETHYLENE
- STEEL STRENGTH MEMBERS
- THERMAL WRAP
- POLYETHYLENE TUBE
- FIBER RIBBONS

PLASTIC-CLAD SILICA CABLE

- HYTREL OUTER JACKET
- KEVLAR YARN
- HYTREL BUFFER
- CLADDING
- CORE

Losses in Fibers

Transmission losses in optical fiber cables are one of the most important characteristics of the fiber. Losses in the fiber result in a reduction in the light power and thus reduce the system bandwidth, information transmission rate, efficiency and overall system capacity. The predominant fiber losses result from:

- Optical fiber losses
- Microbending losses
- Connector losses
- Splice losses
- Coupling losses

The information carrying capacity, better known as bandwidth multimode fiber, is specified by the term "bandwidth distance product" (BWDP). The BWDP is a measure of the capacity of the fiber when measured under precisely defined conditions. The BWDP is inversely related to pulse spreading – hence the lower the pulse spreading, the higher the BWDP. The BWDP is stated in units of MHz-km. The fiber must show a BWDP as a minimum.

- These conditions include equilibrium modal distribution and zero chromatic dispersion. These conditions do not exist in operating systems. Because of the difference between these conditions, the effective, or real bandwidth of a link is not equal to the BWDP divided by the distance. The effective bandwidth will be less, often significantly less, than the result of such division.

Refer to the "bit rate distance product" below (typical estimates) since the spectral width of the light source was not stated in the base document. In addition, the theoretical capacity of a singlemode fiber is so high that most people refer to it as being essentially unlimited.

Optical Fiber Losses

Light is an electromagnetic wave of a vibrating nature. Just as the speed of light slows when traveling in transparent materials, each wavelength is transmitted differently in the fiber. Therefore, attenuation or optical power loss must be measured in specific wavelengths for each fiber type. Losses of optical power at the different wavelengths occur in the fiber due to absorption, reflection and scattering. These occur over distances depending on the specific fiber, its size, purity and refraction indexes.

Absorption

Absorption loss in optical fibers is analogous to power dissipation in copper cables. Impurities in the fiber absorb the light and convert it to heat (Figure 2.16). The ultrapure glass used to manufacture optical fiber is approximately 99.999 percent pure. There are three factors that contribute to the absorption losses in optical fibers:

- Ultraviolet absorption
- Infrared absorption
- Ion resonance absorption

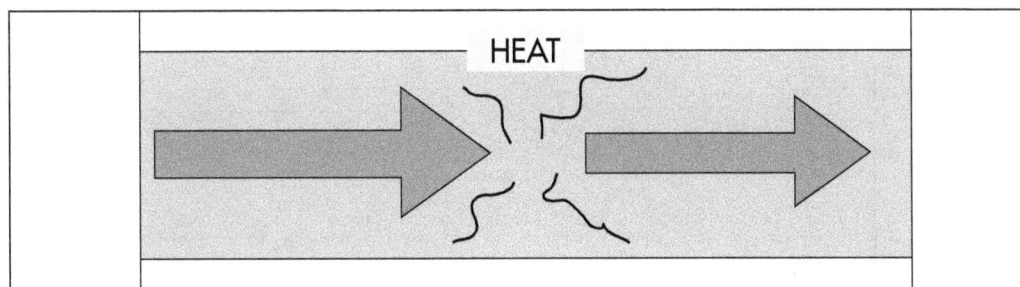

HEAT

Figure 2.16

Ultraviolet Absorption

Ultraviolet absorption is caused by valence electrons in the silica material from which fibers are manufactured. Light ionizes the valence electrons into conduction. The ionization is equivalent to a loss in the total light field and contributes to the transmission losses of the fiber.

Infrared Absorption

Infrared absorption is a result of photons of light that are absorbed by the atoms of the glass core molecules. The absorbed photons are converted to random mechanical vibrations typical of heating.

Ion Resonance Absorption

Ion resonance absorption is caused by OH-ions in the material. The source of the OH-ions is water molecules that have been trapped in the glass during the manufacturing process. Iron, copper and chromium molecules also cause ion absorption.

Scattering

During the manufacturing process, glass is extruded (drawn into long fibers of very small diameters). During this process, the glass is in a plastic state (not liquid and not solid). The tension applied to the glass during this process causes the cooling glass to develop submicroscopic irregularities that are permanently formed in the fiber. When light rays propagating down the fiber strike one of these impurities, they are diffracted. Diffraction causes the light to disperse or spread out in many directions. (See Figure 2.17.) Some of the diffracted light continues down the fiber and some of it escapes through the cladding. The light rays that escape represent a loss in light power. This is called Rayleigh scattering loss.

Spectral Dispersion

Spectral dispersion occurs when portions of a signal cause signal spreading, which were launched together, do not arrive together. This can occur because of chromatic or modal dispersion.

Chromatic Dispersion

The refractive index of a material is wavelength dependent. Light emitting diodes (LEDs) contain a combination of wavelengths. Each wavelength within the composite light signal travels at a different velocity. Light rays that are simultaneously emitted from an LED and propagated down an optical fiber do not arrive at the far end of the fiber at the same time. The result is a distorted receive signal and is called chromatic distortion. Chromatic distortion can be eliminated using a monochromatic source such as an injection laser diode (ILD).

Figure 2.17

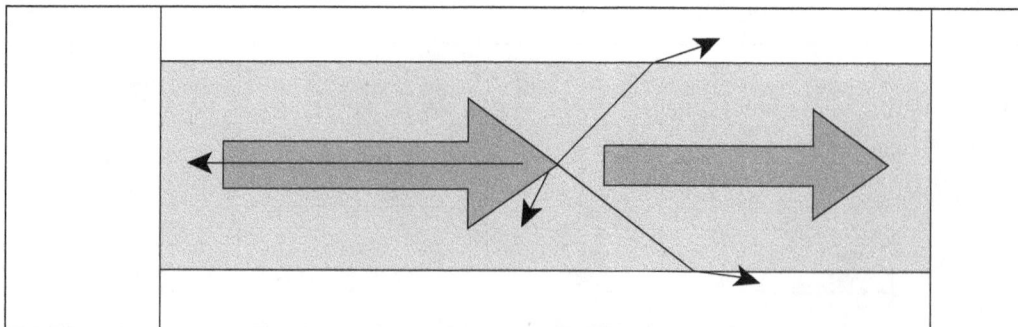

Modal Dispersion

Modal dispersion, or pulse spreading, is caused by the difference in the propagation times of light rays that take different paths down a fiber. Modal dispersion can occur only in multimode fibers. It can be reduced considerably by using graded-index fibers and almost entirely eliminated by using singlemode step-index fibers.

Modal dispersion can cause a pulse of light energy to spread out as it propagates down a fiber. In a multimode step-index fiber, a light ray that propagates straight down the axis of the fiber takes the least amount of time to travel the length of the fiber. A light ray that strikes the core/cladding interface at the critical angle will undergo the largest number of internal reflections and take the longest time to travel the length of the fiber.

Figure 2.18 shows three rays of light propagating down a multimode step-index fiber. The lowest-order mode travels in a path parallel to the axis of the fiber. The middle-order mode bounces several times at the interface before traveling the length of the fiber. The highest-order mode makes many trips back and forth across the fiber as it propagates the entire length. If the three rays of light were emitted into the fiber at the same time and represented a pulse of light energy, the three rays would reach the far end of the fiber at different times and result in a spreading out of the light energy in respect to time. This is called modal dispersion and results in a stretched pulse, which is also reduced in amplitude at the output of the fiber.

Pulse Spreading

Light rays entering a fiber simultaneously that travel separate paths and arrive at the output of the fiber at different times is known as pulse spreading. The light source emits rays of light both passed to the axis of the fiber and at the angles to the respective axis. All of these rays are part of the same pulse, which carries the information. Since these rays will travel different paths, they will arrive at the end of the fiber at different times. The longer the length of the fiber, or the larger the NA of the fiber, the larger will be the difference in time of arrival of the rays. The difference in time of arrival will be the amount of pulse spreading. See Figure 2.18.

There are three major areas of pulse spreading:
a. Modal dispersion is the largest cause of pulse spreading.
b. Chromatic dispersion results in a fiber when different wavelengths of light travel at different speeds in all materials.

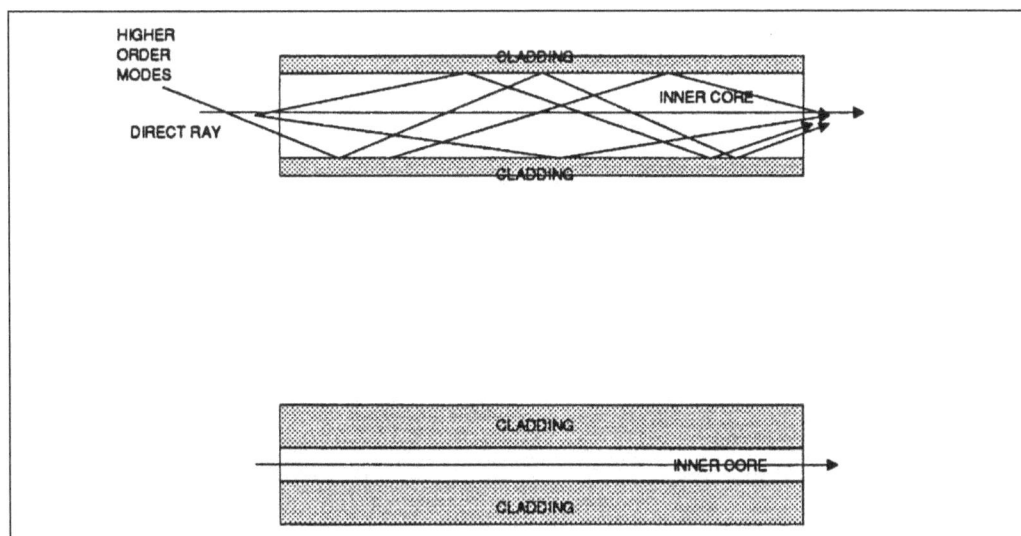

Figure 2.18

c. Material dispersion occurs in fibers when the atomic level structure of the fiber is not constant. As a result, rays of light traveling in different areas of the core and traveling at different speeds, will arrive at different times.

Microbending Losses

Without support, an optical fiber is subject to losses of optical power. Small bends and kinks in the fiber cause radiation losses. Microbends and constant-radius bends are two types of bends. Microbending occurs as a result of differences in the thermal contraction rates between the core and cladding material. A microbend represents a discontinuity in the fiber where Rayleigh scattering can occur. Constant-radius bends occur when fibers are bent during handling or installation.

Connector and Splice Losses

Fibers may be joined permanently by fusion, welding, chemical bonding or mechanical joining. Junction losses are most often caused by one of the following alignment problems; lateral misalignment, end separation, angular misalignment and imperfect surface finishes. (See Figure 2.19.)

Lateral Misalignment

This is the lateral or axial displacement between two pieces of adjoining fiber cables. (See Figure 2.19.) The amount of loss can be from a couple of tenths of a decibel to several decibels. This loss is generally negligible if the fiber axes are aligned to within 5 percent of the smaller fiber's diameter.

Figure 2.19

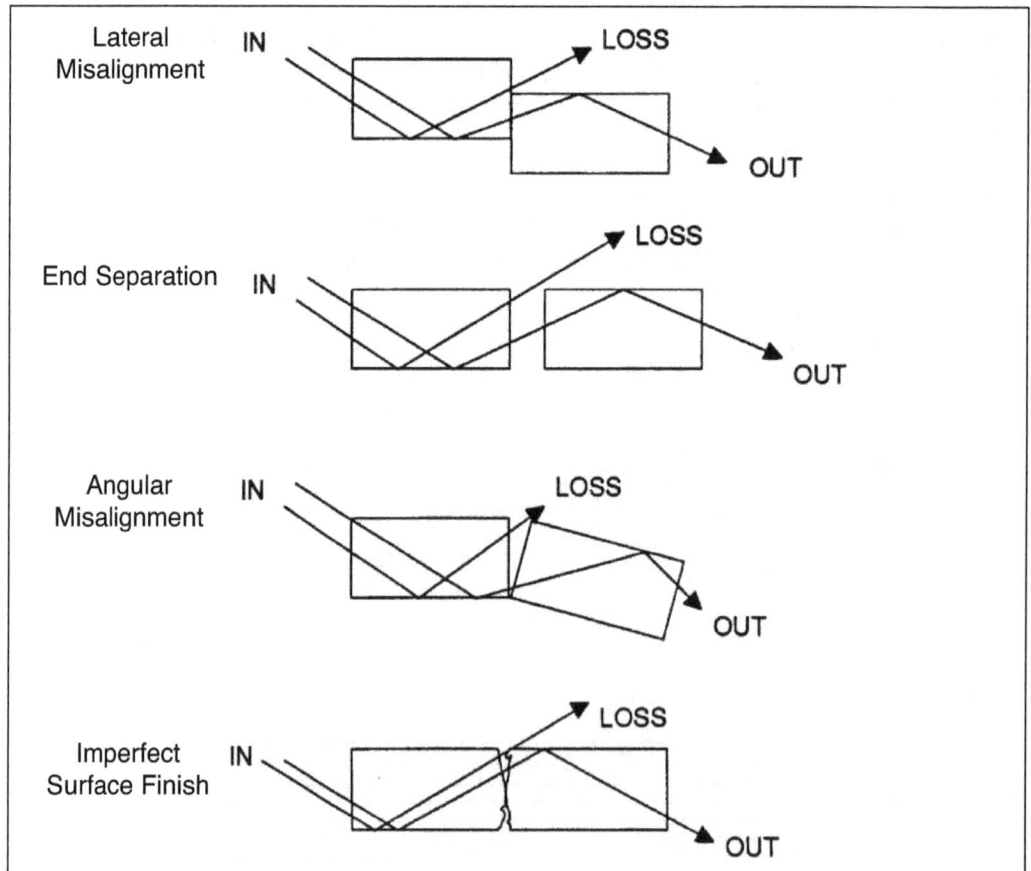

End Separation

When splices are made in optical fibers, the fibers should actually touch. The farther apart the fibers, the greater the loss of light. If two fibers are joined with a connector, the ends should not touch. This is because the two ends rubbing against each other in the connector could cause damage to either or both fibers.

Angular Misalignment (See Figure 2.19.)

This is sometimes called axial displacement. If the axial displacement is less than 2 degrees, the loss will be less than 0.5 dB.

Imperfect Surface Finish (See Figure 2.19.)

The ends of the two adjoining fibers should be highly polished and fit together squarely.

Attenuation

Attenuation is defined as the loss of optical power over distance in a fiber optic cable. Attenuation is measured in dB/Km. For example, there can be up to 300 dB/Km for plastic fibers and .21 dB/Km for singlemode fibers. It will vary with wavelength but not with frequency.

Windows

Windows are low loss regions where the fiber carries light with minimal attenuation.

- Window #1 820 to 850 nm
- Window #2 zero dispersion range - 1300 nm
- Window #3 1550 nm A 50/125 multimode graded-index fiber may have an anticipated loss of 4 dB/Km at 850 nm wavelength and 2.5 dB/Km at 1300 nm wavelength. That's a 30 percent increase in efficiency.

High Loss Regions

The source should be selected to ensure that the emitting light is in the low loss region. High regions that should be avoided are 730 nm, 950 nm, 1250 nm and 1380 nm. Plastic fiber operates better in a visible light range around 650 nm.

Coupling Losses

In fiber cables, coupling losses can occur at light source-to-fiber connections and fiber-to-light detector connections. Loss of optical power between the fiber and the light source or the light detector is a function of both the device and the type of fiber used. For example, LEDs emit light in a broad spectral pattern when compared to laser diodes. Therefore, LEDs will couple more light when a larger core fiber is used. Laser diodes can be used more effectively with smaller core diameters. The optical index of refraction difference between core and cladding also determines the approach that light can take to the core and be accepted for transmission. Using a light source not matched to a particular fiber's numerical aperture and core size will cause less than optimum light coupling for the system.

Optic Sources

There are two devices commonly used to generate light for fiber optic communications systems: LEDs and injection laser diodes (ILDs). Although the LED provides less power and operates at slower speeds, it is amply suited to applications requiring speeds to several hundred megabits and transmission distances of several kilometers. For higher speeds and longer distances, the laser must be considered.

Light Emitting Diodes

A light emitting diode is simply a p-n junction diode. An LED converts electrical energy to light energy. LEDs emit light by spontaneous emission. Light is emitted as a result of the recombination of electrons and holes. When forward biased, minority carriers are injected across the p-n junction. Once across the junction, these minority carriers recombine with majority carriers and give up energy in the form of light. This process is the same in a conventional diode except in LEDs, certain semiconductor materials and dopants are chosen so the process is radiative and a photon is produced. A photon is a quantum of electromagnetic wave energy. Photons are particles that travel at the speed of light but at rest, have no mass. In conventional semiconductor diodes, the process is primarily nonradiative and photons are not generated. The energy gap of the material used to construct an LED determines whether the light emitted is invisible or visible and the color of the light.

Numerical Aperture (NA)
The Light-Gathering Ability of a Fiber

Simply put, NA is the light-gathering ability of the fiber. Only light injected into the fiber at angles greater than the critical angle will be propagated. The material NA is a dimensionless number that relates to 1) the refractive indices of the core and cladding and 2) the acceptance cone, which is the maximum angle at which a cable will propagate rays entering the cable.

The NA provides an indication of how well the fiber accepts and propagates light. A large NA accepts light well. A small NA requires highly directional light. Fibers with high bandwidth have a low NA and accept fewer modes of light. As a result, we have less dispersion and greater bandwidth. Typical NAs range from 0.50 for plastic fibers to 0.20 for glass graded-index fibers. The NA is not usually specified for singlemode fibers, but it is approximately 0.11. It is not really a critical parameter for a system designer.

The NA is affected by distance. Higher order modes (the ones that are carried nearer the critical angle) are often lost. When a graded-index fiber reaches EMD, the NA can be reduced by up to 50 percent. Thus, the light exiting the fiber does so at angles much less than those defined by the acceptance cone. The spot diameter emerging from the fiber can also be reduced.

Sources and detectors also have an NA. Especially for sources, it is important to match the NA of the source to the NA of the fiber to ensure efficient propagation. Mismatches in NA are sources of loss when coupling from a higher NA to a lower NA.

TEST MODULE 2

Multiple Choice: Circle the appropriate answer to the following questions.

1. The wavelength of light is typically measured in nanometers, micrometers and angstroms. Which unit is most often used in fiber optics?
   a) angstroms
   b) nanometers
   c) micrometers

2. What is a specific path light takes in an optical fiber?
   a) mode
   b) core
   c) splice
   d) buffer

3. What is the light-gathering ability of the fiber?
   a) modal dispersion
   b) light emitting diodes (LED)
   c) attenuation
   d) numerical aperture (NA)

4. What is the center layer of the fiber?
   a) cladding
   b) core
   c) buffer
   d) plastic

5. What is the mathematical unit used to describe the attenuation of a fiber optic cable?
   a) dBm
   b) Snell's Law
   c) dispersion
   d) dB/Km

6. What is the inner light-carrying section of fiber called?
   a) cladding
   b) buffer
   c) core
   d) coating

7. How is multimode step-index fiber different from the singlemode configuration?
   a) The center core is much larger.
   b) It has a larger light-to-fiber aperture.
   c) It allows more light to enter the cable.
   d) All of the above.

Section 3.0
Construction and Application of Cables

Objective

Discuss the types and uses of fiber cables.

Outline

- *Types of Cables*
- *Loose Buffer Cables*
- *Tight Buffer Cables*
- *Simplex and Duplex Cables*
- *Color Coding*

Learning Activities

Assessment: *Test Module 3*

LabExercise: *Familiarization of types of indoor and outdoor cables.*

Types of Cables

Fiber cables are offered in many different configurations. When choosing a fiber cable, keep in mind the installation environment temperature, extremes, flexibility, durability and tensile strength. It is obvious that exterior cabling must be much more robust than cabling installed inside a temperature-controlled building. Exterior cabling must be able to withstand such things as ice, extreme temperatures, high winds, rodents and other hazards.

There are probably 100 configurations available from a fiber optical cable manufacturer. Most have excellent guides with good technical explanations of the types of cables used for specific requirements. There are at least 40 specific choices that are required in order to purchase a fiber optic cable. (See Figure 3.1.)

Figure 3.1

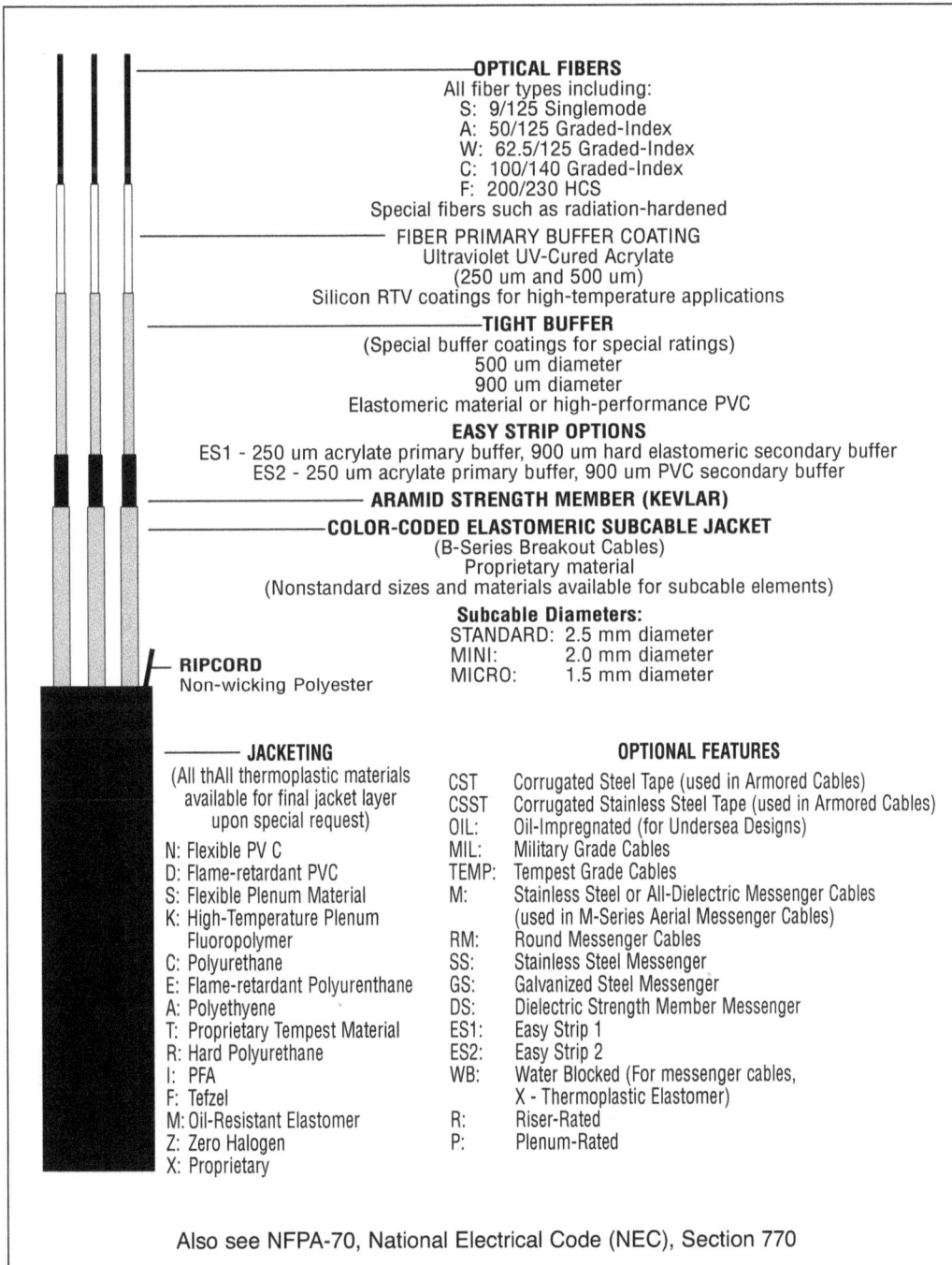

OPTICAL FIBERS
All fiber types including:
S: 9/125 Singlemode
A: 50/125 Graded-Index
W: 62.5/125 Graded-Index
C: 100/140 Graded-Index
F: 200/230 HCS
Special fibers such as radiation-hardened

FIBER PRIMARY BUFFER COATING
Ultraviolet UV-Cured Acrylate
(250 um and 500 um)
Silicon RTV coatings for high-temperature applications

TIGHT BUFFER
(Special buffer coatings for special ratings)
500 um diameter
900 um diameter
Elastomeric material or high-performance PVC

EASY STRIP OPTIONS
ES1 - 250 um acrylate primary buffer, 900 um hard elastomeric secondary buffer
ES2 - 250 um acrylate primary buffer, 900 um PVC secondary buffer

ARAMID STRENGTH MEMBER (KEVLAR)

COLOR-CODED ELASTOMERIC SUBCABLE JACKET
(B-Series Breakout Cables)
Proprietary material
(Nonstandard sizes and materials available for subcable elements)

Subcable Diameters:
STANDARD: 2.5 mm diameter
MINI: 2.0 mm diameter
MICRO: 1.5 mm diameter

RIPCORD
Non-wicking Polyester

JACKETING
(All thAll thermoplastic materials available for final jacket layer upon special request)
N: Flexible PV C
D: Flame-retardant PVC
S: Flexible Plenum Material
K: High-Temperature Plenum Fluoropolymer
C: Polyurethane
E: Flame-retardant Polyurenthane
A: Polyethyene
T: Proprietary Tempest Material
R: Hard Polyurethane
I: PFA
F: Tefzel
M: Oil-Resistant Elastomer
Z: Zero Halogen
X: Proprietary

OPTIONAL FEATURES
CST   Corrugated Steel Tape (used in Armored Cables)
CSST   Corrugated Stainless Steel Tape (used in Armored Cables)
OIL: Oil-Impregnated (for Undersea Designs)
MIL: Military Grade Cables
TEMP: Tempest Grade Cables
M: Stainless Steel or All-Dielectric Messenger Cables
        (used in M-Series Aerial Messenger Cables)
RM: Round Messenger Cables
SS: Stainless Steel Messenger
GS: Galvanized Steel Messenger
DS: Dielectric Strength Member Messenger
ES1:   Easy Strip 1
ES2:   Easy Strip 2
WB: Water Blocked (For messenger cables,
        X - Thermoplastic Elastomer)
R: Riser-Rated
P: Plenum-Rated

Also see NFPA-70, National Electrical Code (NEC), Section 770

There are two types of basic fiber optic cables: loose buffer (often called loose tube)(Figure 3.2) and tight buffer (Figure 3.3).

Loose Buffer Cable

The important thing to remember about loose buffer cable is that it is designed to be used outside or in an environment where broad temperature changes are likely to take place. Loose buffer cable is ideal for long-distance telecommunication lines connecting large cities over varying terrain and weather conditions. Interbuilding backbone cables, on the outside plant cable between buildings, may also be exposed to adverse weather conditions. Loose buffer cable can be used in other applications, but its most common use is in exterior applications. (See Figure 3.4.)

Depending on the manufacturer and type of cable, the buffer tube and/or the primary coating on the fibers may be color-coded for identification. The area around the buffer types, referred to as the interstitial cable core, is often filled with a floating compound to help prevent water penetration into the cable. This is surrounded by a strength member such as Kevlar and a polyethylene sheath. This may be the outside sheath or, in the case of an armored cable, there may be a corrugated steel tape wrapped around the cable that is then covered with a second polyethylene outside sheath. (See Figure 3.5.) The fibers are loose inside the buffer tubes, thereby allowing the tube and other cable parts to expand and contract with temperature changes without putting stress on the fibers. Some of the jacketing materials used are polyethylene, polypropylene, polyurethane and Teflon. The strength members can be steel, fiberglass or Kevlar. (See Figures 3.6 and 3.7.)

Figure 3.2

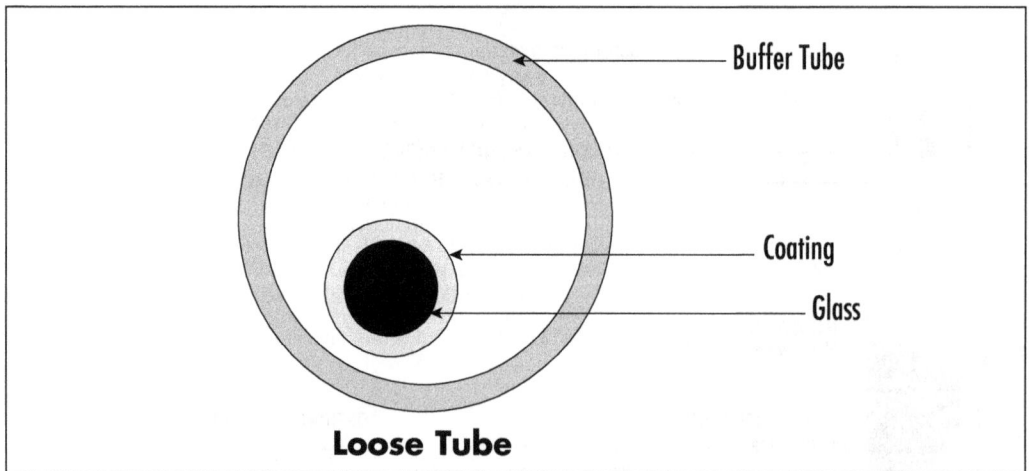

Buffer Tube

Coating

Glass

Loose Tube

Figure 3.3

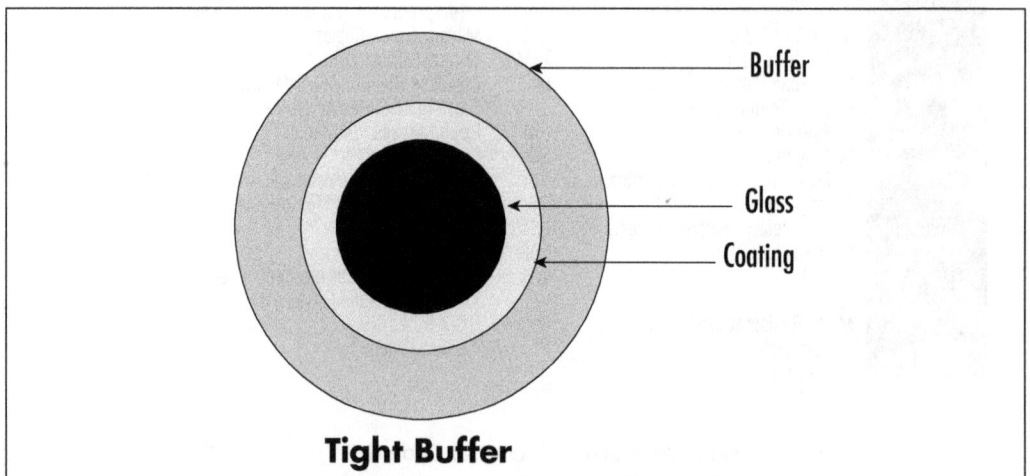

Buffer

Glass

Coating

Tight Buffer

3.4

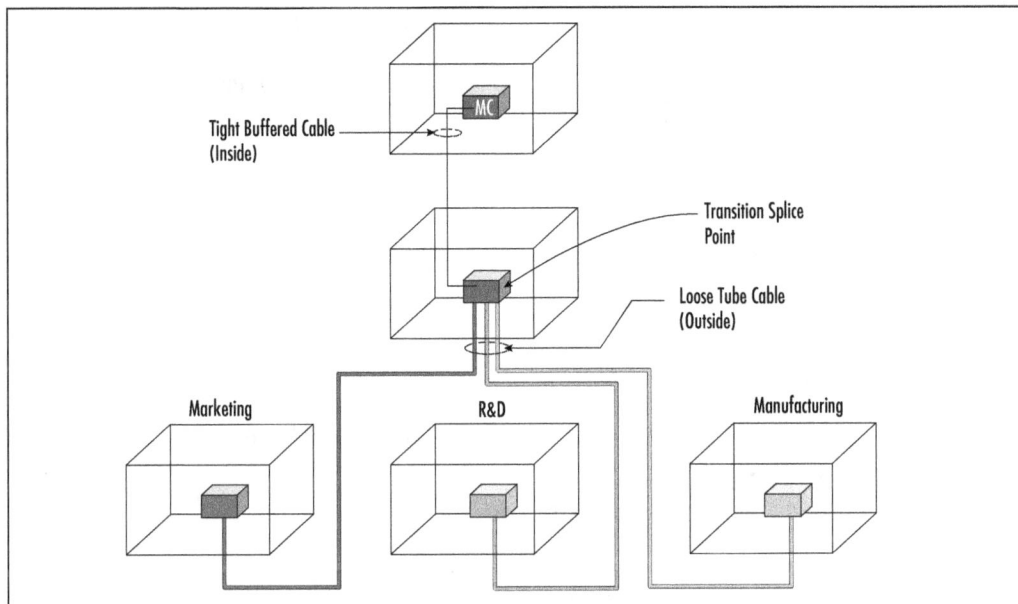

Figure 3.4

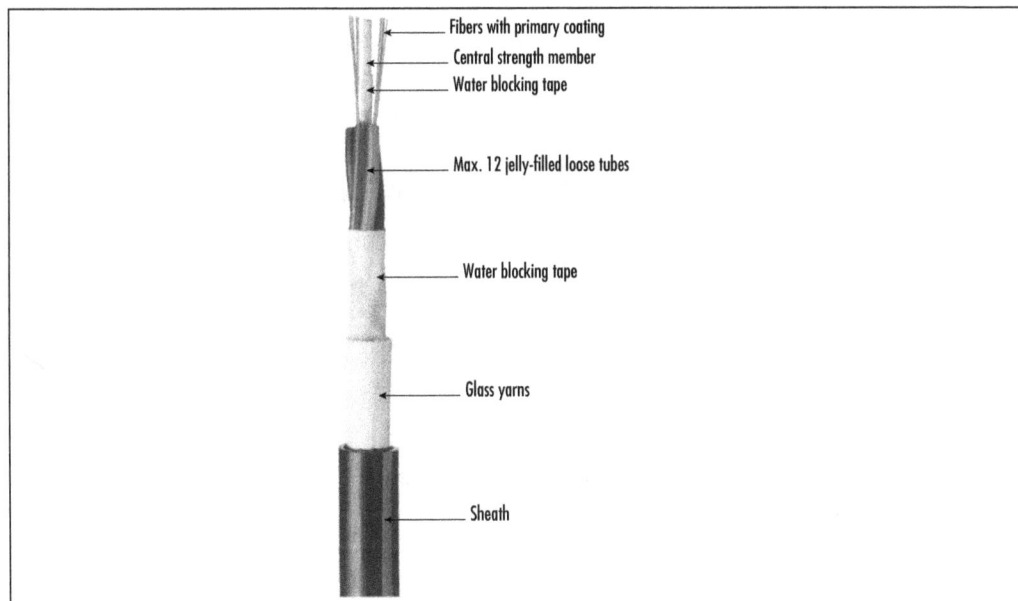

Figure 3.5

Nonmetallic Strength Members
and Glass Yarn Rodent Protection

Loose buffer cables for special uses are also available, for example, a self-supporting aerial cable is shown in Figure 3.8.

Tight Buffer Cables

Tight buffer cables (Figure 3.9) are designed for intrabuilding, horizontal wiring cross-connect/interconnect applications. In some applications where temperature variations are minimal and cable runs are short, tight buffer cabling can be used reliably as a backbone between buildings. Burying this type of cable in a conduit below the frost line is recommended. Tight buffer cables contain fibers that have been coated with a thermoplastic approximately 900 microns in diameter, much like the insulation on copper wires. This tight buffer is applied over the primary coat. It is not just an extra thick primary coat.

Figure 3.6

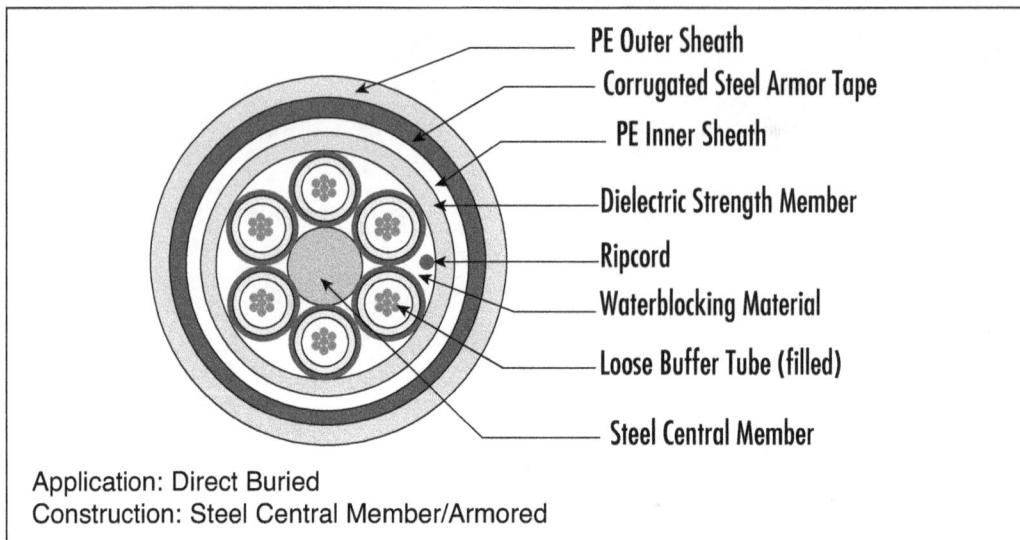

- PE Outer Sheath
- Corrugated Steel Armor Tape
- PE Inner Sheath
- Dielectric Strength Member
- Ripcord
- Waterblocking Material
- Loose Buffer Tube (filled)
- Steel Central Member

Application: Direct Buried
Construction: Steel Central Member/Armored

Figure 3.7

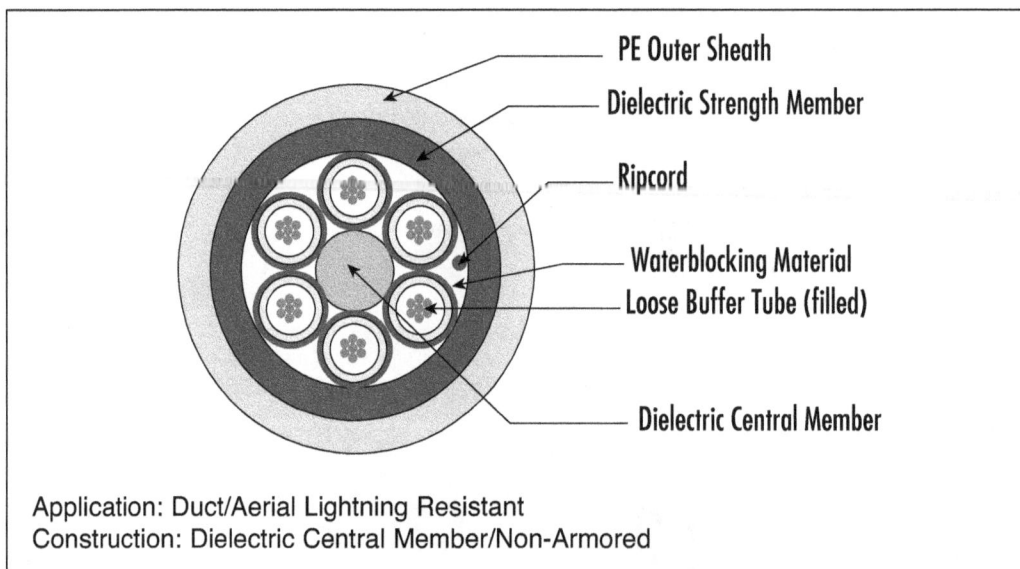

- PE Outer Sheath
- Dielectric Strength Member
- Ripcord
- Waterblocking Material
- Loose Buffer Tube (filled)
- Dielectric Central Member

Application: Duct/Aerial Lightning Resistant
Construction: Dielectric Central Member/Non-Armored

Figure 3.8

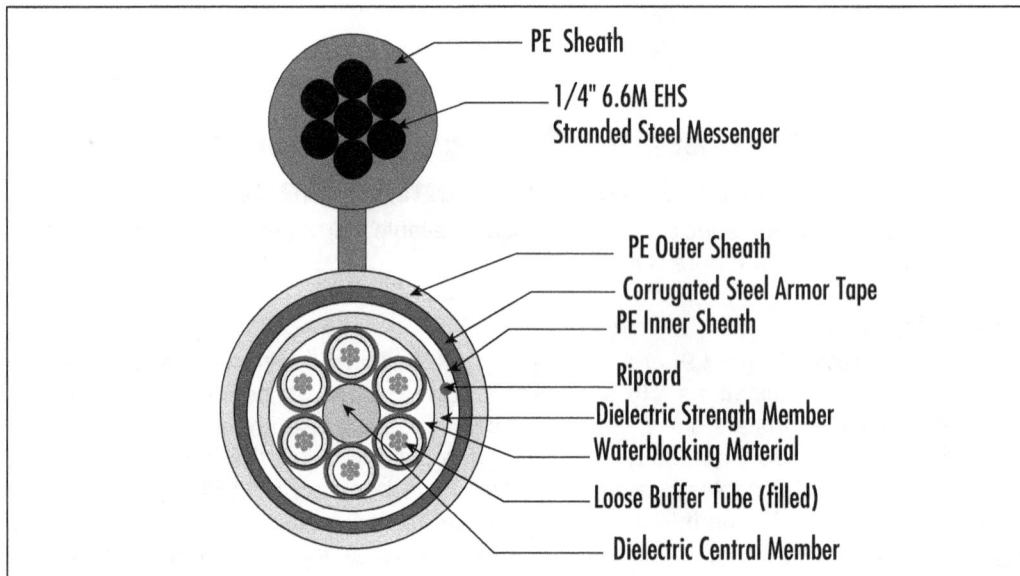

- PE Sheath
- 1/4" 6.6M EHS Stranded Steel Messenger
- PE Outer Sheath
- Corrugated Steel Armor Tape
- PE Inner Sheath
- Ripcord
- Dielectric Strength Member
- Waterblocking Material
- Loose Buffer Tube (filled)
- Dielectric Central Member

Figure 3.9

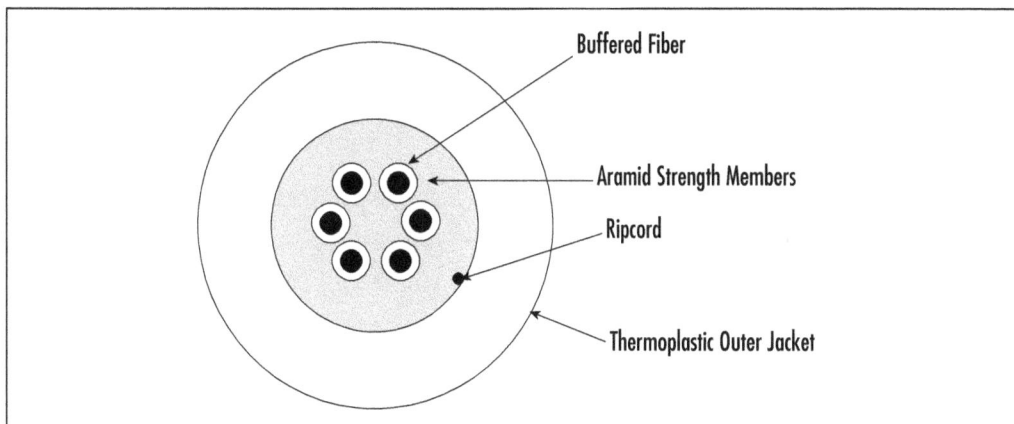

A stripping tool is necessary to remove the buffer. Although the primary coat will usually come off when stripping the fiber, it doesn't always when splicing or assembling a connector. On these occasions, it is obvious that they are two different coats. These cables are more sensitive to temperature variations and are used primarily inside buildings where the temperature is controlled. Tight buffer construction has better impact and crush-resistance. The simplest example of this type of fiber cable is the common cross-connect jumper.

Some of the most common indoor cables include simplex cables, duplex cables, plenum cables and undercarpet cables.

Simplex and Duplex Cables
A simplex cable is your everyday jumper. It refers to a cable with only one fiber and it is generally used to send information in only one direction. (See Figure 3.10.)

Duplex cables, also called zipcords, contain two fibers usually for two-way communication. Duplex cables are actually two simplex cables with a common jacket. (See Figure 3.11.) Some are made with ripcord construction, allowing the two fibers to be easily separated. Two simplex cables could be used, but the duplex cable is just easier to work with than two separate cables.

Plenum Cables
Plenum cables have a special jacket material that does not give off noxious fumes when burned. The National Electrical Code requires that cables run in plenum be contained in fireproof conduit or the cable itself be jacketed with low-smoke, fire-retardant materials. Halar and Teflon are two jacket materials that are now offered with a plenum rating.

Figure 3.10

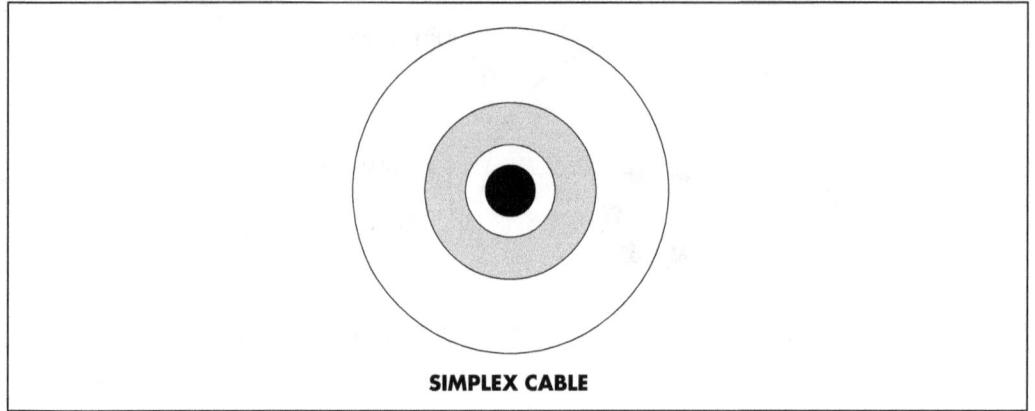

SIMPLEX CABLE

Figure 3.11

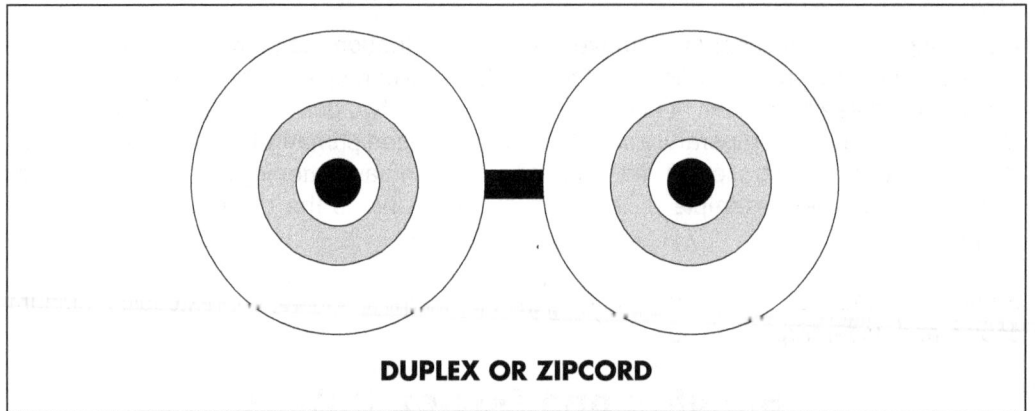

DUPLEX OR ZIPCORD

Color-Coding

Cables, unless customized, follow the industry standard color code system for easy identification. Cables with 12 or fewer individual components will follow the color sequence: blue, orange, green, brown, slate, white, red, black, yellow, violet, rose and aqua.

For cables having more than 12 fibers, grouping is done following the same sequence for the subgroup and for the fibers within it. Example: 24 fiber loose tube cable with six fibers in each of four tubes - tube colors will be blue, orange, green and brown. Each tube will have one each of fibers in the first six colors (blue, orange, green, brown, slate and white). Fibers are then identified by tube color/fiber color - blue/white being the white fiber in the blue tube.

When cables have more than 144 fibers, a black stripe is added to each of the first six colors in order to make 18 recognizable subgroups.

TEST MODULE 3

1. The maximum number of fibers used in a fiber optic cable using the standard color code is 12.

   True   False

2. A strength member in a fiber optic cable can be fiberglass, steel or stainless steel.

   True   False

3. There are two types of basic fiber optic cable; loose tube and tight buffer.

   True   False

4. A stripping tool is necessary to remove the core.

   True   False

5. Simplex cable is called zipcord and contains two fibers usually for two-way communication.

   True   False

LAB EXERCISE

Familiarization of types of indoor and outdoor cables.

Section 4.0
Cable Preparations

Objective

Perform the process of removing the jacket off fiber optic cable and placing the cable in the cabinet connected to other hardware, know the items inside the cable that must be secured, trimmed and stored once the jacket is removed.

Outline

- *Tools Required*
- *Cable Stripping*
- *Sheath Removal*
- *Preparation Guide*
- *Cleaning the Cable*
- *Central Member Preparation*
- *Placing Cable In a Closure*
- *Grounding the Strength Member*

Learning Activity

Assessment: Test Module 4

Lab Exercise: Students participate by getting familiar with the tools used to place cables in cabinets, removing cables, stripping cables, cleaning cables and grounding the strength member.

Tools Required

Sheath removal basically consists of removing the jacket (outer cover) of a fiber cable. Fiber optic cables can have jackets ranging from a simple zipcord type to metal, PVC or other material.

Listed below are the basic tools and materials required when removing most jackets. (See Figure 4.1.)

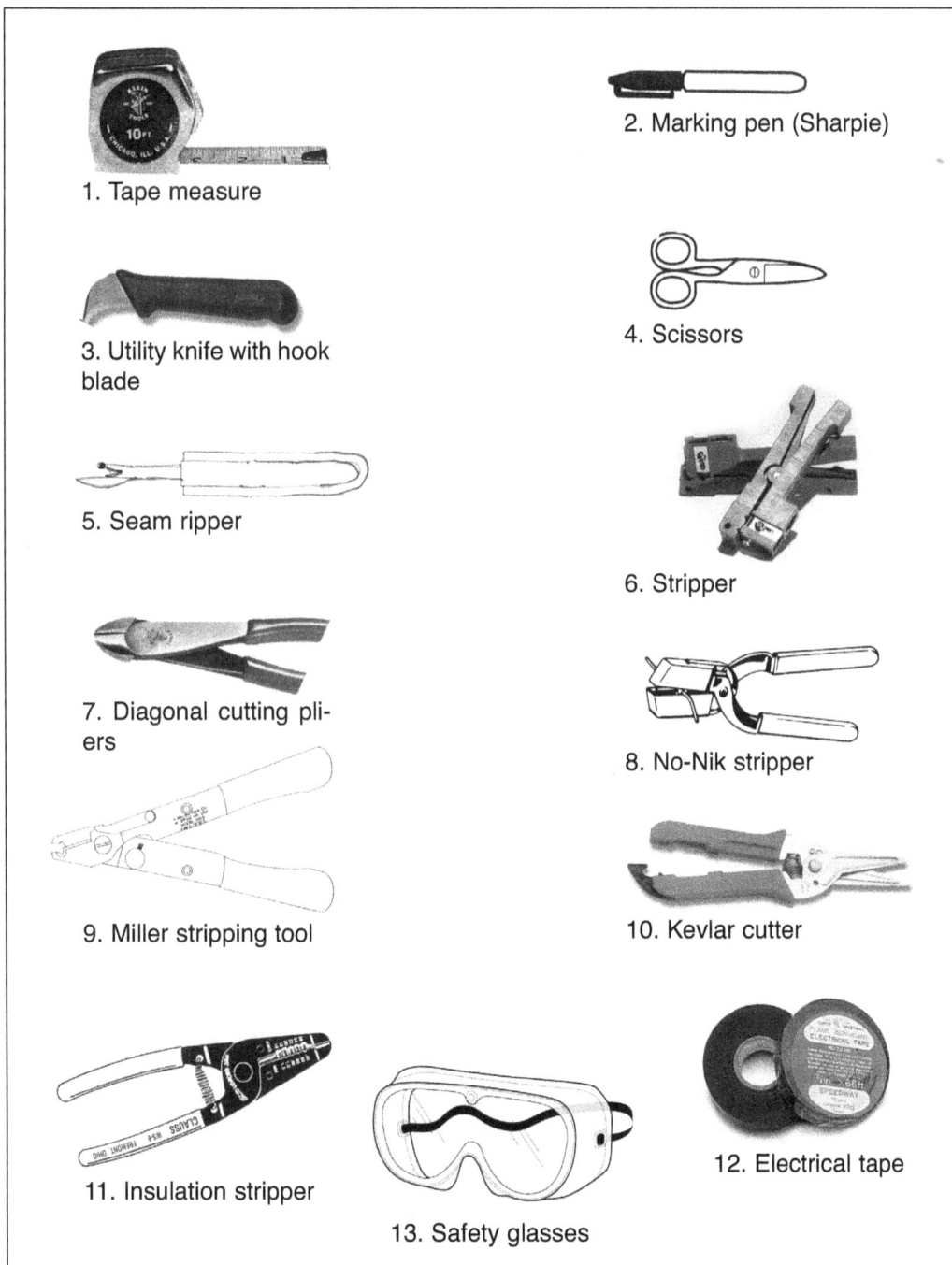

Figure 4.1

1. Tape measure

2. Marking pen (Sharpie)

3. Utility knife with hook blade

4. Scissors

5. Seam ripper

6. Stripper

7. Diagonal cutting pliers

8. No-Nik stripper

9. Miller stripping tool

10. Kevlar cutter

11. Insulation stripper

12. Electrical tape

13. Safety glasses

Cable Stripping Types

There are four methods of stripping cables. (See Figure 4.2.)

Sheath Removal

PVC Sheath

The difference between loose and tight buffer cables and their construction was explained in previous chapters. This first cable preparation will be a simple jacketed tight buffer cable. This typical cable is called an interconnect cable. (See Figure 4.3.)

Tight-buffer refers to the way the optical fiber is held by its applied coating. Rather than allowing the fiber to float inside a stiff buffer tube, as in loose-tube designs, the fiber is held inside a tightly extruded, flexible buffer compound.

Preparation Guide

Most manufacturers of connectors and splices will provide a preparation guide or template indicating the lengths of jacket, yarn and fiber buffer to be removed. (See Figure 4.4.)

Optimax ST Fiber Preparation Guide

Prior to making a connector or a splice, you must follow the specific requirements of the manufacturer.

Figure 4.2

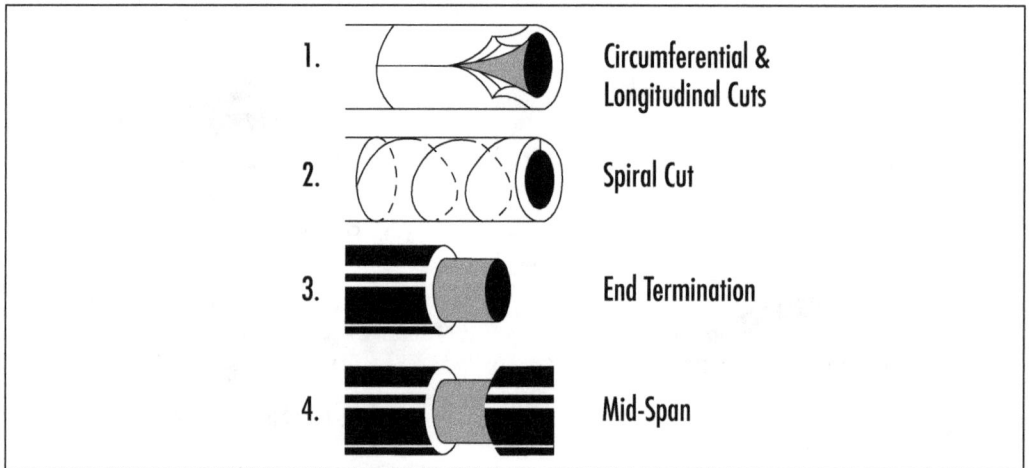

1. Circumferential & Longitudinal Cuts
2. Spiral Cut
3. End Termination
4. Mid-Span

Figure 4.3

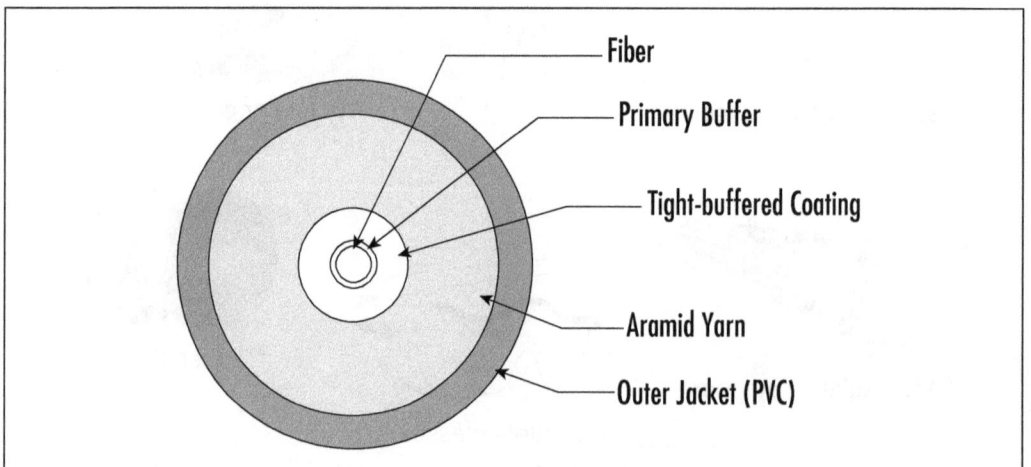

Fiber

Primary Buffer

Tight-buffered Coating

Aramid Yarn

Outer Jacket (PVC)

Some examples of requirements include:
a. Length of jacket materials to be removed
b. Yarn or Kevlar to be removed
c. Buffer coating removal

- Begin by measuring the length of jacket to be removed by marking the distances on the outer jacket with a Sharpie or marker. (See Figure 4.5.)
- The connector manufacturer usually suggests a boot (packaged with the connector) be slipped onto the cable, narrow end first until it's out of your way. (See Figure 4.6.)
- Using the wire stripper, remove the length of described jacket, beginning at the previously marked distances. (See Figure 4.7.)
- For best results, hold the stripping tool perpendicular to the cable, ensuring the cable is in the correct notch. (See Figure 4.8.)
- Cut the yarn to a specific length recommended by the connector's manufacturer. (See Figure 4.9.)
- Each connector's manufacturer has different requirements for the amount of jacket, yarn and buffer to be removed. Before the buffer is stripped, the connectors should be slipped on the fiber. (See Figure 4.10.)
- Place the fiber end to the fiber template and mark the length of buffer to be removed. (See Figure 4.11.)

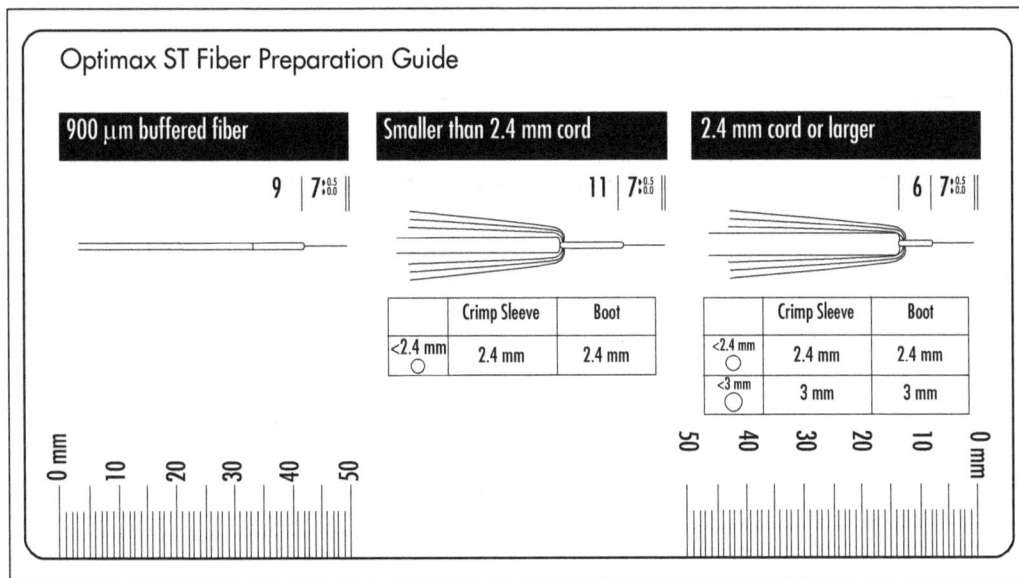

Figure 4.4

Optimax ST Fiber Preparation Guide

| 900 μm buffered fiber | Smaller than 2.4 mm cord | 2.4 mm cord or larger |

	Crimp Sleeve	Boot
<2.4 mm ○	2.4 mm	2.4 mm

	Crimp Sleeve	Boot
<2.4 mm ○	2.4 mm	2.4 mm
<3 mm ○	3 mm	3 mm

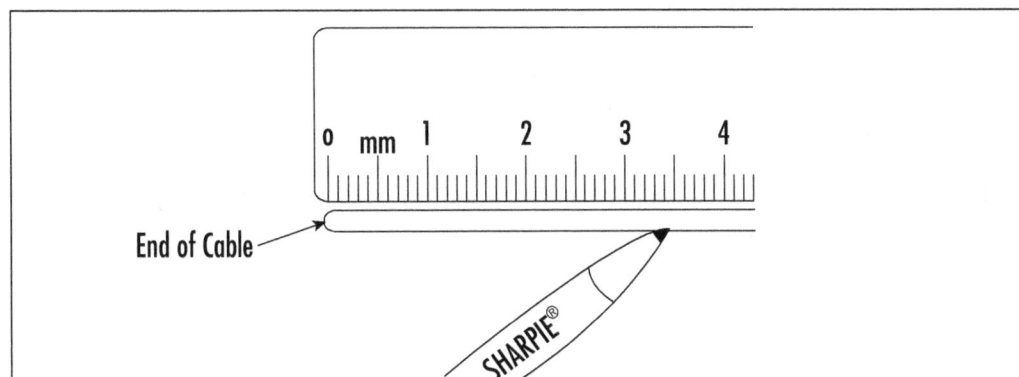

Figure 4.5

End of Cable

SHARPIE®

4.5

Figure 4.6

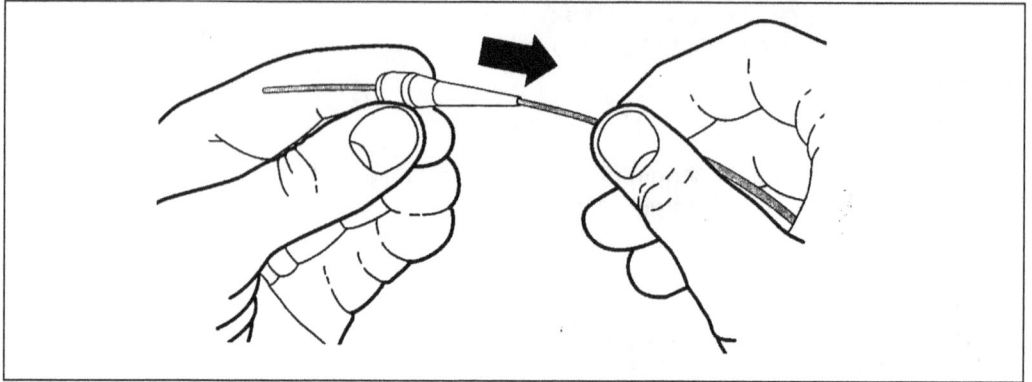

Figure 4.7

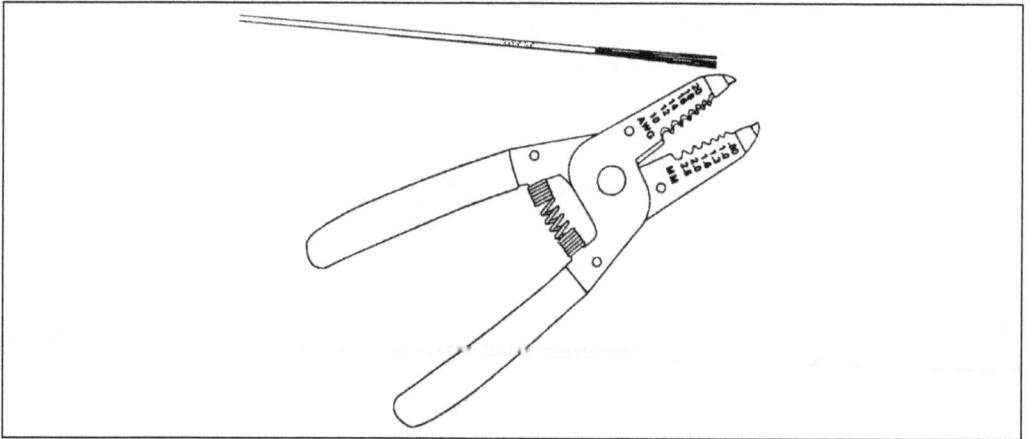

Buffer Removal

Using a No-Nik tool, remove the fiber buffer in 1/4 to 5/16 inch increments at a time until you reach your mark. (See Figure 4.12.)

The final step is to clean the stripped fiber using a lint-free tissue soaked in alcohol. (See Figure 4.13.)

Zipcord

This type of jacket removal is a two-fiber zipcord cable. (See Figure 4.14.) This is the same type of cable as previously described except it has two fibers and can be used for patch panels, jumper assemblies, closet to the wall outlets ceiling installations and fiber to the workstation.

Tools required are similar to those required in single cordage except a razor knife and a coaxial cable cutter are also needed.
- Begin by separating the subunits by placing the cable on a worktable or surface. Place the razor knife about eight inches from the end of the cable and cut into the web to the end of the cable. (See Figure 4.15.)
- The purpose of this starter's cut will provide a controlled tear. By grasping both cable subunits, one in each hand, pull one away from you and one toward you. Determine the length of cable to be pulled apart from the connector or splice manufacturer's recommendation.
- Next, remove the sheath.

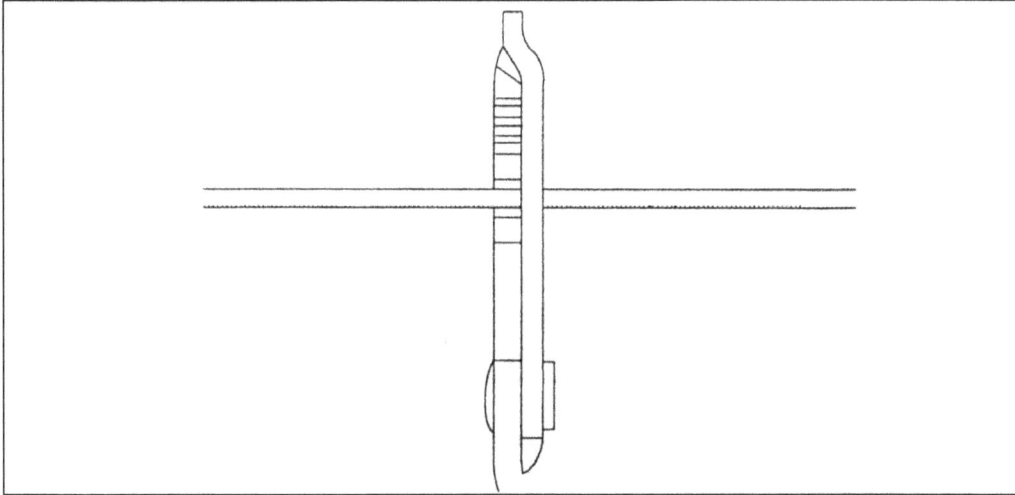
Figure 4.8

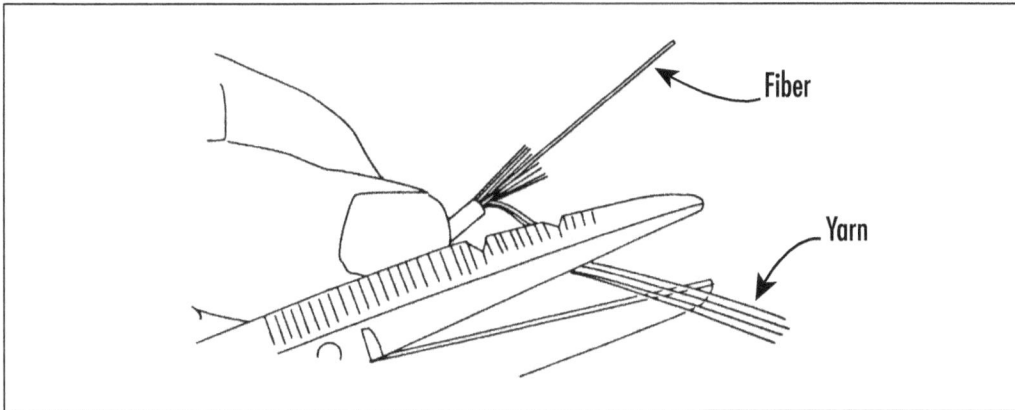
Figure 4.9

Fiber

Yarn

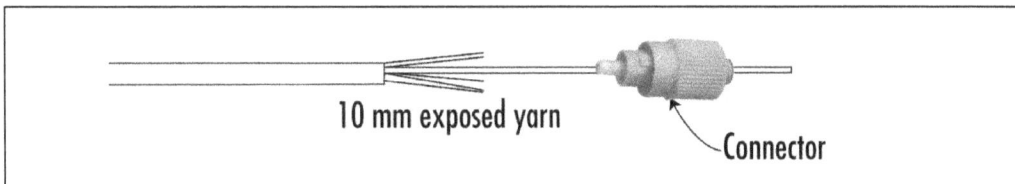
Figure 4.10

10 mm exposed yarn

Connector

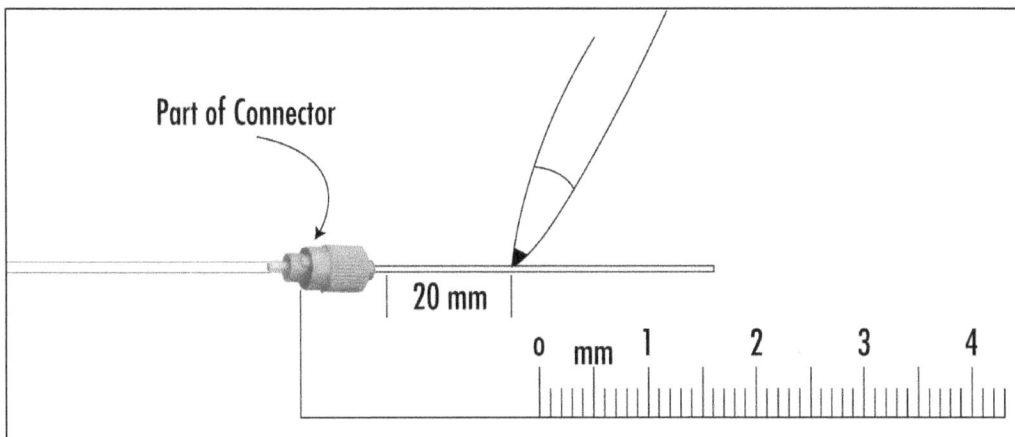
Figure 4.11

Part of Connector

20 mm

0 mm 1 2 3 4

Figure 4.12

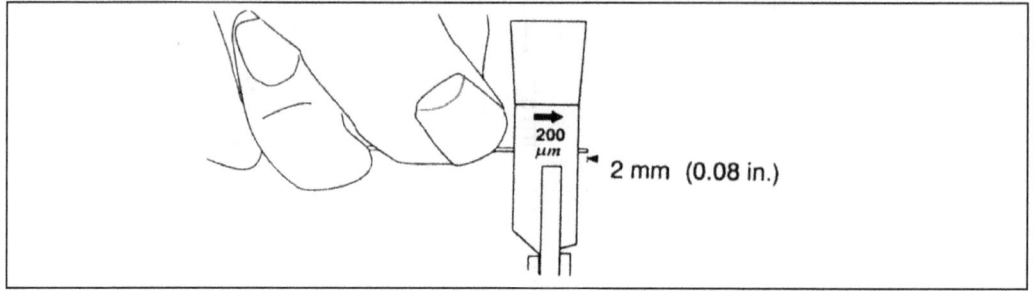

Figure 4.13

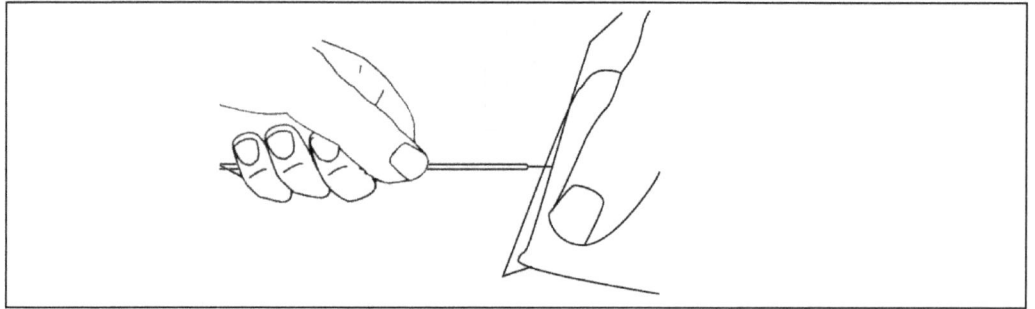

Figure 4.14

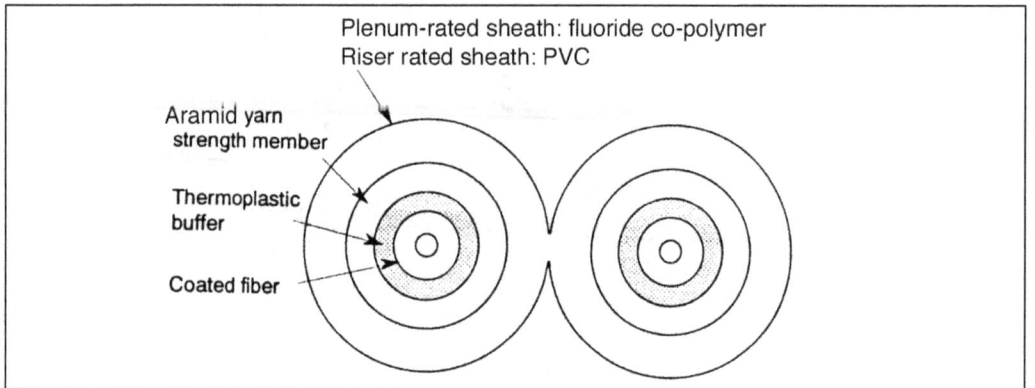

Figure 4.15

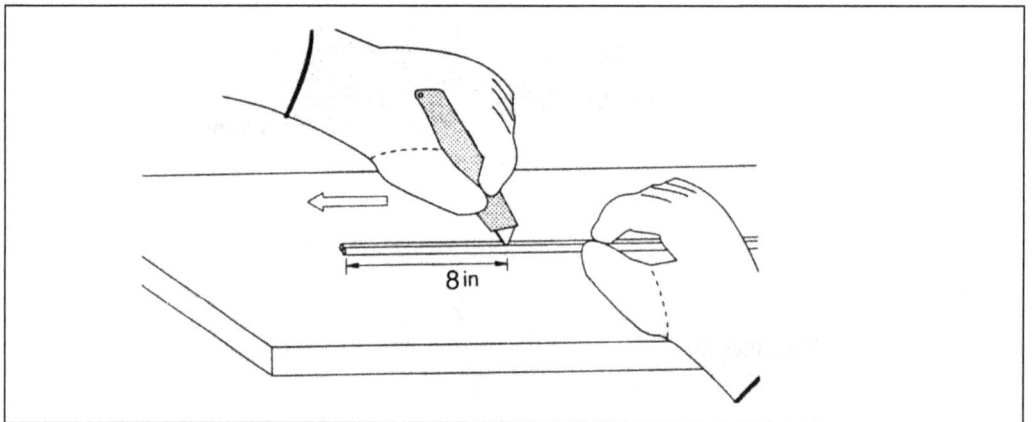

In the previous section, a wire stripper was used to remove the sheath. In this section, a coaxial cable stripper will be used.

- Before using the cutter, ensure it is properly adjusted. Use a small slotted screwdriver (Figure 4.16.) to adjust one of the blades on the side of the cutter so it sits against the lower jaw but does not force the jaw open.
- The next step is to ring-cut or make a cut around the cable. Begin by measuring the desired length of jacket removed.
- Slide the boots in the subunits.
- Open the stripper by squeezing the handles together and placing the stripper blade on the subunit at the desired strip point length. (See Figure 4.17.)
- As shown in Figure 4.17, hold the cable firmly and twist until the sheath is cut (usually two turns). Now you can slide the sheath off the cable subunit and remove the cutters. Follow the same instructions for the other subunit.
- Follow the procedures in the previous section doing the following:
  a. Slip the connector onto the subunit as previously described.
  b. Cut and trim yarn per specifications from manufacturer.
  c. Measure and mark the amount of buffer to be removed.
  d. Use the no No-Nik tool remover to remove the buffer. (See Figure 4.18.)
  e. Clean the bare fiber with a lint-free tissue soaked in alcohol.

The cable is now ready for connectorization.

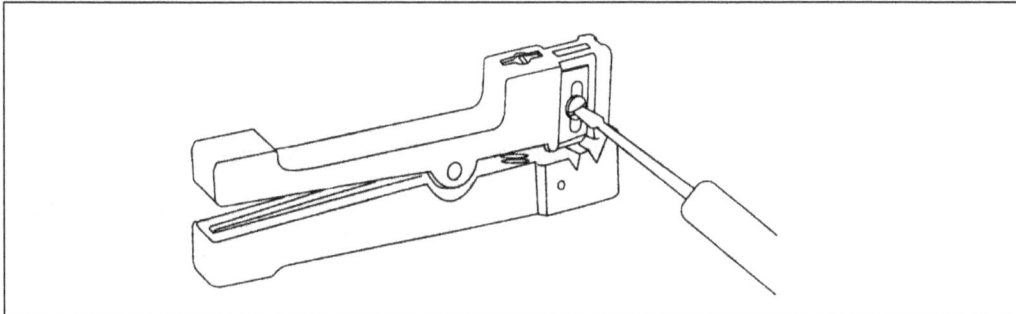

Figure 4.16

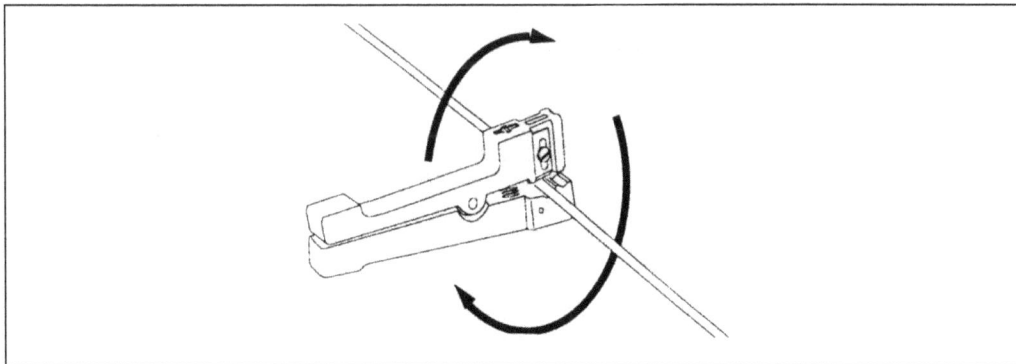

Figure 4.17

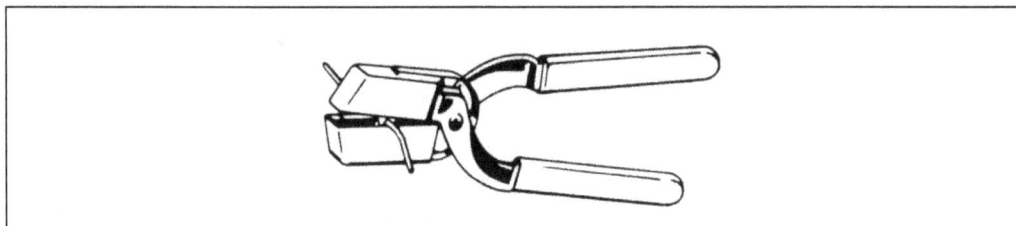

Figure 4.18

4.9

Rubber Heavy-Duty PVC - Armored

Before removing any sheath, check the manufacturer's recommended policy for the amount of sheath removal required. Some manufacturers suggest removing a minimum of 50 inches of sheath.

Armored Protected Cable

Begin by marking the length of sheath to be removed and place a wrap of electrical cable at the end mark. Today, electrical tape comes in a variety of colors other than black. (See Figure 4.19.)

- Using a hook knife, (Figure 4.20) make a ring cut through the outer sheath of the cable at the tape mark. Be careful and don't cut into the inner sheath. Always look at the end of the cable to determine how many layers of rubber, PVC, foil or armor make up the cable.

One of the tools replacing the hook knife is a cable slitter. A cable slitter is capable of 1) adjusting for ringing (cutting around the cable) the cable and 2) slitting the length of cable.

- Gently bend the cable at the slice to break the armor. Don't overdo it and violate the bend radius of the cable.

No Ripcord Type of Armored Cable

- Push the blade of either the hook or the slitter so it can travel down the cable between the armor and the inner sheath toward the end. (See Figure 4.21.)

Tape Marker

- Hold the slitter or knife out straight and begin pulling the cable on the upper lip of the knife. (See Figure 4.22.)
- Start at the end of the cable and carefully peel the outer jacket and armor away from the inner sheath. Continue peeling until you have reached the point where you made the ring cut.

Figure 4.19

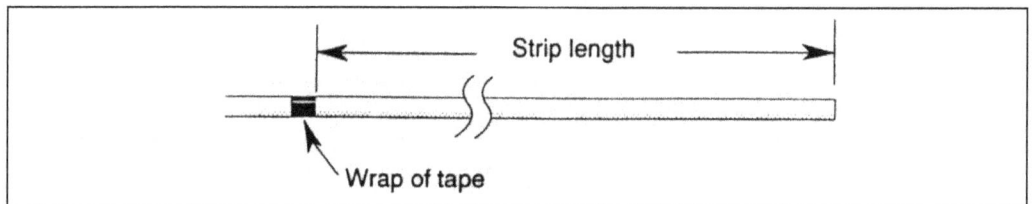

Strip length

Wrap of tape

Figure 4.20

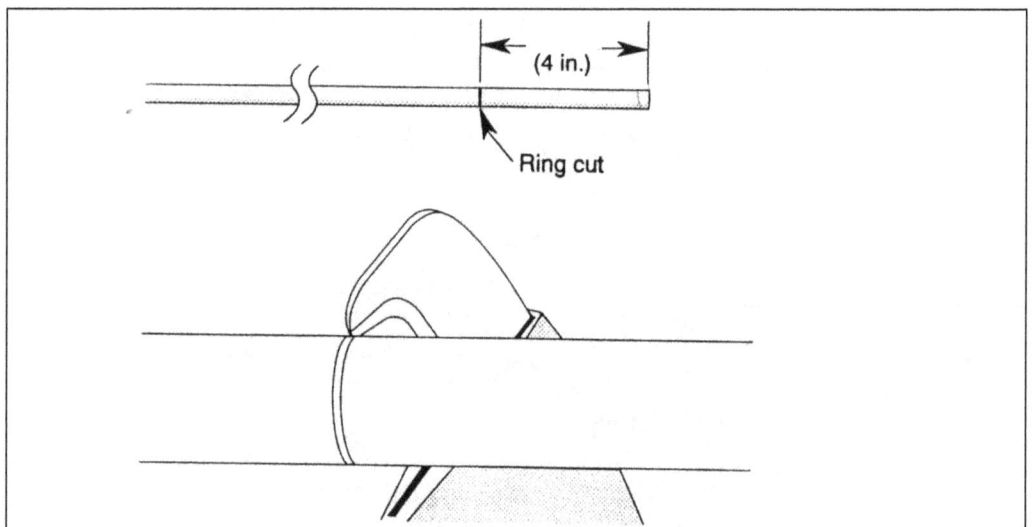

(4 in.)

Ring cut

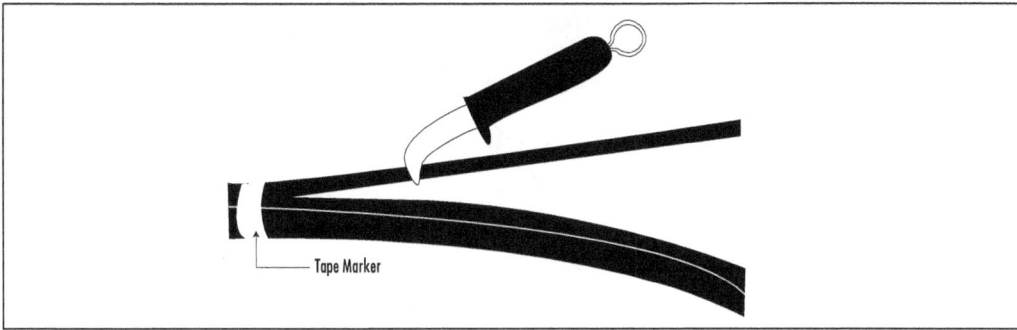

Figure 4.21

Tape Marker

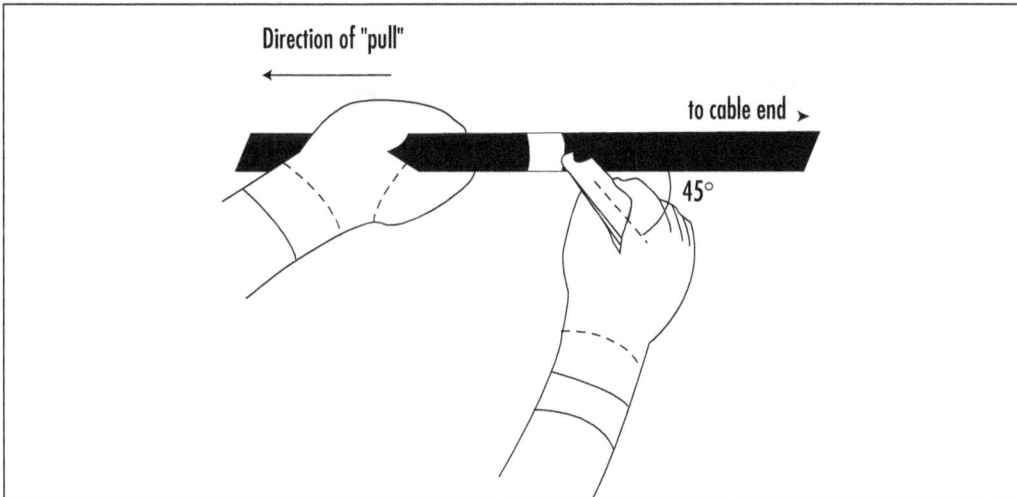

Figure 4.22

Direction of "pull"

to cable end

45°

- If your cable has a ripcord on it, as in Figure 4.23, begin the same procedures except only make a few inch notches to locate the ripcord. Grab and wrap the ripcord around a screwdriver (See Figure 4.24.) and pull the ripcord through the sheath until you reach the tape.
- Cut and trim the ripcord and begin pulling the sheath and armor away from the inner sheath. Cut and trim excess materials. If you have additional armored sheath, follow the previous procedures.

Non Armored with Ripcord

- Remember to check with the hardware manufacturer to determine the amount of sheath that is recommend to remove. Remove sheath without armor about four inches at a time. Using a slitter, adjust the depth of the blade being careful not to penetrate beyond the sheath and damage the buffer tubes. Ring the cable about four inches (Figure 4.25) from the end of the cable and slide the sheath off the end of the cables. (See Figure 4.26.)
- Locate and separate the ripcord (Figure 4.27) from yarns in the cable. Cut a notch in the cable sheath to start the ripcord.
- Pull the ripcord through the sheath until you reach the taped area. Begin peeling down the sheath from the core of the yard. (See Figure 4.28.)
- Trim off the splice section at the tape mark using scissors or cutters. (See Figure 4.29.)

Figure 4.23

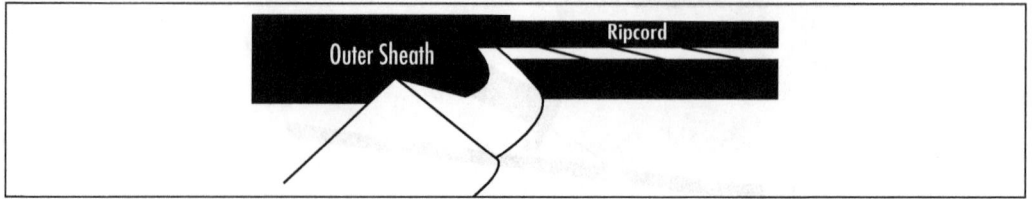

Figure 4.24

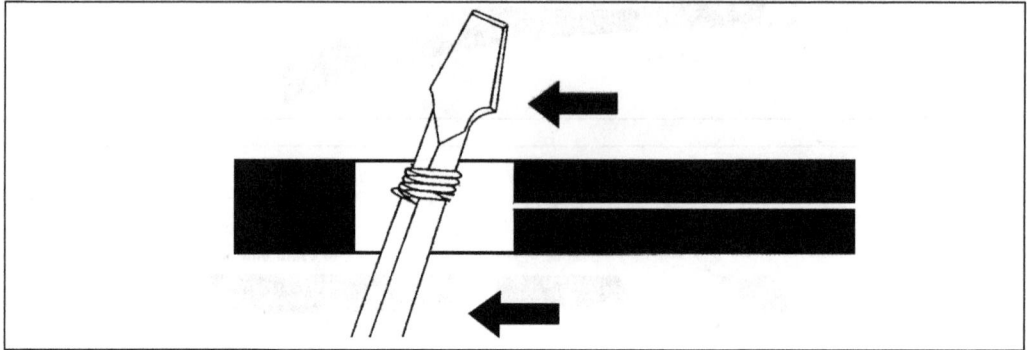

Figure 4.25

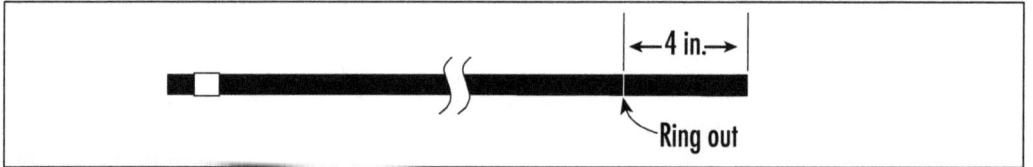

Figure 4.26

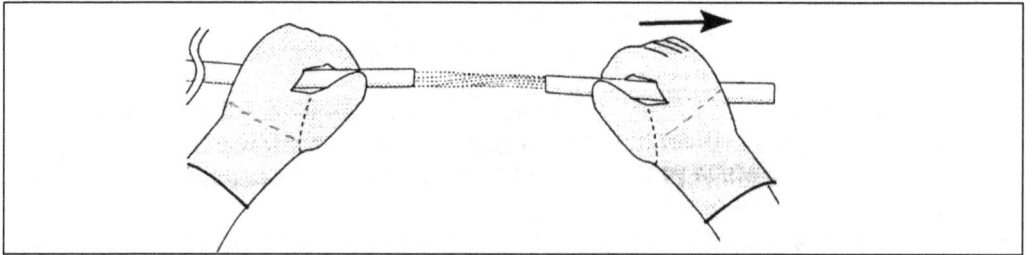

Figure 4.27

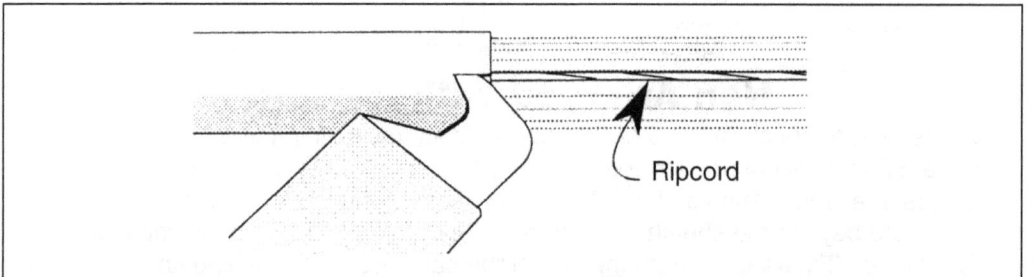

Figure 4.28

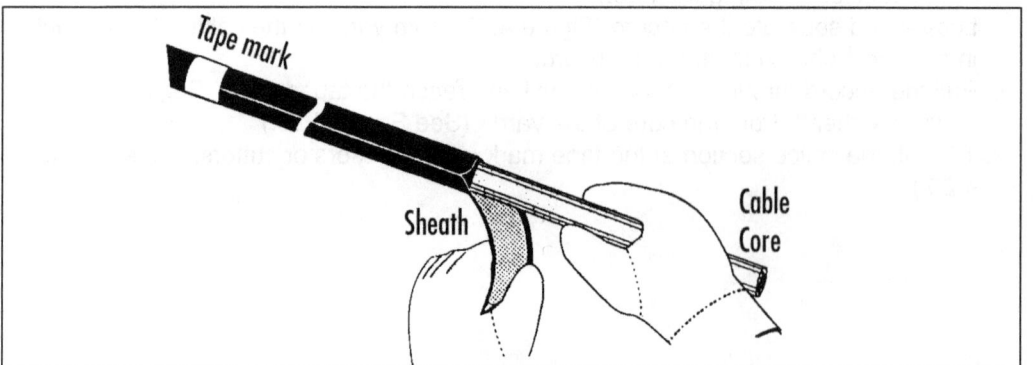

Cable Preparation

Once the sheath and armor have been removed, prepare the fiber for connectorization of splicing. The buffer tubes will be wrapped with yarn, which first must be removed. Start by pulling the yarn away from the fiber tubes until about six inches is exposed. Carefully cut the yarn at the six-inch point away from the tape mark. (See Figure 4.30.)

- At the cut point, slide the yarn to the end of the cable. The yarn will bunch up but will slide to the end of the cable.
- Using a seam ripper, cut the binding tapes wherever they may overlap. (See Figure 4.31.)
- Using a cloth soaked in filling compound remover, push the birding tapes away from the tape mark toward the end of the cable.
- Clean and inspect the buffer tubes for damage.

Cleaning the Cable

Carefully unwind the buffer tubes from around the central member of the cable (Figure 4.32) and wipe the filling compound off the buffer tubes and yarn with a cloth socked in filling compound remover. For final cleaning, use a dry cloth.

Central Member Preparation

The fiber cable will contain either a dielectric or a steel central member inside the cable. Either of the two must be secured to the housing for holding the cable and/or grounding.

Dielectric Central Members

Use side cutters to cut the dielectric central members to about 6 inches or the amount specified by the manufacturer. (See Figure 4.33.)

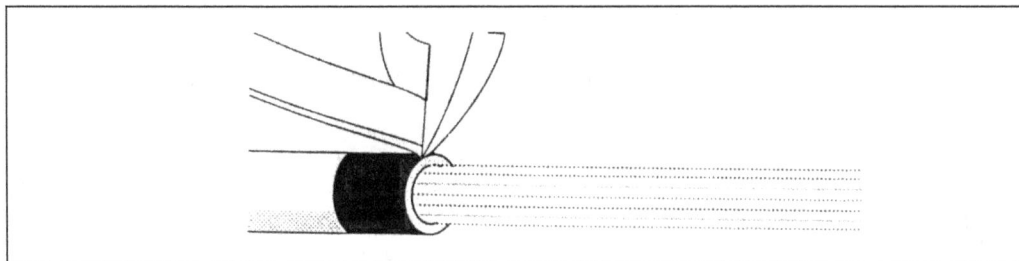

Figure 4.29

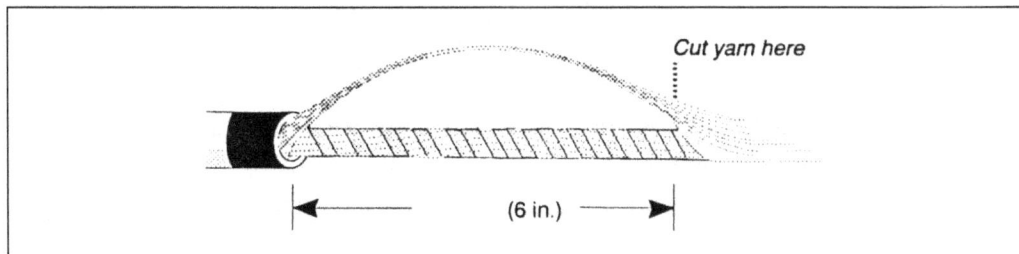

Cut yarn here

(6 in.)

Figure 4.30

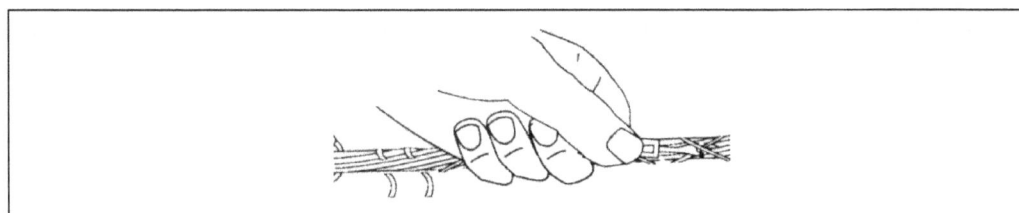

Figure 4.31

Figure 4.32

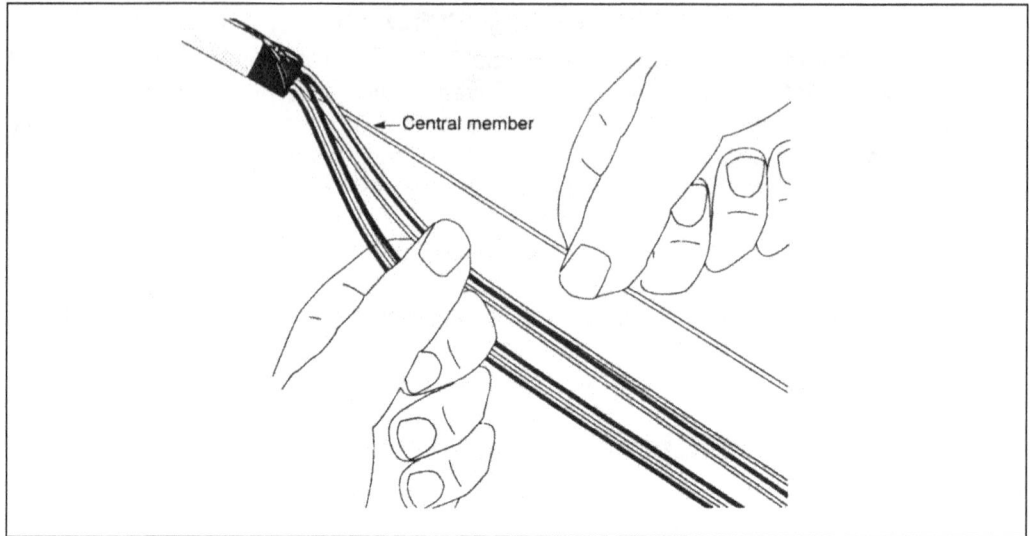

Central member

Figure 4.33

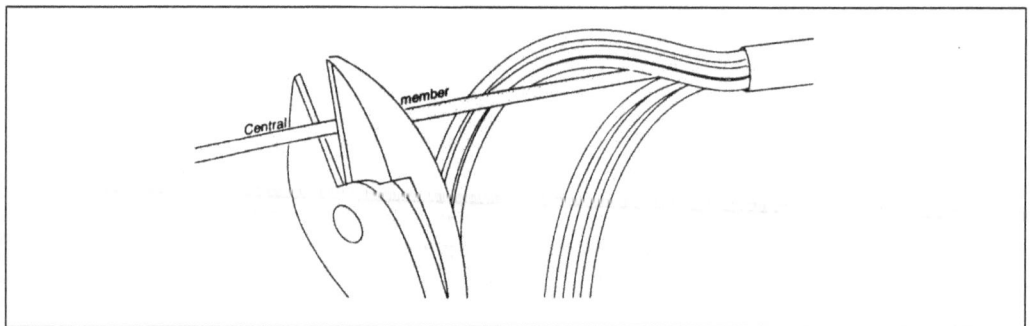

Central member

Steel Central Members

Beginning approximately two inches away from the tape, strip about four inches of coating from the steel central member. Cut the steel central member a total length of six inches (Figure 4.34) or the amount specified by the closure manufacturer.

Placing Cable in a Closure or Housing

Figure 4.35 shows how the manufacturer suggests the cable be laid, secured and distributed inside the cabinet. Starting with (A) the cable is brought into the housing and secured by the wraps (B). The manufacturer recommended 50" of stripped fiber, which is coiled all the way around the cabinet splicer case (C). The cable is secured to the strain refill bracket (D) with tie wraps. Next, an anchoring screen (E) shown in a close-up in Figure 4.36 secures the strength member and yarn.

Grounding the Strength Member

Fiber optic cable with a metallic central member can be grounded directly to a housing. (See Figure 4.37.) Both the aramid or fiberglass yarn and the central member are secured with a u-shaped washer and a bolt. Insert the metallic central member between the u-shaped washer and the flat washer. Wrap yarn between the u-shaped washer and the side of the housing. Tighten the nut. Be sure to ground the utility rack to a central ground bus.

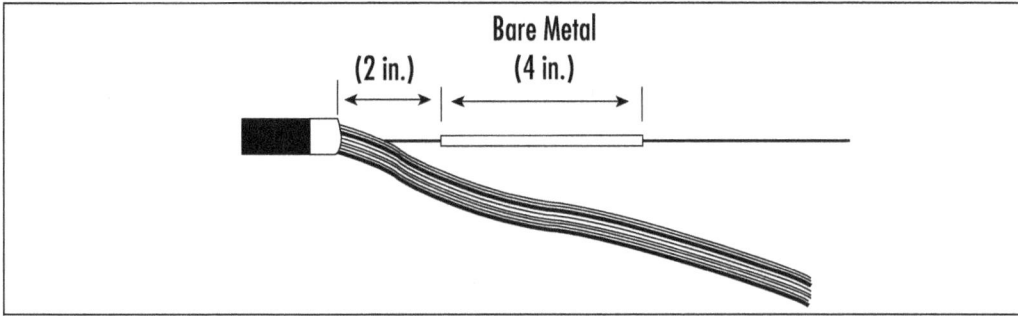

Figure 4.34

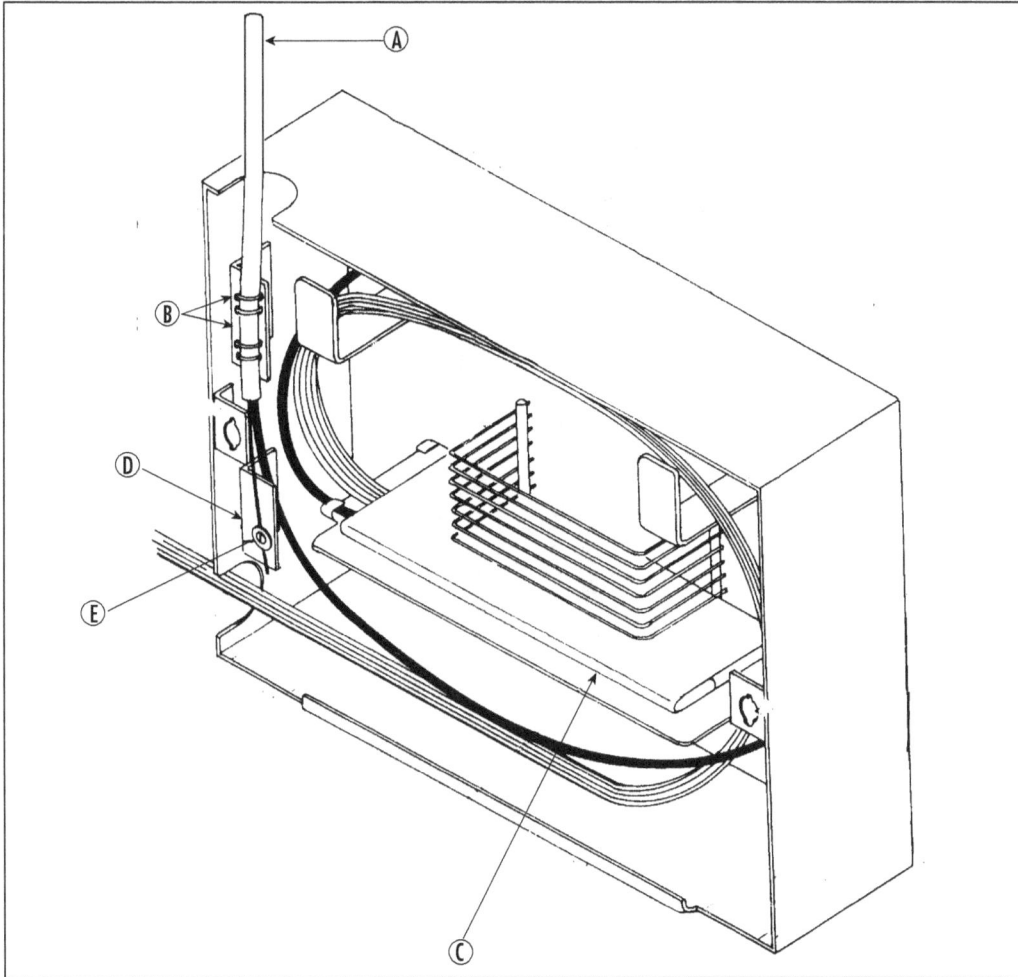

Figure 4.35

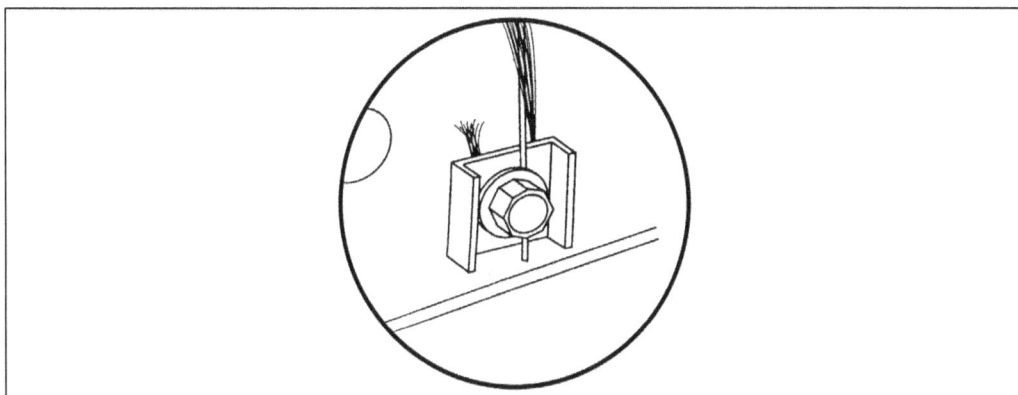

Figure 4.36

Figure 4.37

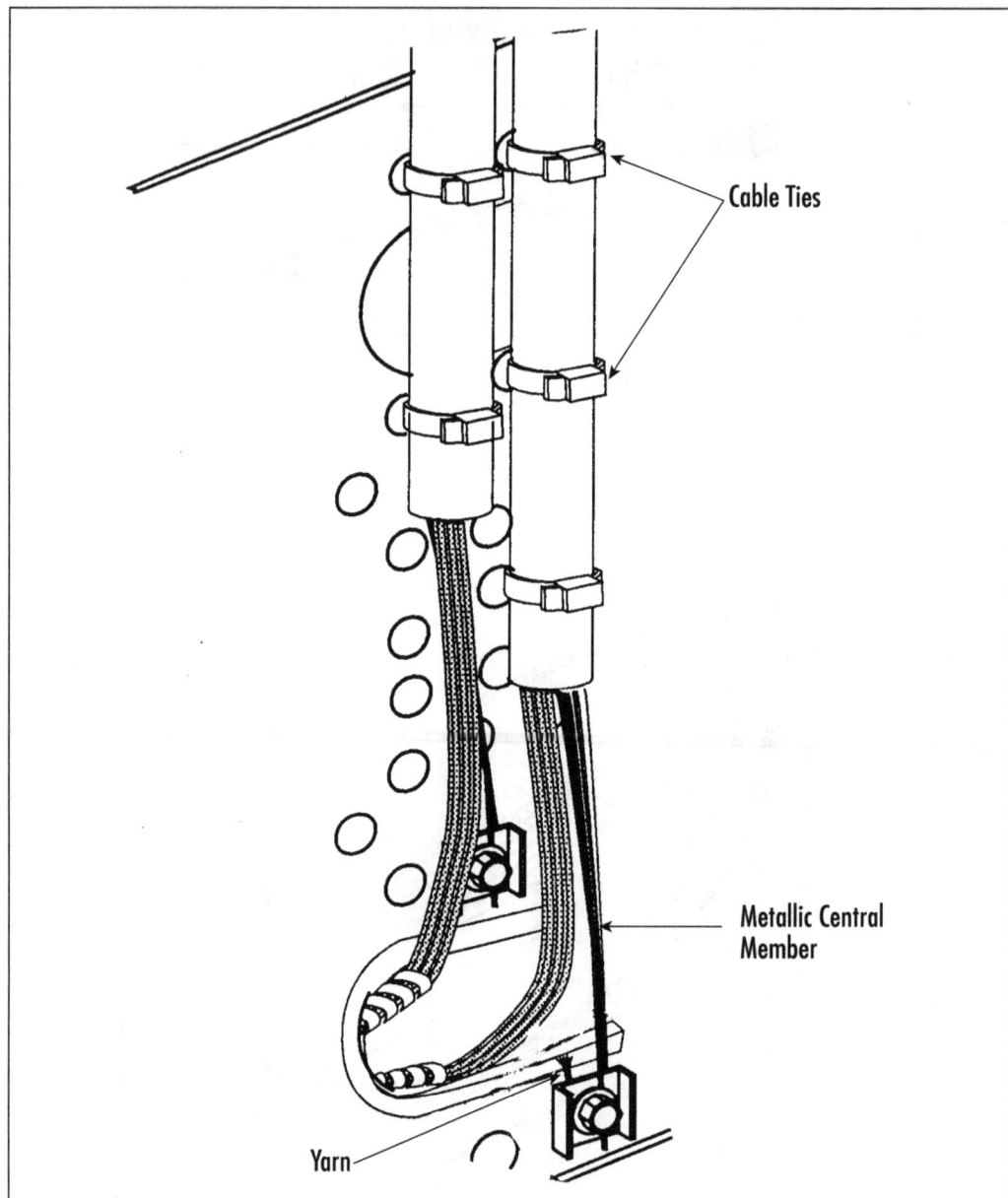

Cable Ties

Metallic Central
Member

Yarn

TEST MODULE 4

1. Sheath removal basically consists of removing the jacket (outer cover) of a fiber cable.

   True   False

2. Using a No-Nik tool, remove the fiber buffer in 1/2 to 5/16 inch increments at a time until you reach your mark.

   True   False

3. Always read the manufacturer's recommended policy for the amount of sheath removal required.

   True   False

4. After the armor and sheath are removed, prepare the fiber for connectorization of splicing.

   True   False

5. The fiber cable contains either a dielectric or a steel member inside the cable, which must be secured to the housing for holding the cable and/or grounding.

   True   False

6. Fiber optic cable with a fiberglass central member can be grounded directly to a housing.

   True   False

7. Loose-buffer refers to the way the optical fiber is held by its applied coating.

   True   False

8. A seam ripper can be used to cut the binding tape.

   True   False

9. Each connector manufacturer has different requirements for the amount of jacket, yarn and buffer to be removed.

   True   False

10. A hook knife is capable of adjusting for ringing (cutting around the cable) the cable and slitting the length of cable.

    True   False

LAB EXERCISE

Students participate by getting familiar with the tools used to place cables in cabinets, removing cables, stripping cables, cleaning cables and grounding the strength member.

Section 5.0
Connector Types

Objective

Understand and identify the many types and style of connectors, the process of alignment and insertion losses. Additionally, understand extrinsic loss and use a microscope to discover the final results in a finished connector.

Outline

- *Introduction*
- *Standard Connector Types and Attributes*
- *Physical Connectors*
- *Inline Splices*

Learning Activity

Assessment: Test Module 5
Lab Exercise: Examine a variety of connectors using a microscope.

Introduction

The purpose of the connector is to provide precision alignment of the fiber core, permanently securing the fiber to the connector and de-coupling the optical fiber for the purpose of patching and cross-connecting. When installing or specifying a connector/adapter system, the following factors are important:

- A connector is often rated by the loss or attenuation of optical signal power, commonly referred to as insertion loss because of the testing method used to rate it. For example, a typical rating could be -0.3 dB or 0.25 dB.
- The cost of connectors is always a consideration. Higher priced connectors are usually easier to install and of good quality, while lower priced connectors may require special tooling or powered assembly equipment and might be of lower quality. A careful cost comparison is prudent. Higher priced units may save labor time.
- Consistency of loss in a connector should be a prime consideration. Every connector installed should have a low dB loss. Variation in loss is a serious impairment and can degrade performance.
- A connector should have the same attenuation loss, regardless of the number of times it is removed and reinserted. This is called the repeatability of a connector. It is usually not specified by the manufacturer of the connector as much as it is with a specific design of connectors being used.
- Today's high-quality manufacturing standards (quality assurance methods) ensure that even the lower priced connectors can deliver the low dB losses that are required. In fact, we should be able to achieve dB loss levels considerably lower than the manufacturer's specifications.
- Properly manufactured and installed fiber optic connectors will minimize losses directly related to fiber interconnection called extrinsic losses. These are losses that can be controlled. Standardization of fiber sizes and tolerances limit these losses. Some connector styles and inferior mating devices as well as human error, contribute to extrinsic losses.
- End surface finish is one of the major causes of loss in a fiber optic connector. Improper polishing methods or dirt and dust contamination of the polishing surfaces creates it. (See Figure 5.1.)

Standard Connector Types and Attributes

Almost every manufacturer of fiber optic connectors in the 1980s seemed to have their own designs, and some of these designs are still in production. Most of the industry has shifted to standardized connector types, with details specified by standards organizations such as the Telecommunications Industry Association, the International Electrotechnical Commission and the Electronic Industries Alliance. Standards groups have developed standards for some two dozen connector types. Most are widely used types and some new types are being developed for emerging needs.

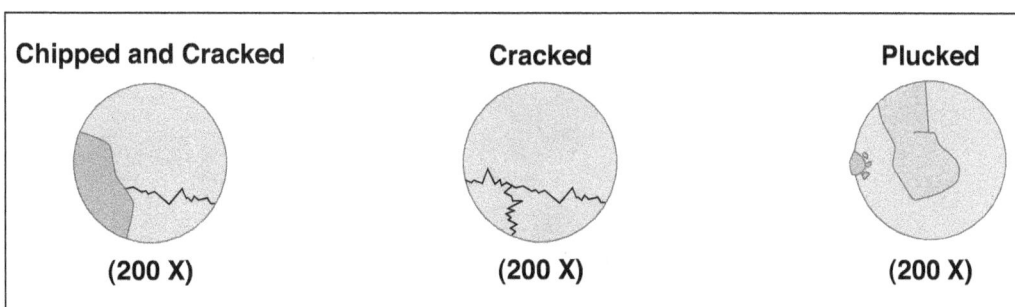

Figure 5.1

Chipped and Cracked	Cracked	Plucked
(200 X)	(200 X)	(200 X)

It is not feasible to cover all varieties of connectors in detail; that's best done by consulting catalogs and product specifications. Some important types used for single and multimode glass fibers are highlighted and other types of connectors may be used for plastic fibers and large-core fibers. These can be divided into single fiber connectors that snap or twist in place, special-purpose connectors, polarizing connectors and multifiber connectors.

There are three commonly used connector form factors: ST, SC and FC (Figure 5.2). The ST is used in LANs primarily with multimode fiber. It is a BNC type "push 'n twist" connection. The SC is used in LANs with singlemode and multimode fiber with a dual strand push-in plug. The dual SC is popular because it assures correct polarity between the Tx and Rx ports. The FC is used in telecommunications/CATV applications with singlemode fiber and has a screw-on termination.

There have been developments with the connectors using factory polished ferrules and crimp-style terminations. The classic styles use epoxy or hot-melt glue to secure the fiber to the connector. Pre-polished connectors significantly reduce installation time but can cost several times as much as the epoxy styles.

Physical Connectors

Mechanical connections are referred to as PC (physical contact) or APC (angled physical contact). (See Figure 5.3.)

Physical contact connectors derive their name from the fact that the faces of the connectors touch each other when mated, although there is always a small air gap involved. This air gap has a much different index of refraction than glass and causes a great amount of reflection. Using index-matching gel in the connector can reduce this by sealing the air gap.

Another alternative is to use APC connectors that are angled so that any reflected light is not sent back "upstream" causing interference, but instead, reflected out of the connector. The result is that there is only an attenuation component of the connection, not reflection.

Figure 5.2

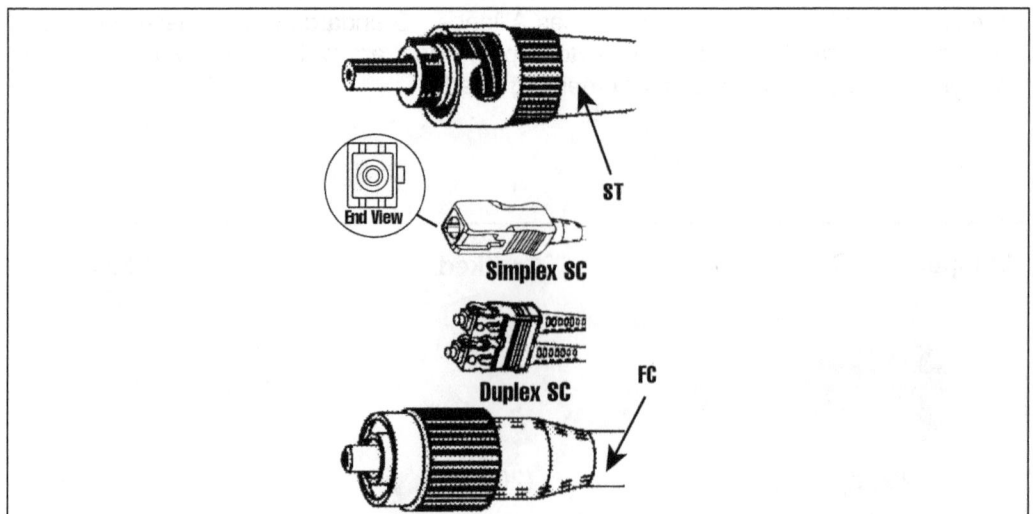

Inline Splices/Connectors

There are two types of splices with fiber optic cable. The first is a mechanical splice. These splices are used for quick restoration of severed lines. The second is a fusion splice that is used during installation for long-term connections where attenuation and longevity is critical. (See Figure 5.4.)

Splices, both mechanical and fusion, are used during initial installation to permanently join sections of fiber. Splices exhibit very little attenuation and usually no detectable reflection. The mechanical splice uses a tube filled with index-matching gel into which the cleaved fiber ends are inserted. The splice is compressed to physically hold the ends of the fibers into position.

Fusion splices, on the other hand, use heat to melt the two strands of glass together into one piece. This is quite durable and adds almost no attenuation. Typically, values of .10 to .00 dB can be achieved.

In the next section, we will look at the practical aspects involved in making a variety of connectors. You will see how important it is to the overall performance of the network that proper techniques are followed.

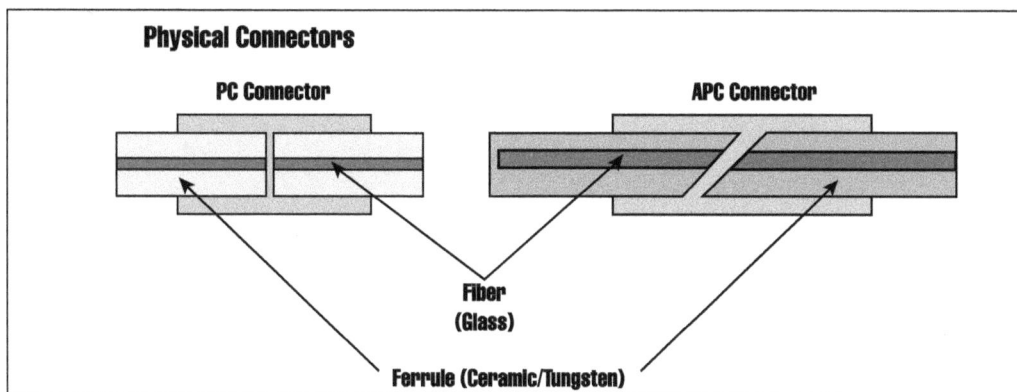

Figure 5.3

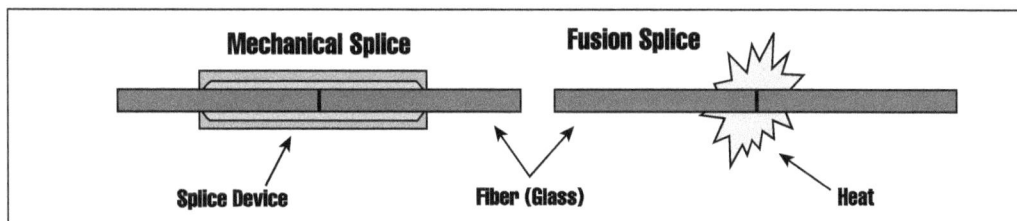

Figure 5.4

TEST MODULE 5

1. What is the purpose of a connector?

2. Every connector installed should have a high dB loss.

   True   False

3. _____ is a common contributor to extrinsic losses.

4. List the three most common connector form factors:

5. The FC is used in telecommunications/CATV applications with multimode fiber.

   True   False

6. Identify one way reflection is reduced when using a physical contact connector.

7. Splices exhibit very little attenuation.

   True   False

LAB EXERCISE

Examine a variety of connectors using a microscope.

Section 6.0
Connectorization

Objective

Demonstrate how to make a variety of connectors, including methods and applications. Adhesives, polishing and required tools will be discussed.

Outline

- *Connectorization Methods*
- *Adhesive and Polishing Connectorization*
- *Epoxyless Connectorization*

Learning Activity

Assessment: *Test Module 6*

Lab Exercise: *Demonstrate the use of required tools to strip cable, prepare a cable for termination, including epoxy and epoxyless methods, as well as inspect the connector for proper cleaving, cleaning and alignment.*

Connectorization Methods

Before you begin, put on safety glasses. In the previous section, the construction and types of connectors were discussed. In this section, the two methods of installing connectors on fiber will be shown.

 a. Fiber optics adhesives - hot and cold
 b. Crimping or epoxyless

Adhesives are used quite extensively in field and in-plant terminations. The proper handling of adhesives is extremely important to ensure that terminations provide the best and most consistent results. By following the proper procedures, every technician can achieve the same results.

The correct adhesive must be chosen, although each type of adhesive has advantages. Epoxies are the most popular and will produce the highest quality terminations. There are three types: heat cure, room temperature cure and fast gelling. The heat-cured adhesives must reach temperatures above 100°C to cure. These systems produce the highest glass transition temperatures and are usually the most rigid of all adhesives. The room temperature curing systems impart less stress and do not require an oven, although moderate temperatures can be used to accelerate the curing process. Fast gelling systems can be processed quickly at room temperature but are usually slightly softer than the other types of epoxies. The fast gelling systems are usually used for field terminations, whereas the other adhesives are used for both in-plant and field requirements.

Other systems also used for field terminations are:
 • anaerobic adhesives, which also offer fast cure times and are relatively rigid and easy to polish
 • hot melts, which process quickly with an oven
 • ultraviolet cured adhesives, which cure quickly under UV light but require a special connector with a glass insert in the ferrule
 • cyanoacrylates, which cure quickly but can be very brittle after curing

Correct processing is essential with whatever system is chosen. With epoxies, it is extremely important to make sure the mix ratio of resin and hardener is correct and both resin and hardener are thoroughly mixed. Inaccurate ratios or insufficient mixing can lead to inconsistent results. For these reasons, it is recommended that epoxies be purchased in packages containing the correct ratio as defined by the manufacturer.

If exposing the epoxy to air is a concern, most room temperature and heat-cured epoxies can have the air removed by either centrifuge or vacuum. Centrifuging is the best process for most fiber optic epoxies. After loading a syringe, either by drawing material through the needle or pouring material into the back end of the syringe, place the capped syringe, needle end up, in the centrifuge at 3.400 rpm for three minutes. In this position, the air is forced to the needle end and can be easily pushed out before applying the adhesive. Only non-filled epoxies that use compatible hardeners and resins can be used in a centrifuge (check with the epoxy manufacturer).

A vacuum chamber should be used to remove the air from filled or viscous epoxies. Pour the mixed epoxy into a vessel in as thin a section as possible, thus creating a large surface area. Place the vessel in the chamber and pull a vacuum of 28 to 30 inches of mercury. A foam head will usually rise and break. Leave the adhesives under vacuum for two minutes after the head breaks and then carefully fill the syringe so as not to reintroduce air. Epoxies should not be left under vacuum for extended periods of time as ingredients may be pulled out that could damage the properties of the finished adhesive.

Room temperature or heat-cured epoxy should be injected into the connector until a small head appears on the outside of the ferrule. This small head should be wiped off before the cleaned and stripped fiber is inserted into the connector. Only the epoxy pulled through with the fiber should be left on the outside of the ferrule. This procedure reduces the stress on the fiber and lessens the risk of fiber cracking, especially with multimode fiber and heat-cured epoxy. Cure the epoxy according to the manufacturer's recommended curing schedule, then cleave and polish.

The fast gelling epoxies are handled in the same manner as the other types. However, transferring to the syringe, injecting the material, and inserting the fiber must be done before the epoxy gels. Some installers will apply material directly from the epoxy package, thus avoiding the transfer to the syringe and saving several minutes of the epoxy's working life.

Application of anaerobic adhesives is easy. These systems are supplied in two parts, the adhesive and the accelerator/activator. The adhesive is injected into the connector and the fiber is dipped or coated with the accelerator. When the fiber is injected into the connector, the accelerator will start the reaction. After one or two minutes, these systems are usually ready for cleaving and polishing. It is extremely important to minimize the amount of adhesive behind the ferrule since excessive amounts will cure slower than the material between the fiber and the ferrule. In some cases, you may want to place a drop of accelerator on the outside of the ferrule to ensure a complete cure.

Hot melt adhesives are supplied preloaded in the connector. The only required step is that the connector be heated to melt the adhesive. The fiber is then inserted and the unit cooled. Care must be taken to protect both the operator and the cable from the high temperatures. These connectors should not be placed in environments that may cycle to higher temperatures as the adhesive may soften causing fiber movement.

UV (ultraviolet) cure connectors require that the adhesive is injected into the connector and the fiber is inserted. The UV light source is positioned in front of the ferrule so the light can reflect through the glass insert. UV adhesives will usually cure in 45 to 60 seconds with most handheld light sources. With usage, the efficiency of the light source may diminish and more time may be needed for curing. The installer must be careful not to shine the light on skin or in the eyes as serious injury could occur.

Cyanoacrylates (CA) are the most finicky of all adhesives. CA is injected into the connector and the fiber is inserted so it protrudes through the ferrule. The protruding fiber is sprayed with an accelerator and the fiber is withdrawn back into the ferrule. The cure is almost instantaneous and therefore the installer has only one opportunity to run the process. If done incorrectly, the connector must be cut off and a new connector installed.

Safety is a primary concern when working with any type of adhesive. It is important not to expose skin, eyes or mucous membranes to any chemical, including adhesives. For this reason, gloves, eyewear and other protective clothing should always be worn when using adhesives. To prevent dermatitis or other reactions, skin exposed to direct contact should be washed immediately with mild soap. Eyes and mucous membranes should be thoroughly rinsed with water. Review the safety data sheet supplied by the adhesive manufacturer for specific information regarding the adhesive to be used.

By following the correct processing techniques with fiber optic adhesives, terminations can be easily performed and yield consistent and dependable results. Each adhesive requires the proper processing steps to ensure that the connector does not fail in the field.

Part I - Connectorization - Adhesive/Polishing Type

ST Type Connector/Adhesive and Polishing

This first connector type requires adhesive and polishing. Keep in mind, there are many versions of this type of connector, and nomenclature changes from connector to connector. It is impossible to list all variations of connectors, however, this general overview will provide a basis or feeling for what is required to install a connector. Always carefully read the manufacturer's installation practices before you begin to install the connector. Once you determine the type of connector to use, buy the tools and types of epoxy recommended by the manufacturer of the connector. We will begin by installing an ST multimode connector.

- Open the package. Lay out all the parts and carefully read the installation procedures. (See Figure 6.1.)
- Prepare the cable as described in the previous section starting with measuring the cable, pulling on the boot, removing the buffer and cleaning. (See Figure 6.2 and Figure 6.3.)
- Next, have UV adhesive already prepared in a syringe. Remove the end cap from the UV adhesive syringe and place a pink needle-nose tip on the syringe and load the connectors. (See Figure 6.4.)
- Hold the connector assembly straight up (Figure 6.5) and insert the syringe tip into the rear until it bottoms on the ferrule.

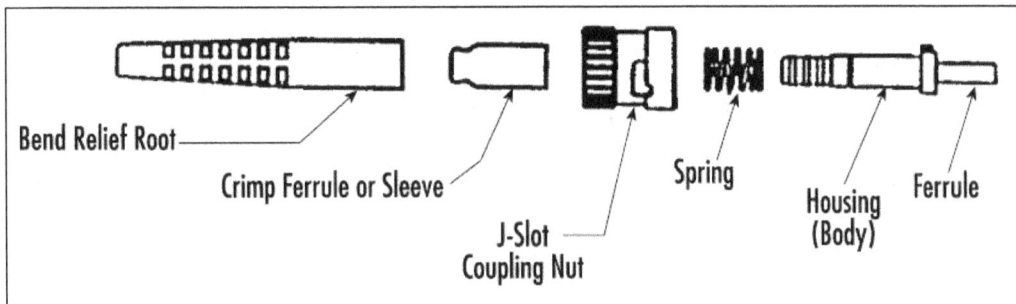

Figure 6.1

Bend Relief Root

Crimp Ferrule or Sleeve

J-Slot
Coupling Nut

Spring

Housing
(Body)

Ferrule

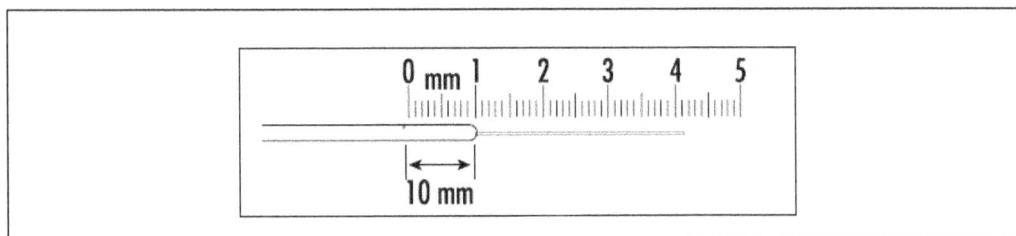

Figure 6.2

0 mm 1   2   3   4   5

10 mm

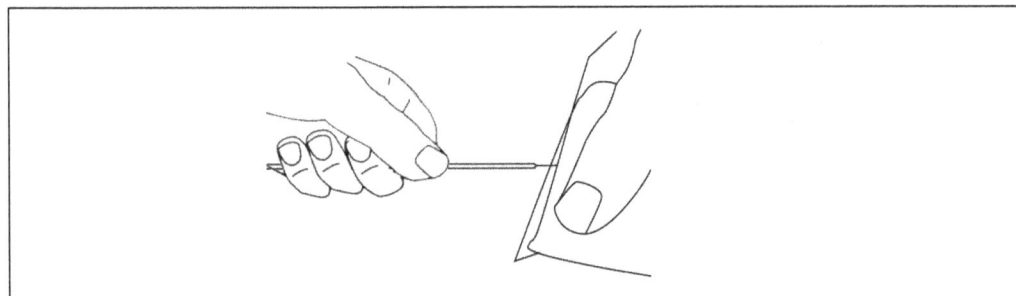

Figure 6.3

6.5

- With the syringe pointed up, hold on to the connector and slowly inject the UV adhesive. When adhesive begins to come out of the tip of the ferrule, let go of the connector.
- Continue to inject the adhesive, allowing the connector to rise on the syringe tip and then remove the syringe tip leaving a tiny bead of UV adhesive. Remove the connector assembly from the syringe tip and wipe off the epoxy bead on the end face with a lint-free tissue. Apply a small amount of adhesive on the Kevlar. Carefully, feeling for the hole in the connector tip, insert the prepared fiber into the connector. (See Figure 6.6.)
- The epoxy will coat the fiber as it enters the tip. When you see the fiber protrude from the connector tip, pull it back gently and watch for movement of the fiber at the tip, which will verify that you have not broken the fiber.

Figure 6.4

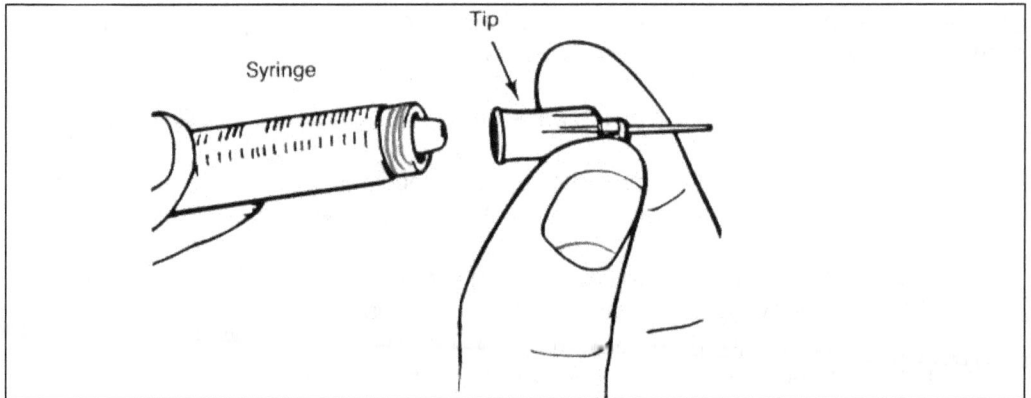

Figure 6.5

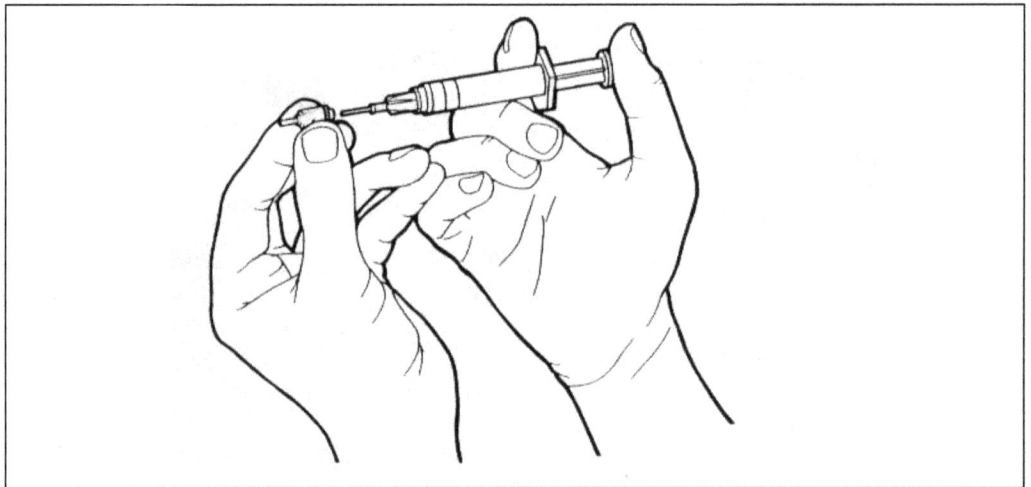

Figure 6.6

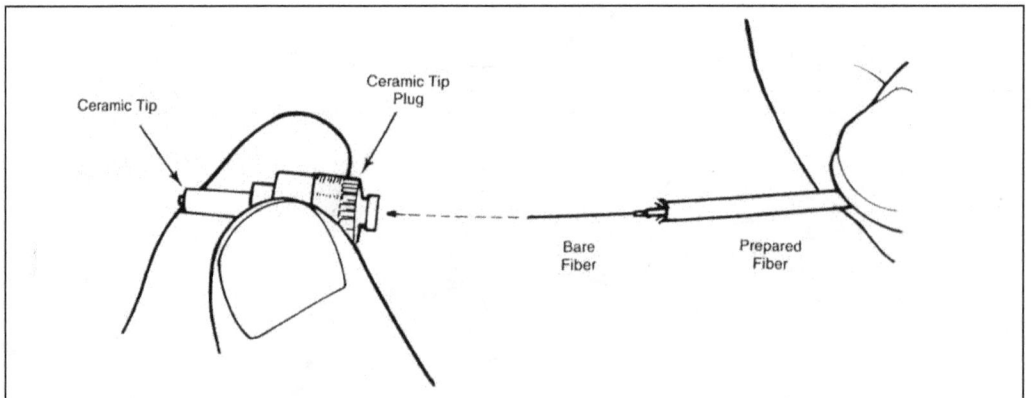

- At this point, depending on which manufacturer's connector is being used, you may be required to crimp the connector to the sleeve or ferrule. (See Figure 6.7.)
- When the connector is properly seated, approximately 3/16" of bare fiber should emerge from the tip of the ferrule. (See Figure 6.8.)
- You may cure one or several connectors at the same time. Pull out the UV light tray and place the connector(s) into the tray. Push the tray forward and leave the connectors in the tray for about 45 seconds.

Depending on the machine, type of connector, type of adhesive and working environment, the length of time for this process varies.

- Once the connectors from the UV machine have been removed, using a scribe tool, score (slice sideways against the base fiber) the fiber as close to the top of the ferrule as possible. (See Figure 6.9 A and B.)

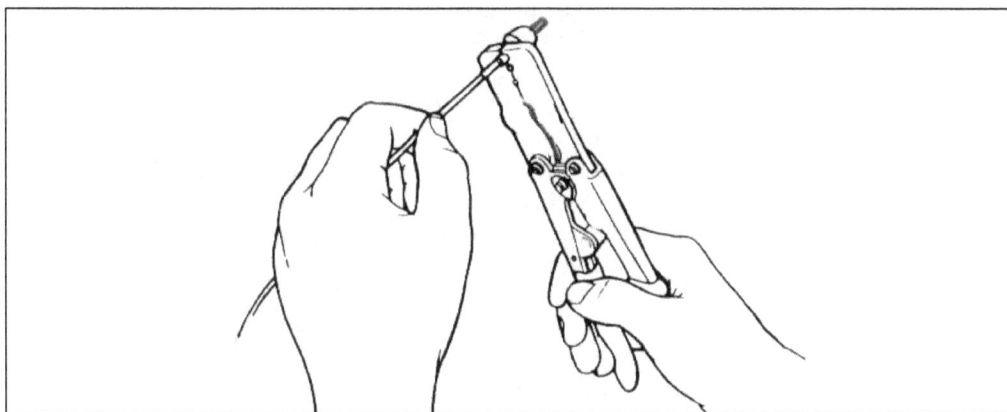

Figure 6.7

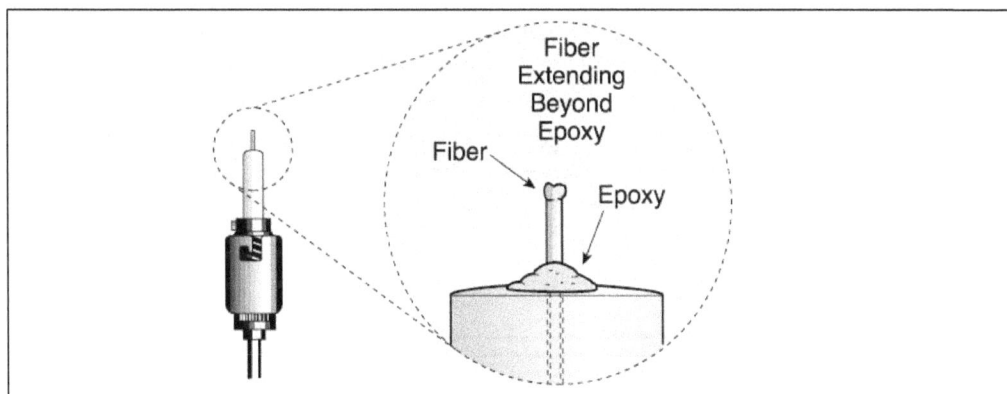

Figure 6.8

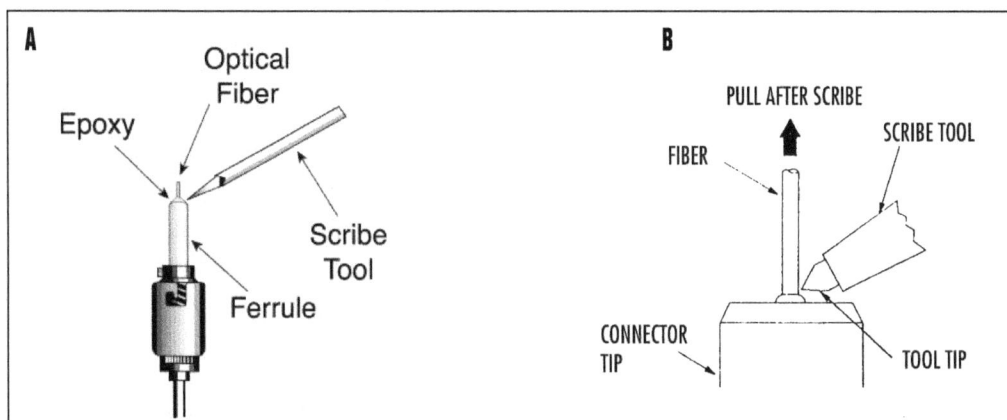

Figure 6.9

- Use tweezers to gently pull the fiber straight up to remove the excess fiber. (See Figure 6.10.)
- Dispose of the detached fiber on a loop of masking tape or a disposal bottle.

Polishing Process

There is a small amount of fiber protruding through the ferrule known as the stinger. Take a piece of 3.0 micron film and form it into a u-shape with the abrasive face inward.

Remove the stinger by making circles on the film. Stop when the fiber stinger no longer scratches the film. Check the condition of the ferrule end face. A small bead of adhesive should be present.

Once the fiber is polished down into the epoxy bead, it is relatively safe from shattering.

The following is the procedure using a single position puck.
- Using the manufacturer suggested polishing paper, lay the paper on the glass plate. The pad acts as a cushion under the fiber and helps prevent shattering the fiber until it is polished down flush with the ferrule. The example shows a sheet of five-micron lapping film (fine sandpaper). (See Figure 6.11.)
- Next, use canned air to clean the film, connector, polishing tool (also called a puck) and polishing platform. (See Figure 6.12.)
- Insert the connector into the puck. (See Figure 6.13.) If the connector does not fit, epoxy may be present on the shaft of the ferrule. Remove any epoxy on the shaft with a razor blade. (See Figure 6.14.)

If you do not remove the fiber burrs that extend through the fiber as previously described, the initial polish may remove it. If this is the case, place the polishing tool on the five-micron paper and push the tip through the tool and onto the paper. Without applying pressure to the plug, slide the fiber end over the film in a figure-eight motion. After doing this three or four times, the fiber should be within the epoxy. (See Figure 6.15.)

Figure 6.10

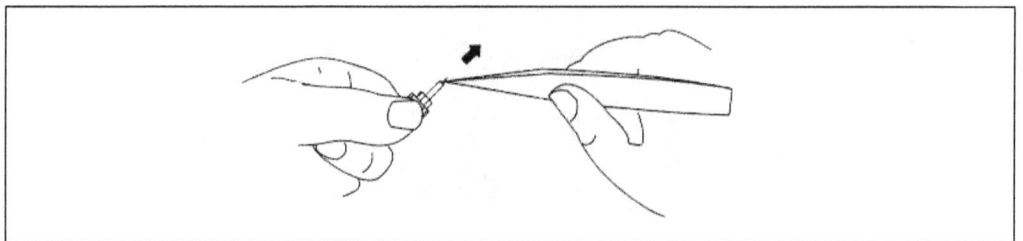

Figure 6.11

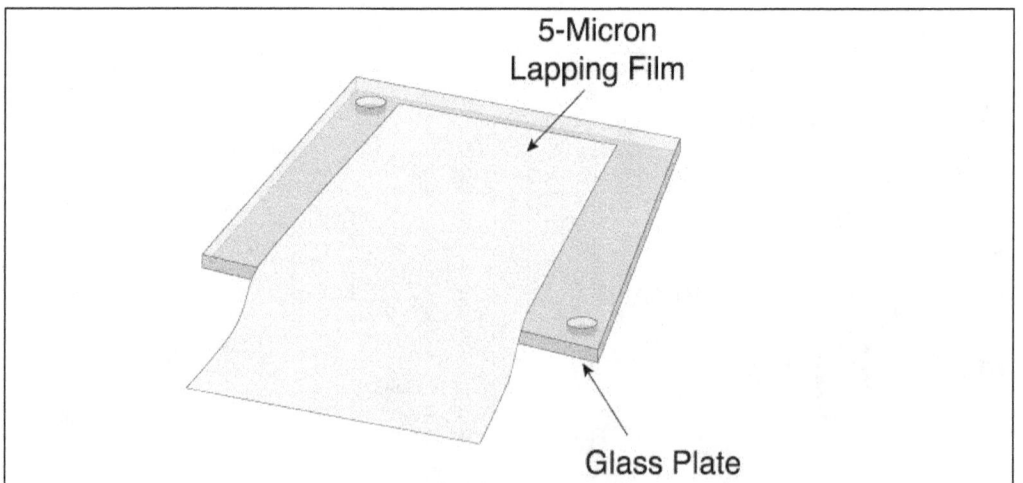

5-Micron
Lapping Film

Glass Plate

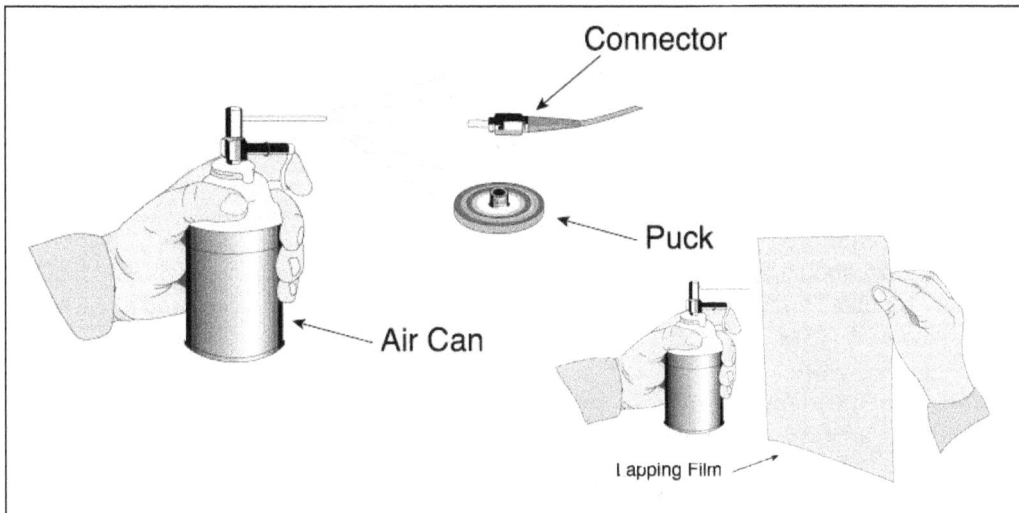

Figure 6.12

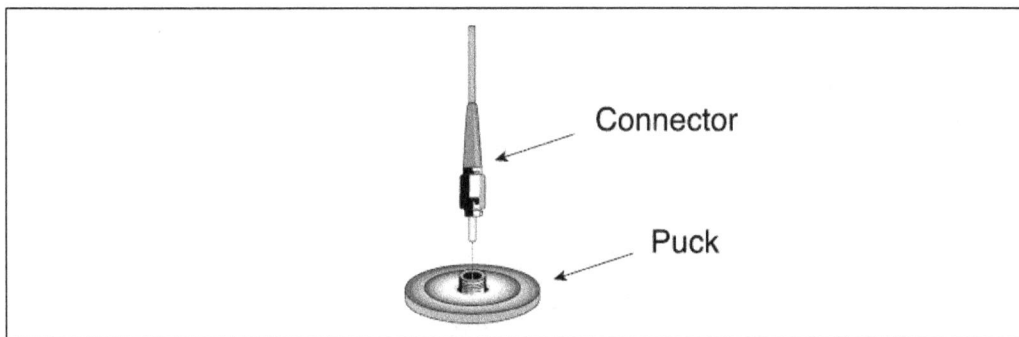

Figure 6.13

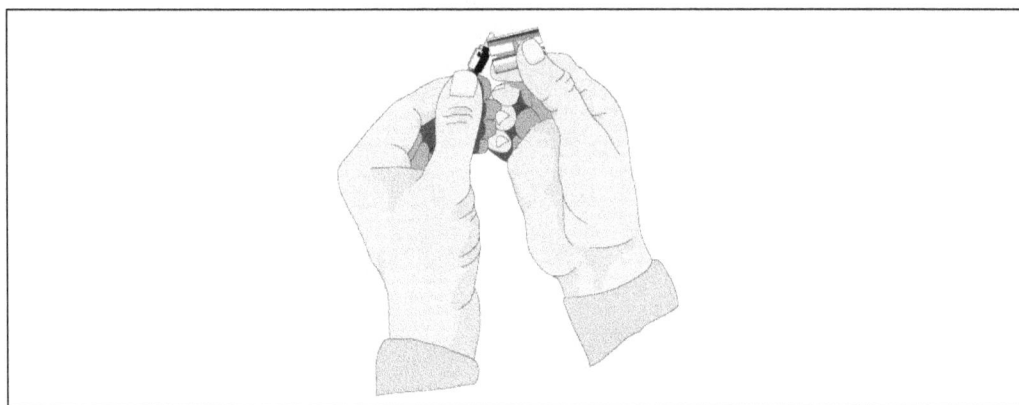

Figure 6.14

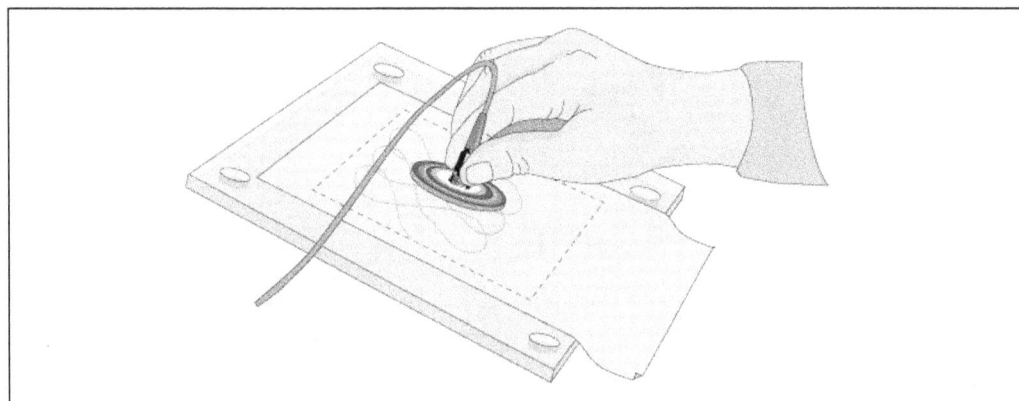

Figure 6.15

- Ensure you do not apply pressure to the fiber end while polishing. Continue to polish the plug, using slight pressure. The first strokes should leave fine, light lines on the film. The lines should get wider as you continue. Use the five-micron paper until a thin layer of epoxy remains.
- Remove the connector from the polishing tool and clean the tip of the connector with an alcohol wipe.
- Next, remove the connector from the puck and examine it with an eye loupe. (See Figure 6.16.)
- The epoxy should look thin but should not be completely removed. As the epoxy thins, an outer ring on the flat part of the ceramic tip will be exposed, and it will have a shine to it. (See Figure 6.17.)

Note: Repeat the same procedure for each connector to be polished. It is very important to clean all materials after each polishing. Clean the glass platform using a dampened alcohol wipe. (See Figure 6.18.)

Remove any dust using canned air. Then, replace the five-micron sheet of lapping film with three-micron film. Use canned air to clean both sides of the five-micron polishing paper. Put the connector back in the puck and repeat the figure-eight motions using medium pressure until the epoxy has been completely polished away.

- Examine the tip with the eye loupe to ensure the epoxy has been removed. If any epoxy remains, polish the connector again using three-micron paper until it is removed. Be careful not to over polish ceramic connectors because excess polishing with any lapping film coarser than one micron can cause pits. You can feel when the epoxy is gone as you are doing the figure-eight.

Figure 6.16

Figure 6.17

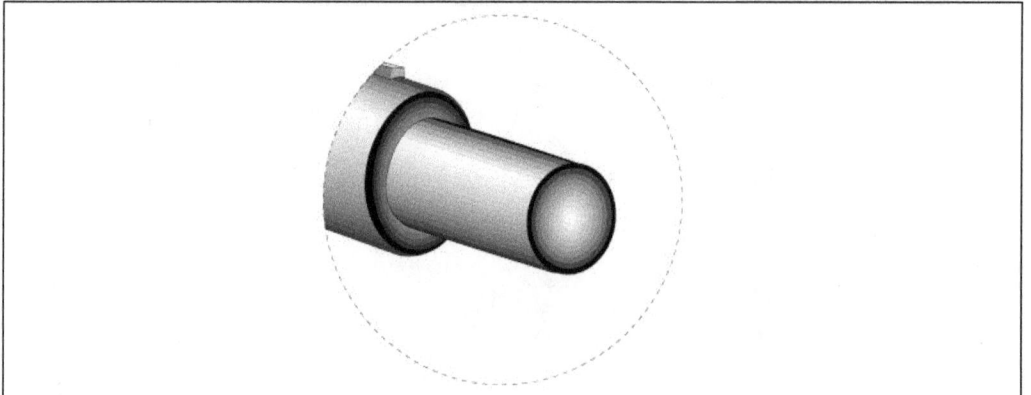

- Examine the tip with a microscope. This is accomplished by placing the tip into the microscope and opening the barrel to illuminate the tip end. (See Figure 6.19.)
- Visually inspect the fiber end. It should appear as shown in Figure 6.20. No cracks, holes or deep scratches should appear in the fiber end. If you cannot polish the fiber end to an acceptable condition, reterminate the fiber end and start again.

Part II - Connectorization - Epoxyless

ST Epoxyless Style Connector (no epoxy)
The epoxyless style connector is easier to install than the type that uses adhesives, both hot or cold, and has become popular in the industry.

Several manufacturers of the connector make use of such items as holding clamps, special molded gigs and type devices, called cams and crimping or clamping tools. Again, read the manufacturer's specification sheets for the type of special tools required. Several manufacturers have their own tool kits in addition to regular hand tools. A typical special tool kit would include items such as ones shown in Figure 6.21.
- Open the package and read the instructions, then prepare the cable to the dimensions set forth by the manufacturer. (See Figure 6.22.)
- Clean fiber with an alcohol soaked tissue.
- Insert the fiber into the connector while slowly rotating the connector back and forth. Slide the connector onto the cable until the connector butts up against the outer jacket.

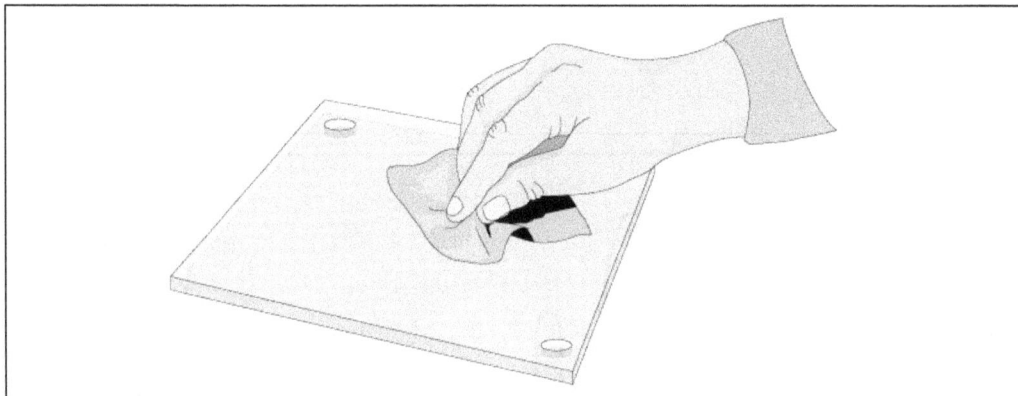

Figure 6.18

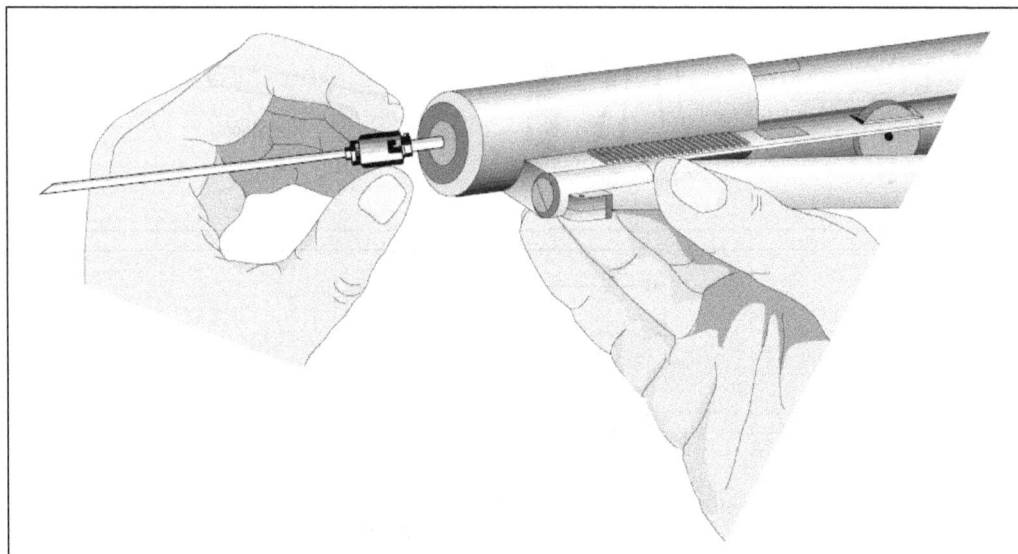

Figure 6.19

- Using the special crimp tool (Figure 6.23), crimp over Area A of the connector with cavity No. 1 of the hex crimp tool. Then crimp over Area B with cavity No. 4. (See Figure 6.24.)
- Slide the crimp sleeve over the Kevlar and connector backpost. Crimp the sleeve over Area C and Area D using hex No. 2 (for plenum 2.5 mm crimp sleeve, use hex

Figure 6.20

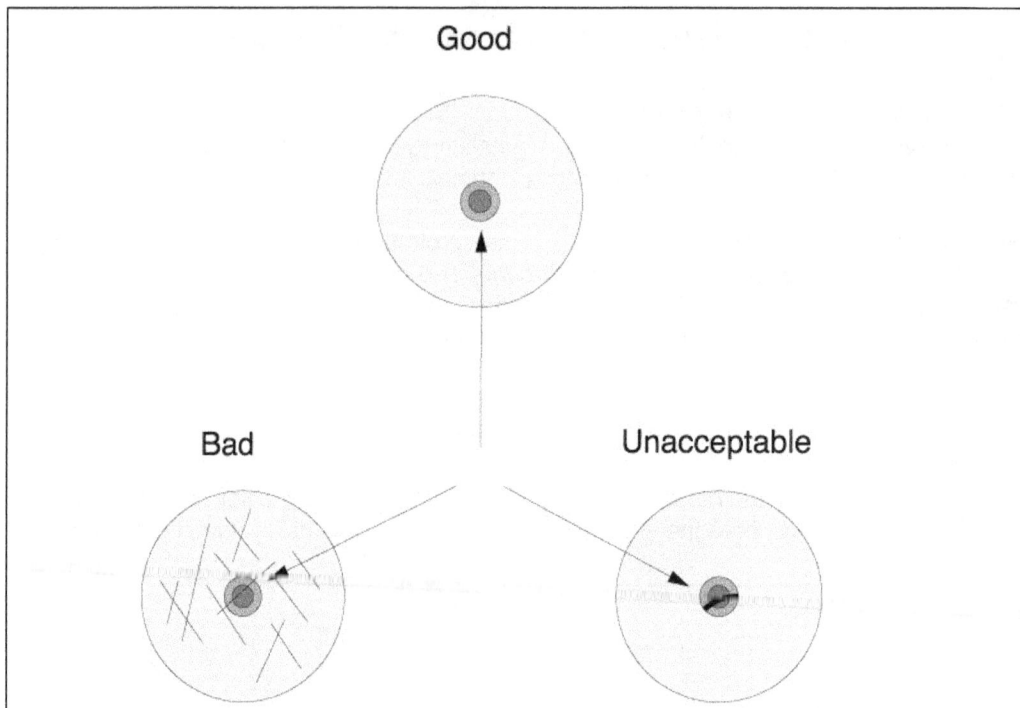

Figure 6.21

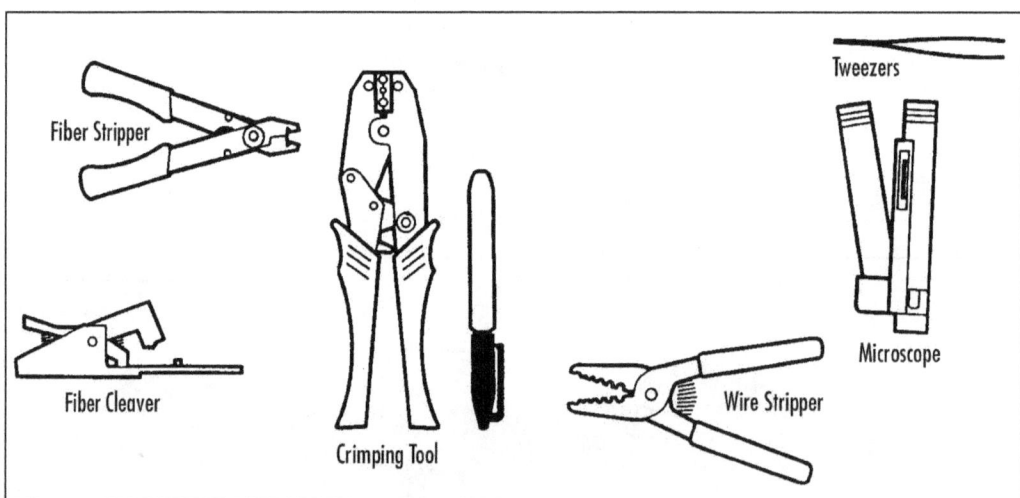

Figure 6.22

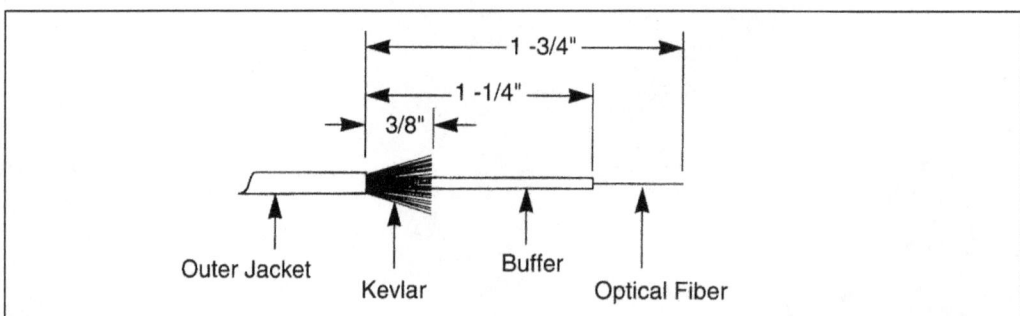

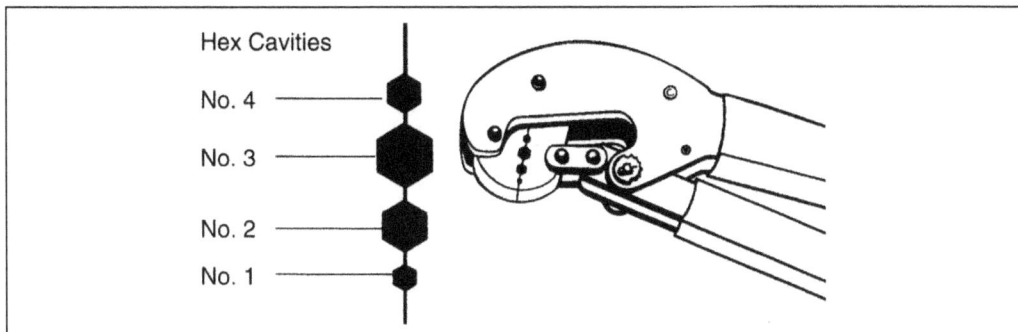

Figure 6.23

Hex Cavities

No. 4

No. 3

No. 2

No. 1

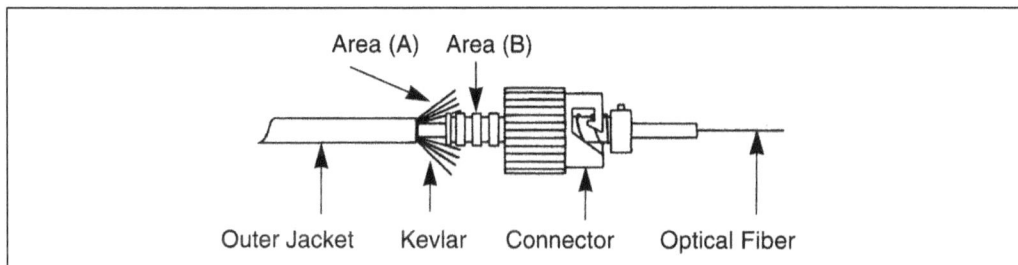

Figure 6.24

Area (A)   Area (B)

Outer Jacket   Kevlar   Connector   Optical Fiber

No. 4) and hex No. 3 respectively. (See Figure 6.25.)
- Slide the strain relief boot over the crimp sleeve and connector. (See Figure 6.26.)
- Using the sapphire scribe, lightly score the fiber at the point where the fiber comes out of the stainless steel ferrule tip. Gently pull up on the fiber until it separates. (See Figure 6.27.)
- Take a piece of 12 micron lapping film and lightly polish the exposed fiber until it is almost flush with the top surface of the stainless steel ferrule. Follow the polishing procedures explained previously and test as required.

Part III - Connectorization - Epoxyless/Crimping Style

Epoxyless Style Connector (no epoxy)
Called the LightCrimp® series, the product manufacturer is AMP Inc. It consists of sin-glemode and multimode SC and ST style connectors that do not require epoxy during the assembly of the connector. The connector kit consists of two cable boots, (either one may be required), a crimp sleeve, a plunger, a connector body, larger-diameter tubing, small tubing and a dust cap. (See Figure 6.28.)
Tooling required:
- Special crimp tool
- Cleaning tool
- Polishing puck
- Snips
- Cable cutters
- Fiber strippers
- Polishing film
- Polishing plate

- Open the connector kit package.   Carefully read the installation practices and ensure all the required pieces are in the package.
- As in previous examples, slide the boot on, mark the cable, strip off the specified amount of cable jacket, trim strength members, remove buffer from end of cable and clean fiber with soaked tissue. (See Figure 6.29.)

Figure 6.25

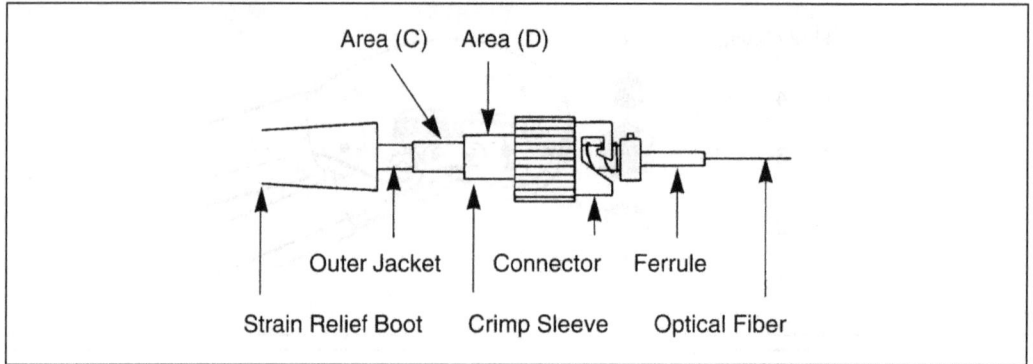

Area (C)   Area (D)

Outer Jacket   Connector   Ferrule

Strain Relief Boot   Crimp Sleeve   Optical Fiber

Figure 6.26

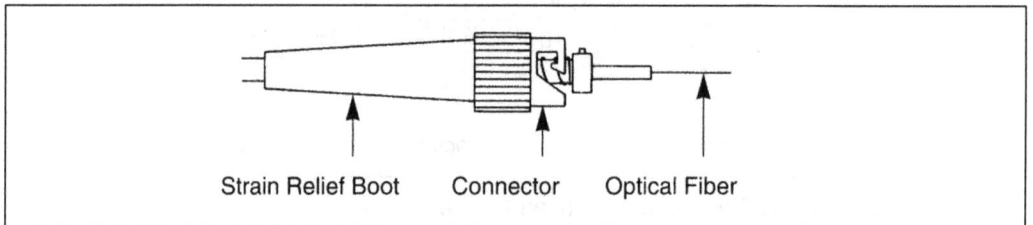

Strain Relief Boot   Connector   Optical Fiber

Figure 6.27

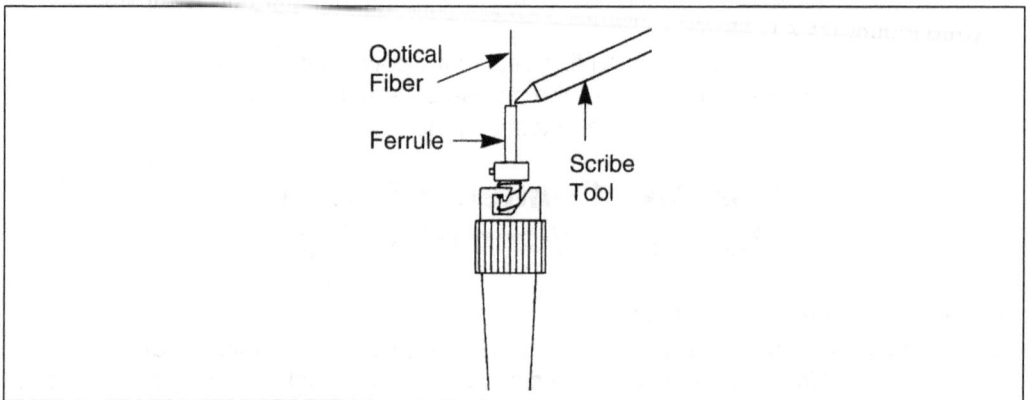

Optical Fiber

Ferrule

Scribe Tool

Figure 6.28

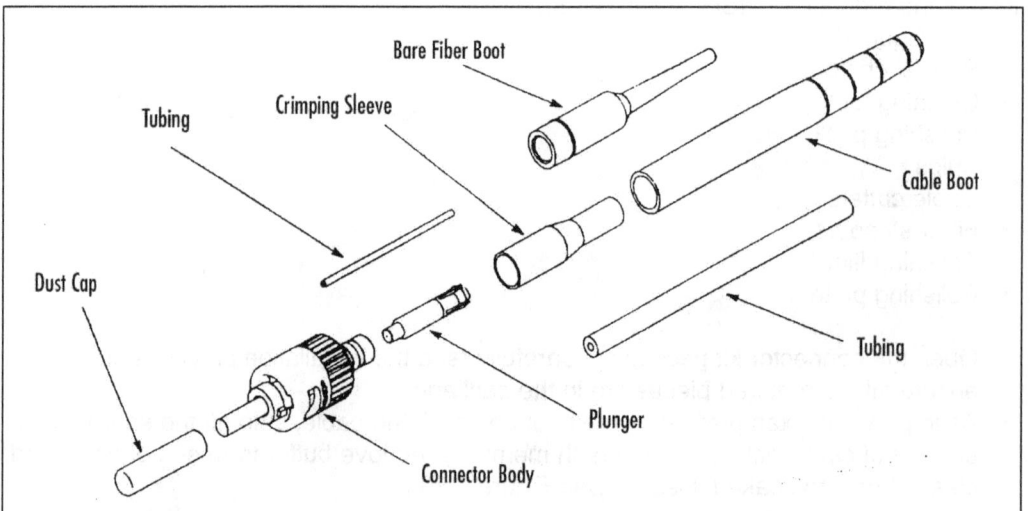

Bare Fiber Boot

Tubing   Crimping Sleeve

Cable Boot

Dust Cap

Tubing

Plunger

Connector Body

Figure 6.29

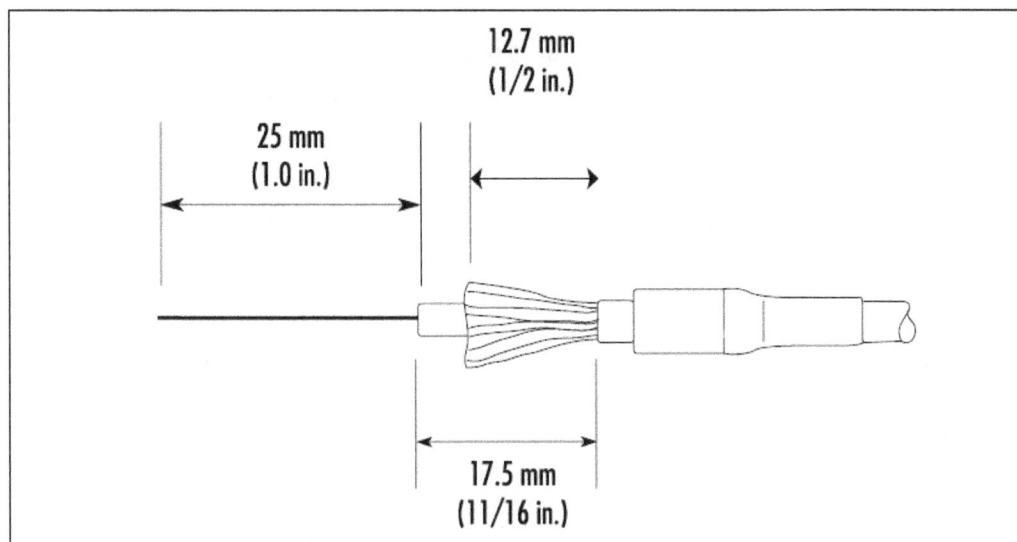

12.7 mm
(1/2 in.)

25 mm
(1.0 in.)

17.5 mm
(11/16 in.)

- Next, slide the crimp sleeve onto the cable and position the sleeve so it captures the strength member in a "bent-back" position. (See Figure 6.30.)
- Insert the plunger into the rear of the connector body until it stops. (See Figure 6.31.)
- Feed the prepared fiber through the rear of the plunger (and connector body) until it stops. (See Figure 6.32.)
- The bare fiber should protrude from the connector's ferrule not more than a half inch.

Using AMP's Crimping Tool

Gently close the tool until you hear one click from the ratchet. Position the connector assembly in the dies so the die set shoulder is located between the plunger and the folded-back strength members. The connector assembly should be aligned with the direction of the arrows on the dies. The eyelet and strength members are positioned in the larger-width pocket. (See Figure 6.33.)

- Slowly squeeze the tool handles together until the ratchet releases.
- Slowly release the tool handles and allow the handles to open fully.
- Carefully remove the crimped connector and its fiber cable from the tool.
- Using the crimping tool, place the larger crimp cavity around the crimp sleeve's larger diameter, closest to the connector body. (See Figure 6.34.)
- Squeeze the crimping tool's handles together until they release.
- Move the crimping tool to the crimp sleeve's smaller section and continue squeezing using the smaller hex cavity on the tool.
- Open the handle.
- Slide the cable boot over the crimp sleeve to complete the connector assembly. (See Figure 6.35.)
- The next step is to score the fiber and prepare for polishing as previously described. Test and put the dust tap on and take a break.

Good job! Hopefully you remembered to wear your safety glasses and discarded broken fiber or scored fiber pieces in the bottle.

Figure 6.30

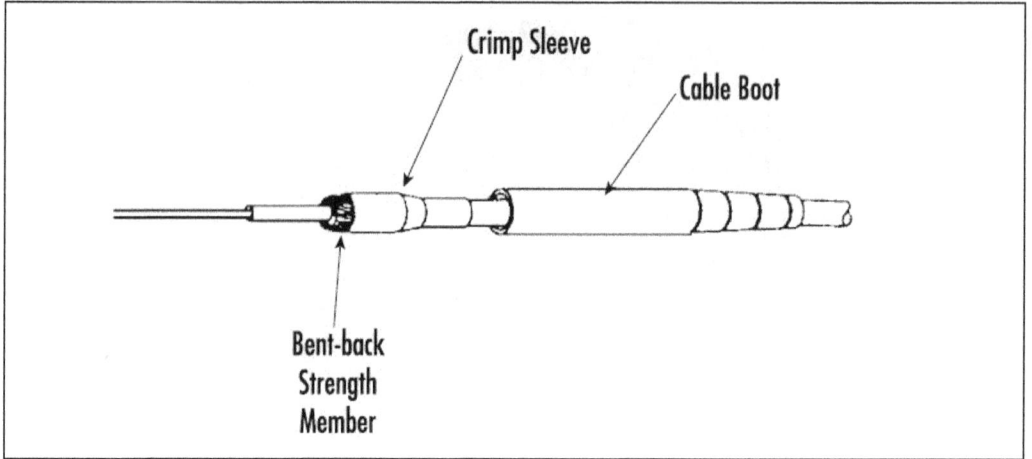

Crimp Sleeve

Cable Boot

Bent-back
Strength
Member

Figure 6.31

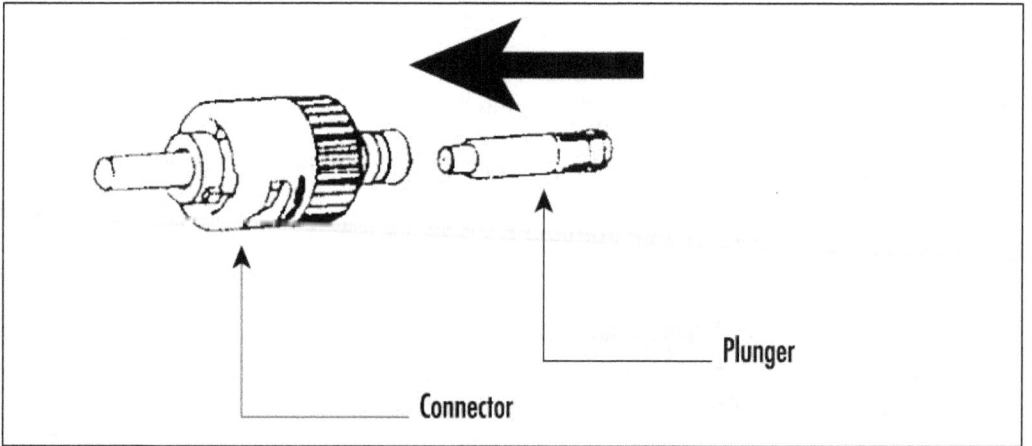

Plunger

Connector

Figure 6.32

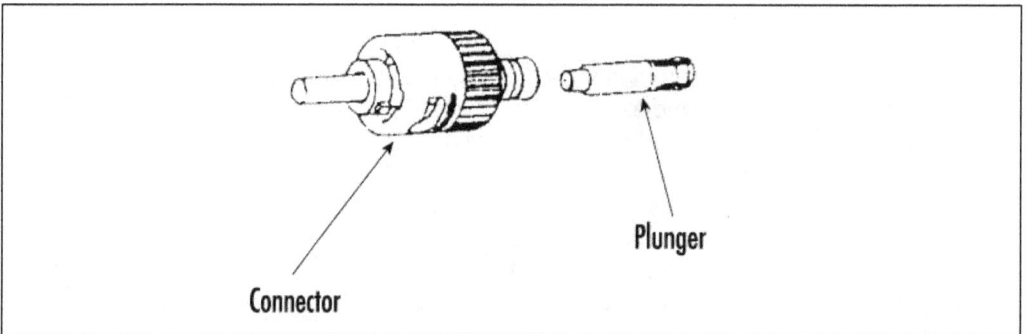

Plunger

Connector

Figure 6.33

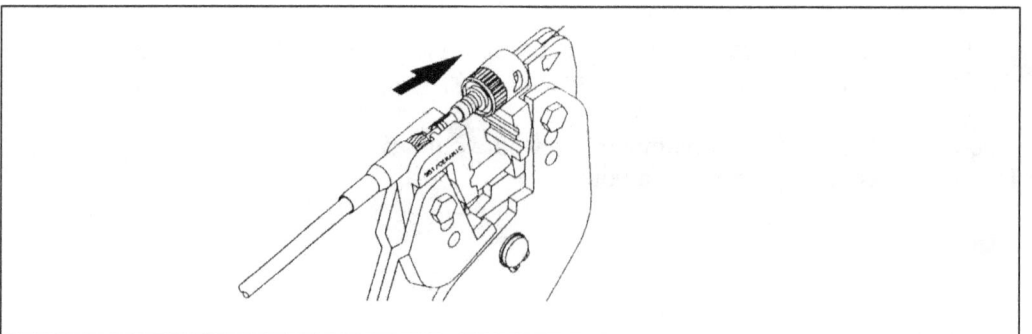

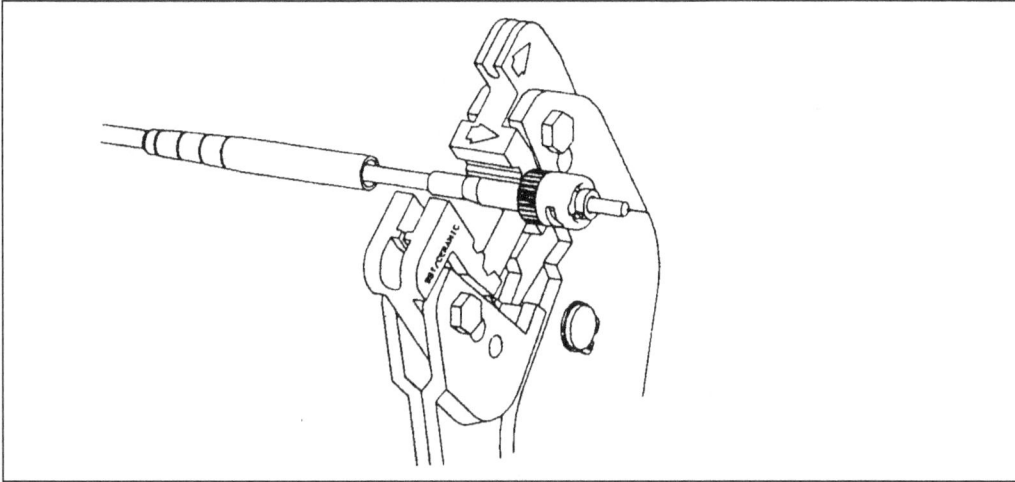

Figure 6.34

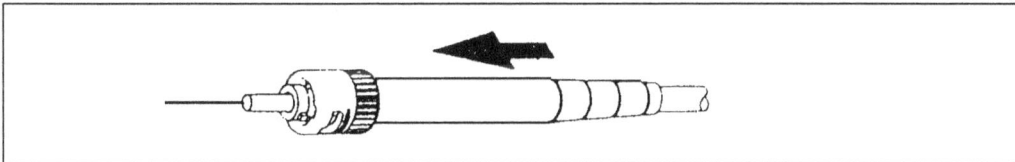

Figure 6.35

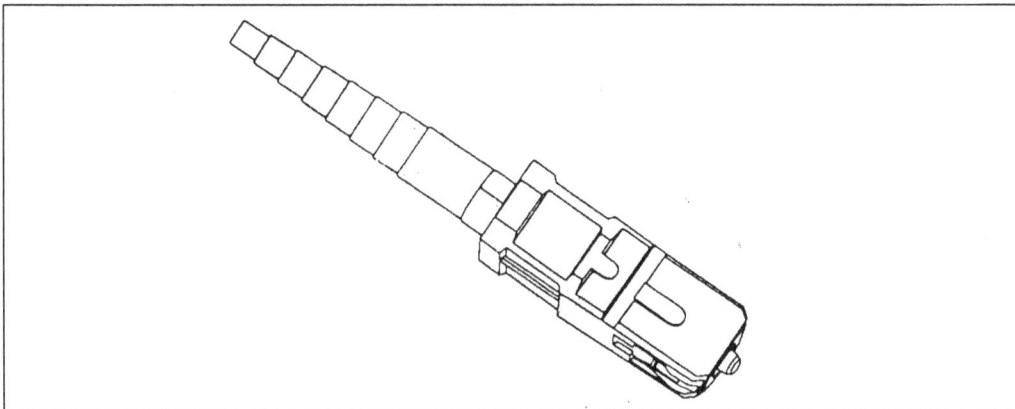

Figure 6.36

Part IV - Connectorization -
Special Tooling/Jig Devices Required

Several manufacturers are making connectorization easier by designing a special tool (jigs), which sometime looks more confusing than it really is. None of these special tools are difficult once you make just one connector. As previously explained, AMP uses a special crimping tool, but Corning Cable Systems (formerly Siecor) uses several types of crimping jigs for both connectors and splices. Thomas and Betts uses a large device that does several tasks at one time and NORDX/CDT uses a plunger method.

Type I - A Corning Cable Systems Product

This connector is manufactured by Corning Cable Systems and is a field-installable connector that does not require epoxy or polishing. The unit incorporates fiber stub that is bonded into a ferrule and polished in the factory and not in the field.

The fiber is cleaved and inserted into the connector so it touches the cleaved end of the fiber stub. (See Figure 6.37.)

When the cam of the boot is rotated, both cleaved ends are pressed into precise alignment inside the connector that is held in place. Once you have chosen the connector, cable size and type, you are ready to begin. Open the connector package, ensure all the pieces and an installation practice card or sheet are enclosed. Examples of parts are shown in Figure 6.38.

The component breakdown of the SC connectors is as follows:
1. Boot - Protects cable interface to fiber and maintains the fiber's minimum bend radius.
2. Strain-Relief - Captures aramid yarn, typically used on patch cords and jumpers for pullout resistance.
3. Spring - Maintains positive force to keep end faces touching.
4. Ferrule and Ferrule Holder - Houses the fiber once it is bonded into it and the end face is polished.
5. Latching Mechanism - Attaches connector to adapter.
6. Dust Cap - Keeps contamination to a minimum and protects polished end face.

The installation tool that will be used is a special tool designed to position the fiber into the UniCam connector. (See Figure 6.40.) Rotate the cam that aligns the fibers, and crimp the fiber in place. A separate crimp tool is required to secure the aramid yarn on the interconnect cable.
 • Begin by flipping the crimp handle open and rotate the wrench handle so it is up. (See Figure 6.41.)
 • Pull back the slider and insert the connector into the tool as far as it will go. The lead-in tube should rest on the crimp platform when the connector is fully seated. The front of the connector should rest in the slider.

Figure 6.37

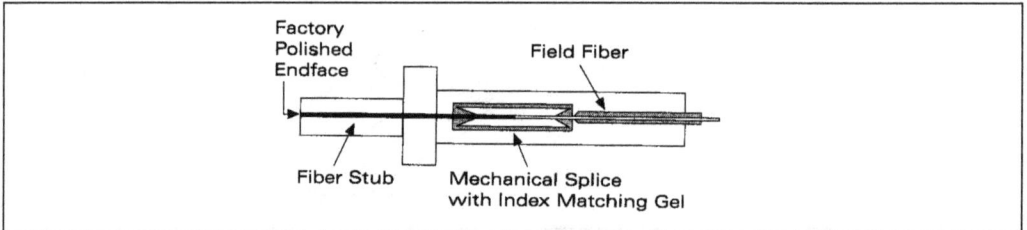

Factory Polished Endface
Field Fiber
Fiber Stub
Mechanical Splice with Index Matching Gel

Figure 6.38

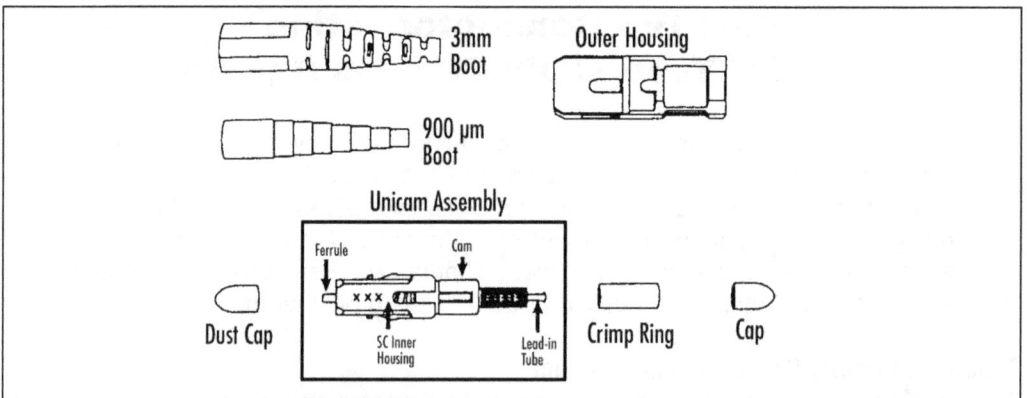

3mm Boot
Outer Housing
900 µm Boot
Unicam Assembly
Ferrule
Cam
Dust Cap
SC Inner Housing
Lead-in Tube
Crimp Ring
Cap

Figure 6.39

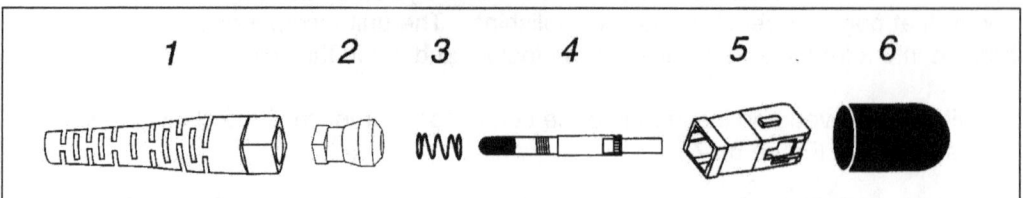

1   2   3   4   5   6

Figure 6.40

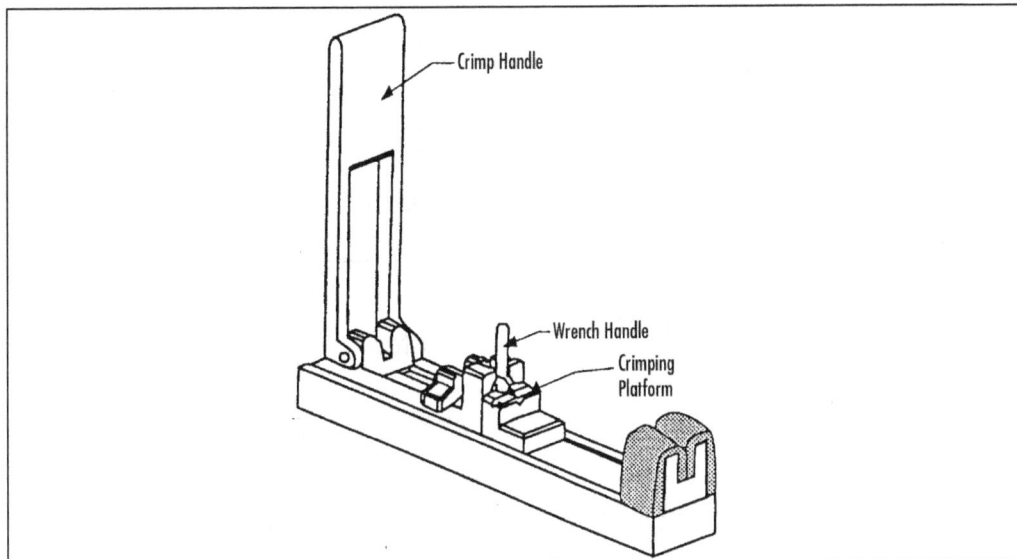

Figure 6.41

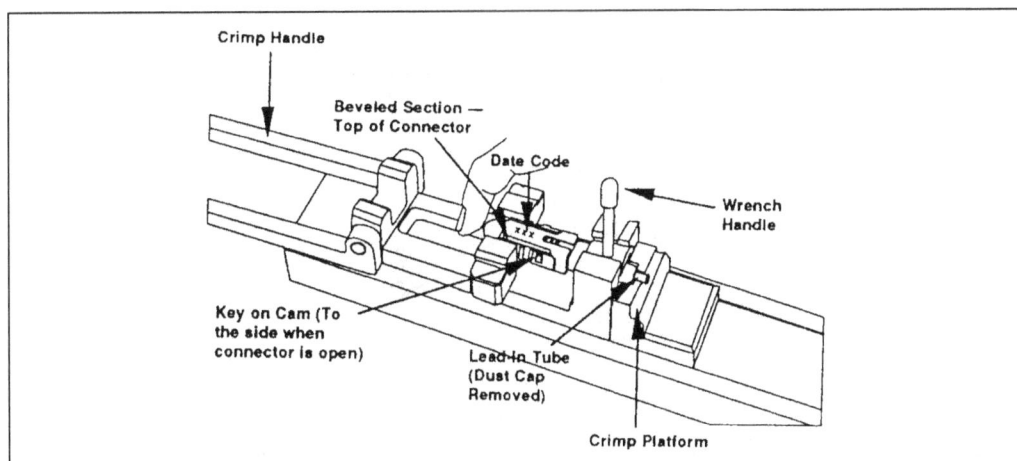

- When your connector is locked into the tool, begin your fiber preparation. In this section, a 900-micron light buffered fiber will be used.
- Prepare the cable as discussed in the previous section.
  a. Slide the boot onto the cable out of the way.
  b. Using the manufacturer's guide sheet, mark your cable for suggested measurements.
  c. Strip off outer jacket and buffer as directed.
  d. Clean the bare fiber with alcohol wipes.
- Next, cleave fiber to the specified dimensions required by the manufacturer (Figure 6.42).

The Cleaving Tool

Corning Cable Systems calls its cleaving tool an Optical Fiber Cleaver (Figure 6.43). It can break the glass by using a score-and-snap method of cutting the fiber known as cleaving. This allows a good method of providing fiber cleaves for fusion or mechanical splicing.

Note: As shown previously, the fiber has already been stripped, cleaned and marked where the manufacturer has recommended cleaving dimensions.
- Press down on the handle to open the cleaver's fiber clamp. (See Figure 6.44.)

Figure 6.42

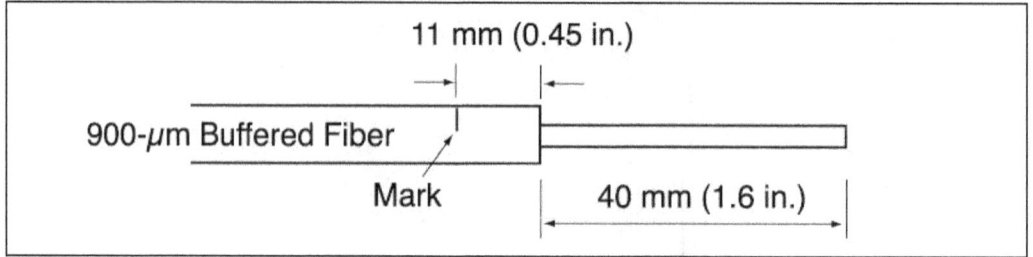

Figure 6.43

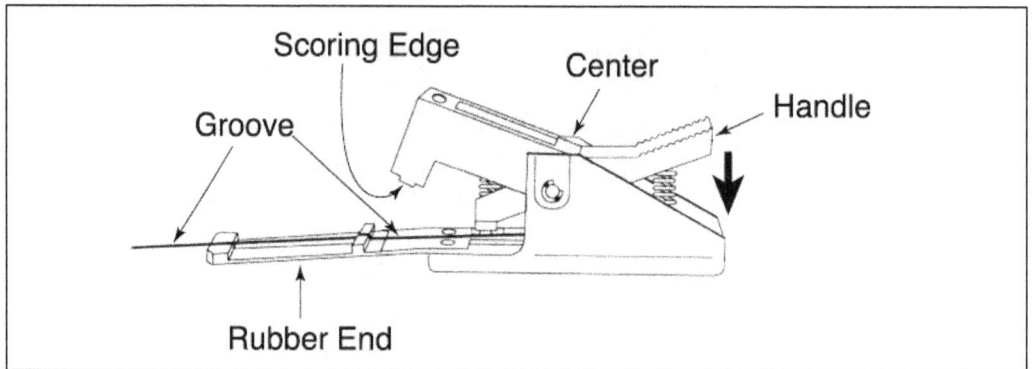

Figure 6.44

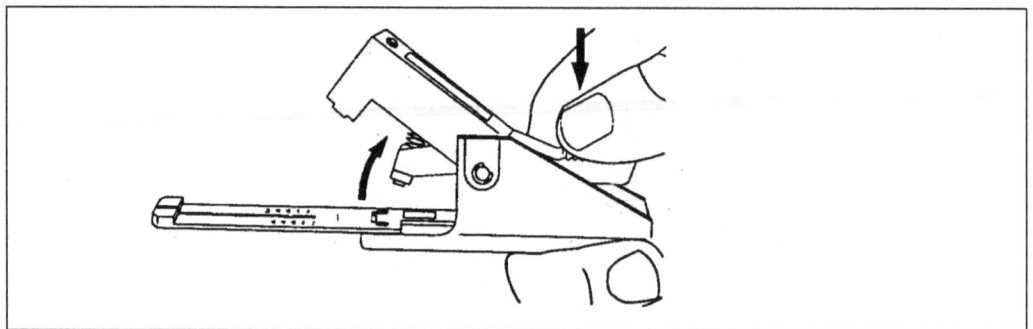

- With your other hand, place the fiber in the cleaver's fiber guide so the end of the fiber is under the fiber clamp and the end of the fiber coating lines up with the desired cleave length marking. (See Figure 6.45.)

DO NOT FLEX THE FIBER GUIDE AT THIS TIME.
- Gently release the handle to lower the clamp onto the bare fiber.
- Press down on the cleaver arm until it just touches the fiber and guide. (See Figure 6.46.) This will apply enough pressure to properly score the fiber.
- Gently release the cleaver arm.
- Flex the fiber guide to snap the fiber as shown in Figure 6.47. The fiber is now ready for splicing.
- Press down on the cleaver's handle to once again lift the fiber clamp. Remove the end piece of fiber with tweezers and place the fiber on a loop of tape or bottle for proper disposal.
- Cleave the fiber to 8 mm.
- As previously accomplished, install the connector in the installation tool. Repeat carefully, inserting the cleaved fiber into the lead-in tube until you feel it firmly stop against the connector's fiber stub. (See Figure 6.48.)
- Guide the fiber in straight. Do not bend or angle it.
- If you feel resistance at the entry tunnel, rotate the fiber back and forth while applying a gentle inward pressure.

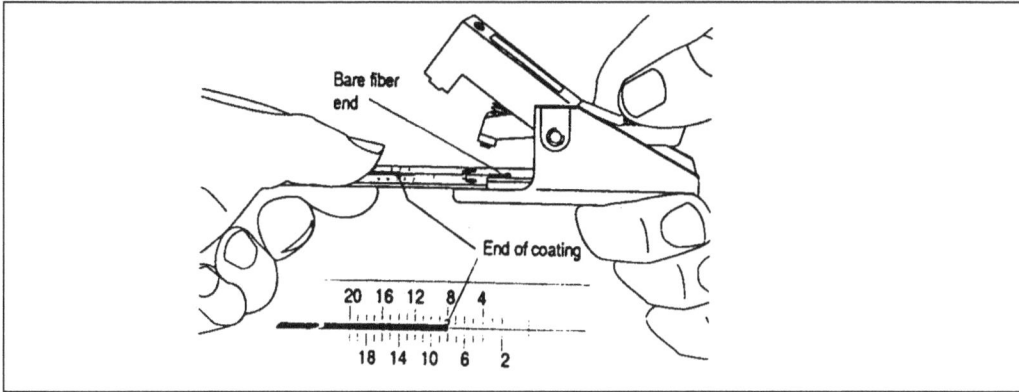
Figure 6.45

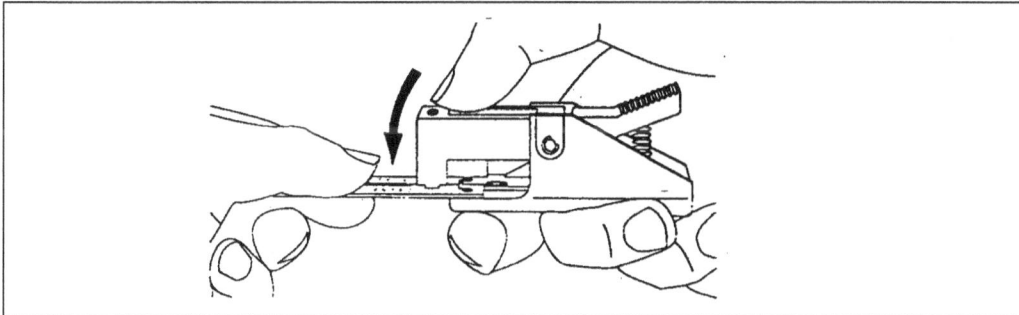
Figure 6.46

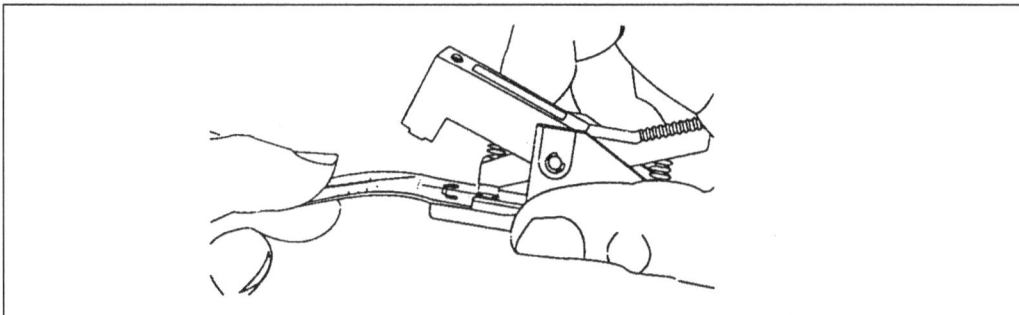
Figure 6.47

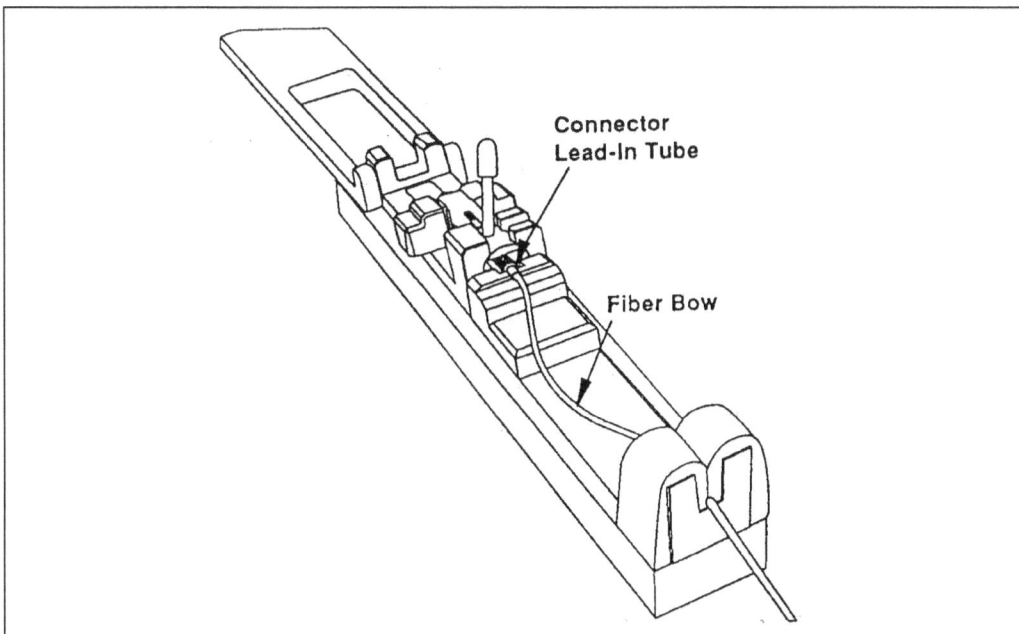
Figure 6.48

Figure 6.49

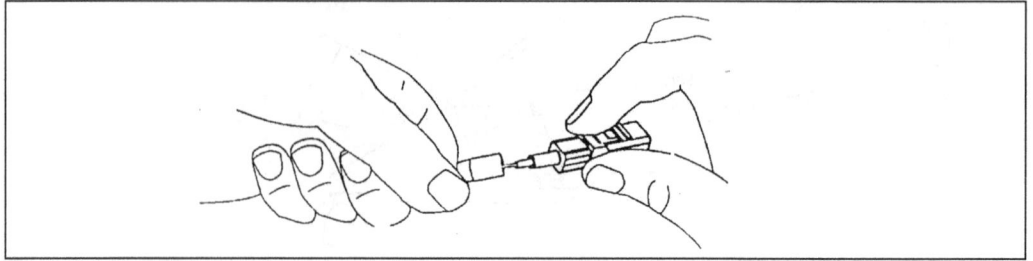

Figure 6.50

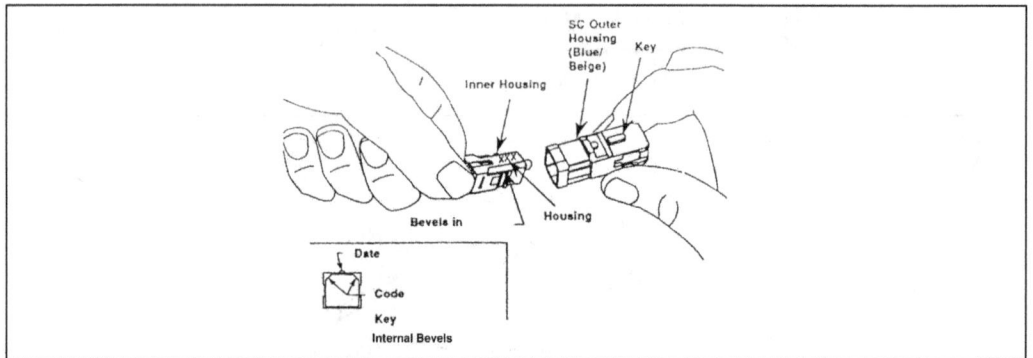

Note: If you have stripped and cleaved the fiber to the correct lengths, the end of the cable jacket or the buffer mark should be within 2 mm (0.08 in) of the lead-in tube.

• Apply a light inward pressure on the fiber to keep it butted against the fiber stub during the next step.
• Rotate the wrench past 90° to cam the connector.

The splice now holds the fiber inside the connector. You no longer need to hold it in place, but be careful not to pull on the fiber.

• Carefully flip the crimp handle 180° until it contacts the crimp tube. Push down firmly to crimp. You should see a flat impression in the crimp tube, indicating a proper crimp. The tool cannot over crimp the connector.
• Flip the crimp handle back. Leave the wrench handle down. Remove the connector by lifting it straight up and out of the tool. Do not pull on the fiber. Handle the connector only.
• Slide the boot up the back of the connector until it reaches the cam.
• A small line of adhesive may be applied around the rear of the connector, just past the metal crimp tube, before putting the boot on. Slide the boot into place quickly. (Figure 6.49).
• To install the UniCam® assembly into the SC outer housing, line up the date code on the inner shroud with the key on the outer housing. Using the boot, push the UniCam assembly into the rear of the outer housing until it snaps into place. You may have to wiggle the parts to make them snap together. (See Figure 6.50.) You have now completed the SC connector.

Type II - A NORDX/CDT Product

This product uses a special installation boot. (See Figure 6.51.) This tool uses a plunger style device that pulls a release and the connector snaps in place. Start installation by opening the package and inspecting all parts.

• Place and tighten down the tool clamp to a tabletop surface. Open the installation tool by rotating the tool body. (See Figure 6.52.)

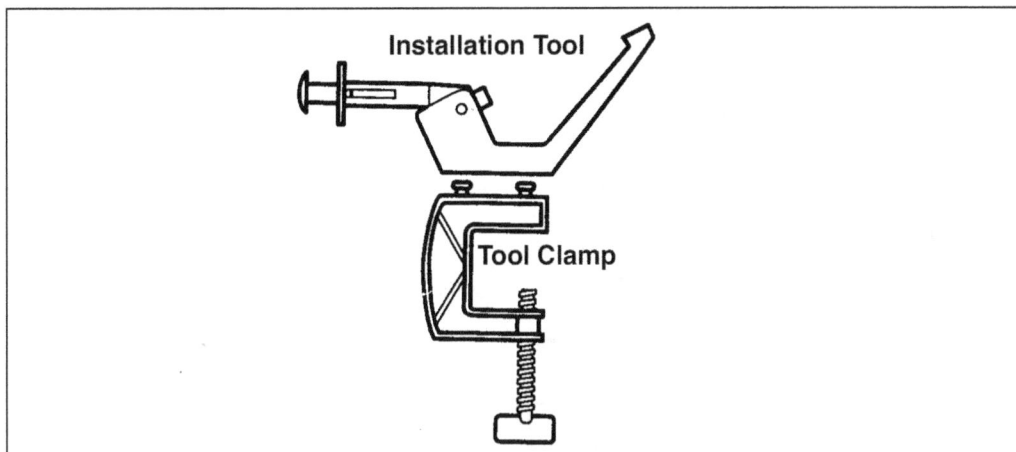

Figure 6.51

Installation Tool

Tool Clamp

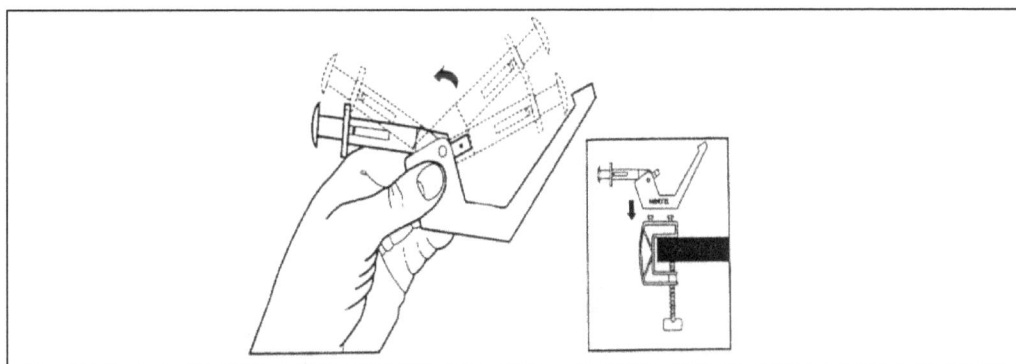

Figure 6.52

- Load the connector body into the tool with the release wire up. (See Figure 6.53.)
- Prepare the cable for connectorization. Measure, put boot on, strip the buffer, clean, cleave and remove dust cap as previously discussed.
- Carefully insert the base fibers into the stem of the connector until it bottoms out. Clamp the fiber into the clamp of the installation tool. (See Figure 6.54.)
- Depress the installation tool plunger and make sure it hooks the release wire. Slowly release the plunger. (See Figure 6.55.)
- Unclamp the fiber and carefully remove the connector from the installation tool.
- Crimp the connector as required. (See Figure 6.56.)
- Slide the boot onto the connector to finish.

Part V - Connectorization - Heat-Curing Method

We have addressed UV-cured connectors, no cure-no polish connectors, and glass insert type connectors. Next, we will address the heat-cured (oven) connector. Heat-cured connectors are very cost-effective but require more time and a little more skill to install compared to UV connectors. This type of connector is excellent if you have several to make because the heaters/ovens have multi-oven ports and can heat up to 24 fiber connectors (Figure 6.57).   Smaller curing ovens can heat as little as four ports.

Type I - A Lucent Product

Before you begin, choose the specific epoxies to use such as high temperature use, fast ambient cure and/or general purpose. Each one of these has a specific curing time ranging from 15 minutes to two hours. Read the instructions and select the proper epoxy for the specific task.

Figure 6.53

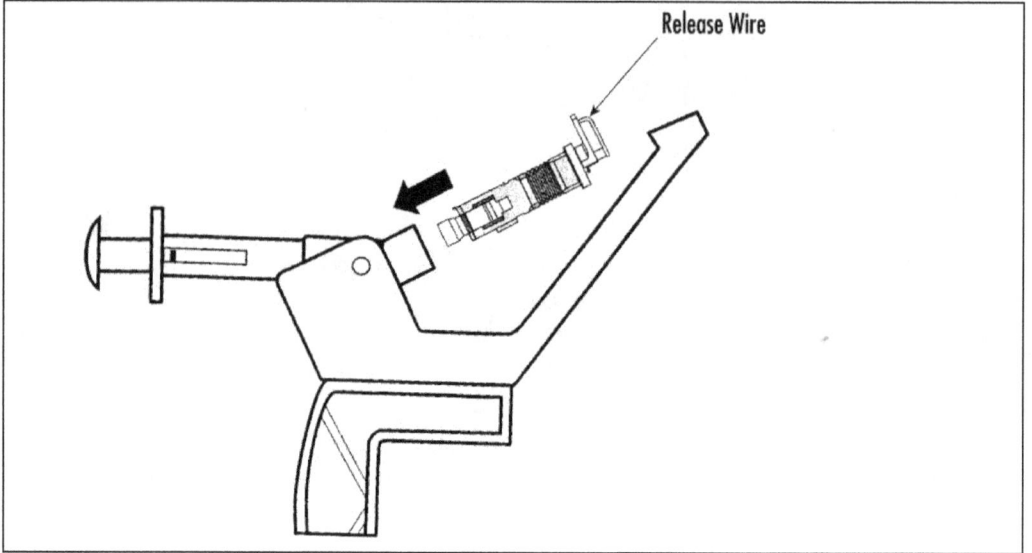

Release Wire

Figure 6.54

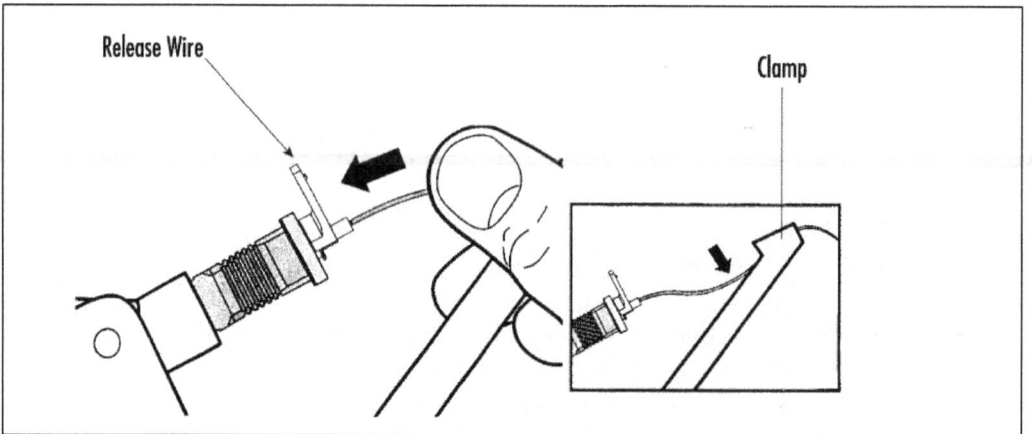

Release Wire

Clamp

Figure 6.55

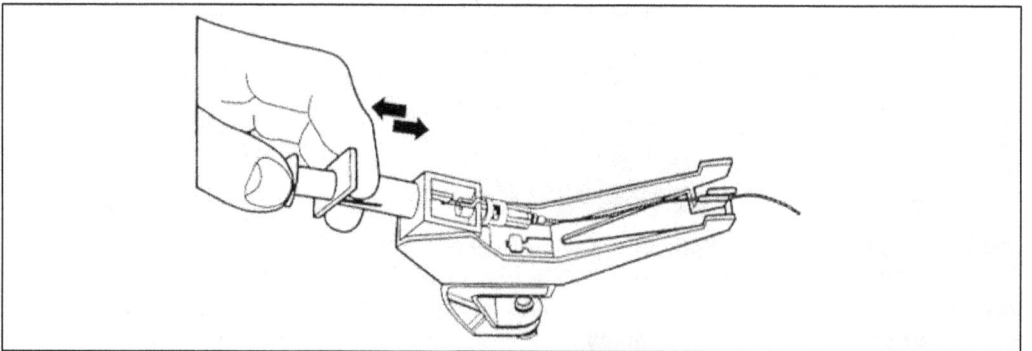

Figure 6.56

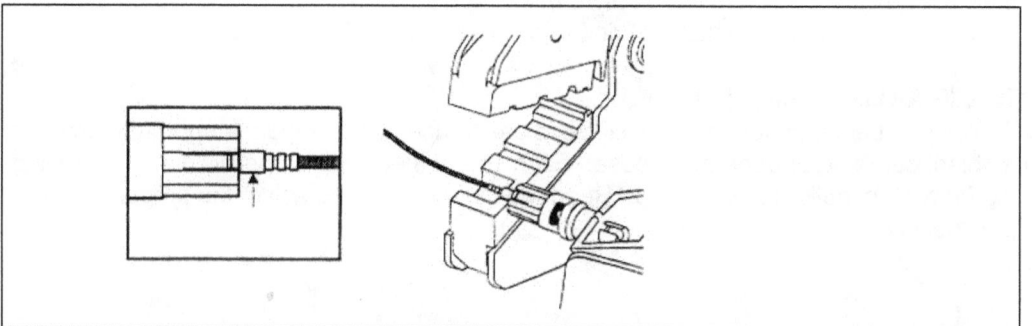

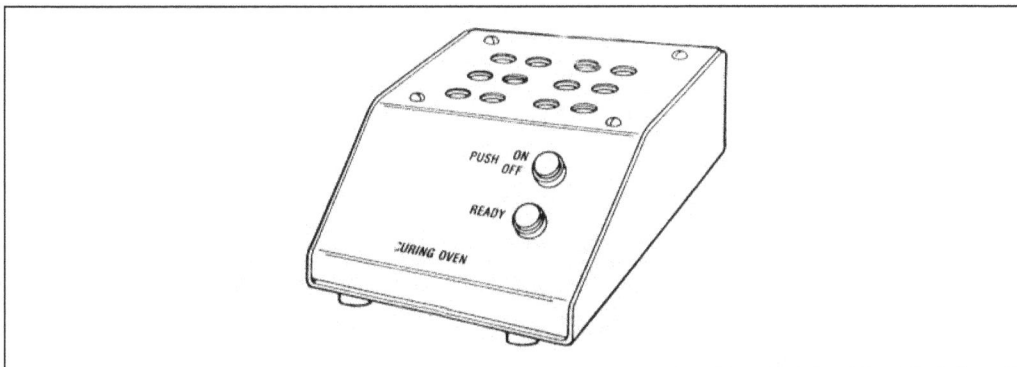

Figure 6.57

Once you select the epoxy to use, determine the number of connectors to be made at one time, since the pot-life (before it hardens) can be as little as 15 minutes. Check the manufacturer's specification sheets. Lucent Technologies manufactures several types of connectors including the one addressed next.

ST™ Connectors Oven-Cure
Epoxy and/or Anaerobic

Note: Anaerobic is a special, almost clean epoxy used exclusively by Lucent.

Once you have finished preparing at least six fibers as discussed in the previous connector examples, place the fiber into a holding block. (See Figure 6.58.)

• Next, prepare the epoxy by squeezing equal lengths of the resin and hardener onto a clean sheet of paper. (See Figure 6.59 and Figure 6.60.) With a small stick, mix thoroughly until the two different colored parts blend into a smooth color as specified by the manufacturer.
• Remove the protective cap from the syringe and then remove the syringe. Using a stick, load the epoxy into the syringe and insert the plunger/handle into the syringe. (See Figure 6.61.)

Remember, it will harden in about 40 minutes.

• Slowly inject epoxy into the plug until a small bead appears on the tip.
• Apply a thin coat of epoxy at the end of the plastic tubing.
• Slide the tubing into the barrel of the connector assembly and rotate to distribute the epoxy. (See Figure 6.62.)
• Apply a thin coat of epoxy to the first 1/4 inch of the buffered part of the fiber. (See Figure 6.63.)
• Insert the fiber into the connector. Be careful and feel for the hole in the connector tip. Don't hurry, and don't use force or you'll break the fiber. Feed the fiber through the connector until you see the fiber protruding through the connector hole (Figure 6.64).
• Place the connector holder on the connector and twist the lock and crimp the barrel at the rear of the connector onto the cable. (See Figure 6.65.)
• Examine the connector and ensure the length is not too long for the oven.
• Curing - insert the connector into one of the oven ports and leave in for about 10 or 15 minutes.

Be careful because the oven is very hot.

Figure 6.58

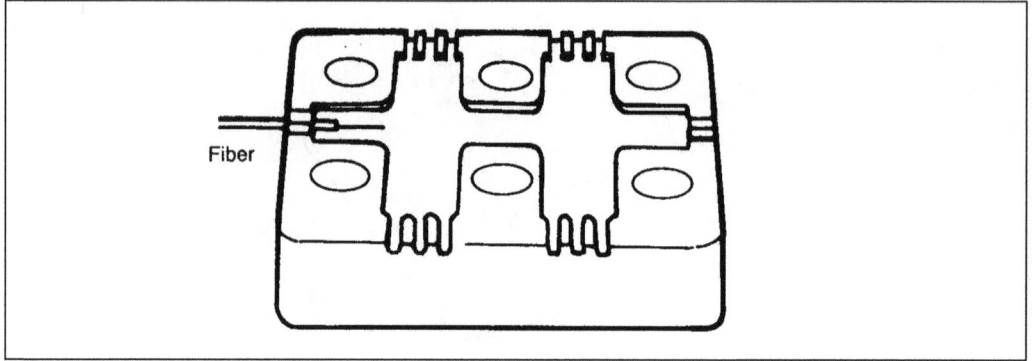

Fiber

Figure 6.59

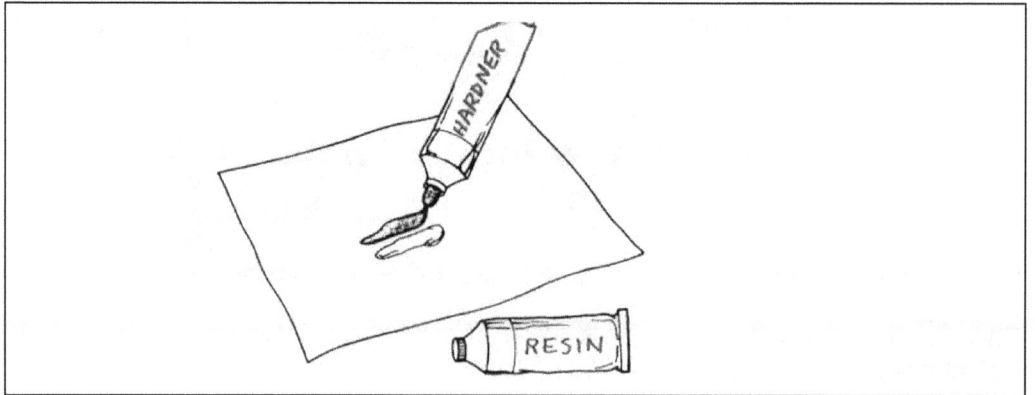

HARDNER

RESIN

Figure 6.60

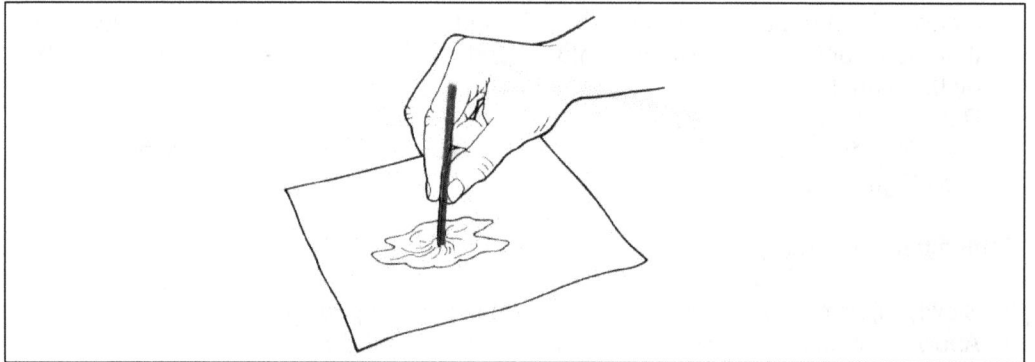

Figure 6.61

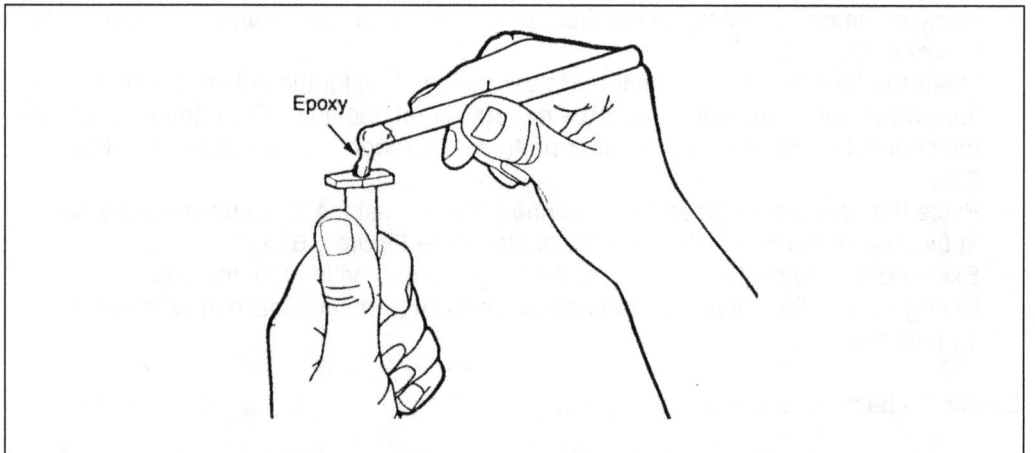

Epoxy

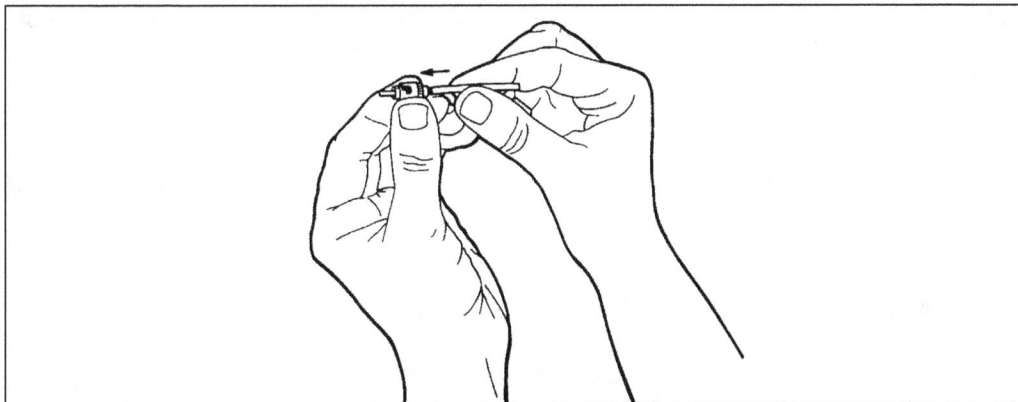

Figure 6.62

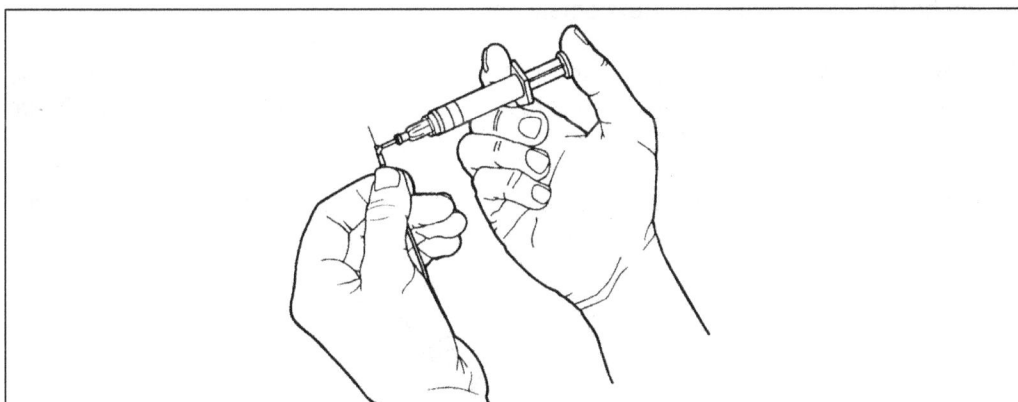

Figure 6.63

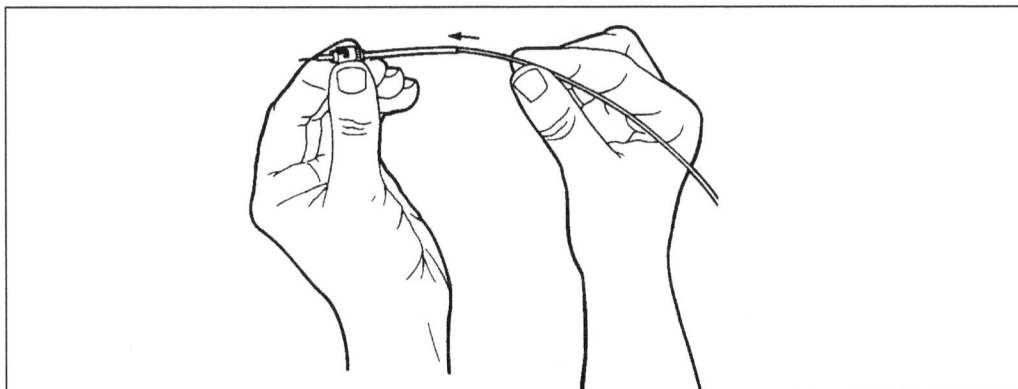

Figure 6.64

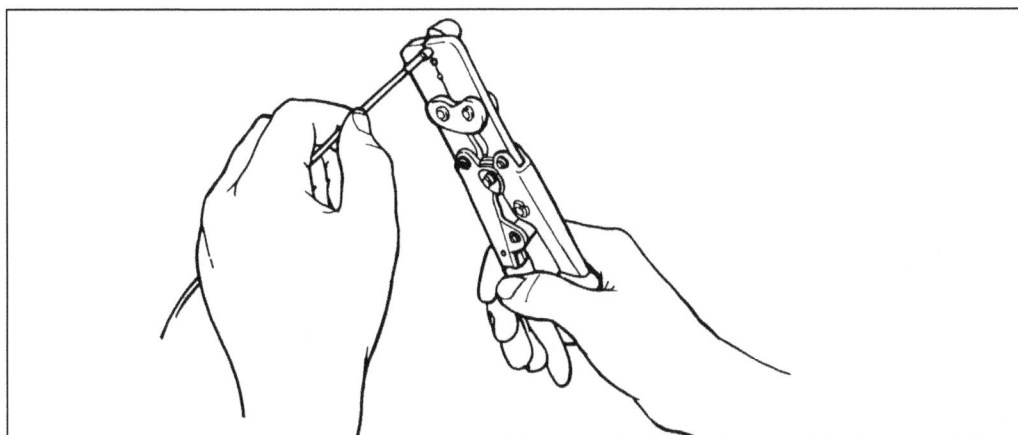

Figure 6.65

Remove the connector from the oven. Put the connector into the holding assembly to cool off. Once it cools off, trim the fiber with a cleave and polish to completion as previously described.

Type II - A 3M Product

Another epoxy curing type connector is the 3M Hot Melt (See Figure 6.66.). The 3M Hot Melt connectors are pre-loaded with epoxy from the factory.

Using a special holder (Figure 6.67), you install the pre-loaded connector into the holder and put it into the oven and allow it to melt the epoxy in the connector.

Once the connector is loaded into the holder, the holder is loaded into a special type of oven. After several minutes of heating, the holder and connector are removed. Be careful, the holder and connector are extremely hot.

With the prepared fiber, hold the holder firmly and insert the fiber into the connector. Do not use force. Feel for the hole and gently rotate the fiber until it bottoms out. (See Figure 6.68.)

Look at the bottom of the holder and ensure your fiber has protruded through the barrel of the connector holder. (See Figure 6.69.)

Figure 6.66

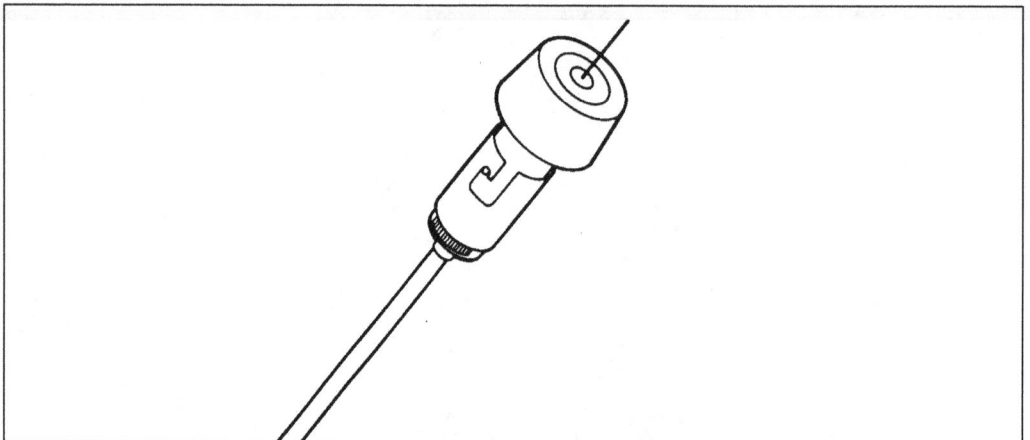

Figure 6.67

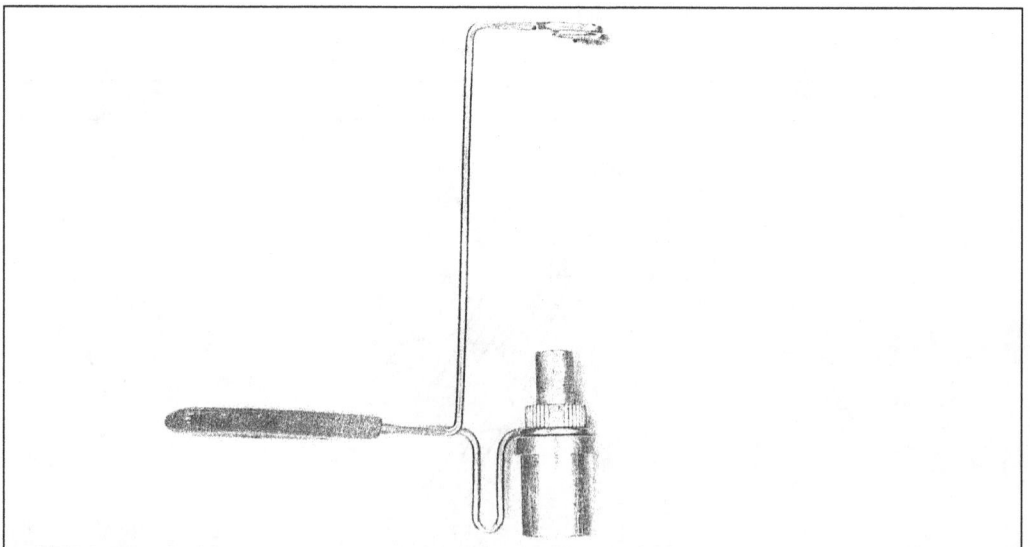

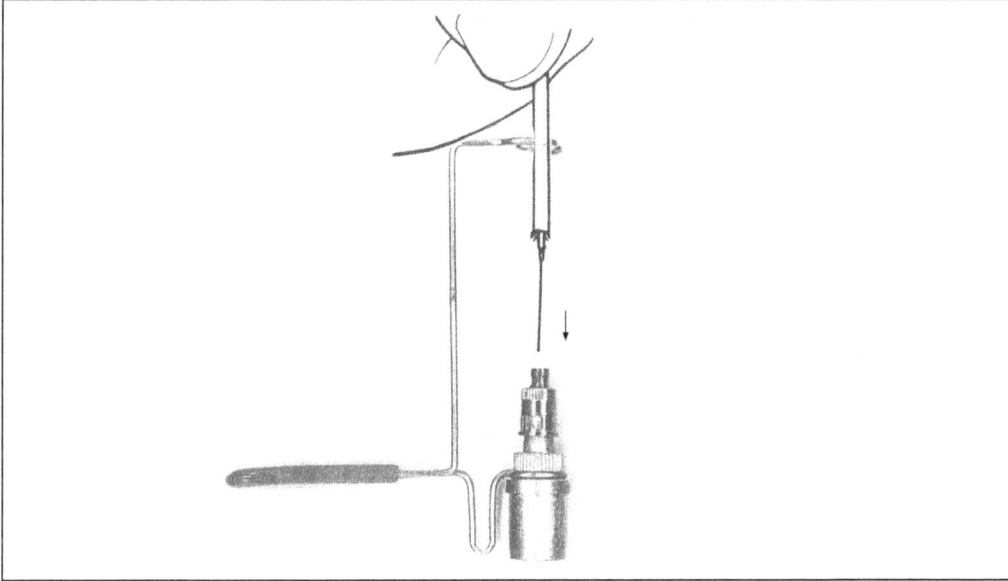

Figure 6.68

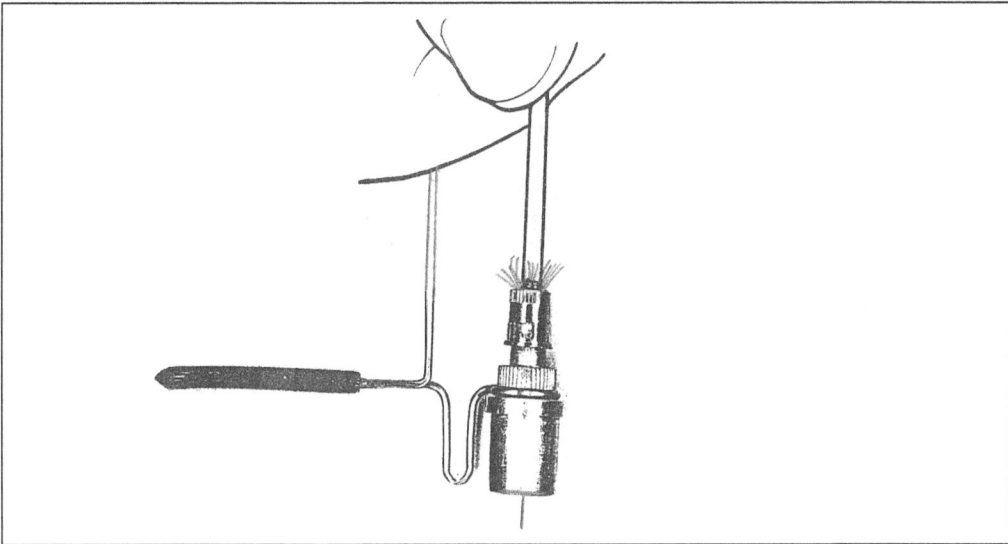

Figure 6.69

Gently insert the holder and connector into a cooling holder and allow to cool for several minutes. Once it has cooled, remove the connector and proceed with crimping, cleaving, polishing and testing as previously explained.

3M Hot Melt kits include cable preparation and polishing tools and materials; a lightweight oven, which is pre-set for the proper operating temperature; a 12 connector cooling fixture; and 12 specially designed connector holders. A replacement connector holder package is also available.

Fewer tools are required for installing the 6100 Hot Melt Connector than with other adhesive methods. 3M conversion kits are available, allowing you to use your existing 3M FC/PC or Biconic Field Termination Kit for Hot Melt installations.

Part VI - Connectorization - Small-Form-Factor

In just the last several years, the Internet and its close cousin, Intranet, have had a major effect on private network data communications. The mix and volume of data and voice traffic being carried on these networks is changing rapidly, with more applications and larger data files than ever before. To accommodate this increased traffic, fiber optic technology is playing a pivotal role in the development of today's private networks.

Increased use of fiber optics results in more electronics, cable and apparatus being installed into smaller and smaller physical spaces. Hence, a primary focus of many development efforts is on the fiber optic connector, that small passive device that connects the fiber optic cable with the optoelectronics driving these networks. These new connectors are known as small-form-factor (SFF) connectors.

A primary goal of all SFF fiber optic connectors is reducing the size to approximately one-half that of the traditional ST and SC connectors. Smaller size is important because, as more fiber is being used in private networks, more electronics are being squeezed into less space. With older connectors, electronics manufacturers could not achieve port densities equivalent to copper on switches, hubs and routers. Now with SFF connectors, the same densities are possible with both types of media. SFF connectors also make practical, smaller fiber network interface cards (NIC) for computer workstations and servers.

Intuitive Operation Similar to RJ-45

When Bell Laboratories invented the first modular connector for copper cable, the RJ-11, in the late 1970s, the designers probably never envisioned the tremendous popularity it would achieve. The compact size, low cost and intuitive operation of these connectors was so attractive that is was used in many successive generations of new modular connectors, culminating in the almost universal RJ-45 8-pin modular connector used presently for most copper private network connections as seen below in a single and duplex fiber connector. (See Figure 6.70.)

Some of today's SFF fiber optic connectors emulate not only the size, but also the intuitive operation of the RJ-45. For example, some are polarized, which ensures that the transmit and receive fibers are always correctly oriented. Connector polarization is required in various standards including TIA 568A (Commercial Building Telecommunication Cabling Standard) and ISO 11801, the international equivalent. In addition, some SFF connectors offer visual verification of proper polarity through the use of A/B markings on the adapter into which the connector is inserted. Audible clicks are another reassuring feature some SFF connectors have adapted from the RJ-45. Color-coding can help users quickly distinguish multimode (beige) and singlemode (blue) connectors, which brings us to our last type of connector, the Opti-Jack® fiber optics connector by Panduit Corp.

Figure 6.70

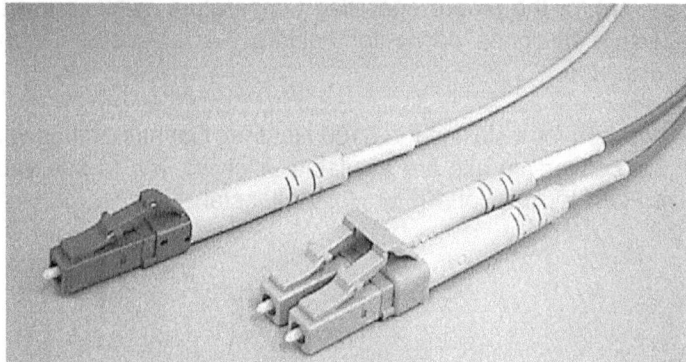

Opti-Jack® Termination Procedures

One major advantage of the Opti-Jack fiber optic connector is that both the plug and the jack can be easily terminated in the field. This allows greater flexibility to make changes on the fly and to maintain connectors and cabling systems for years to come. Plug-to-plug patch cords can be pre-terminated in the factory, or they can be field terminated to length in order to create more aesthetically pleasing and manageable telecommunications closets and workspaces.

Panduit's recommended procedure for fiber terminations employs an anaerobic adhesive and primer. This process secures the fibers within the ferrules in 60 to 90 seconds. (See Figure 6.71 A-H.)

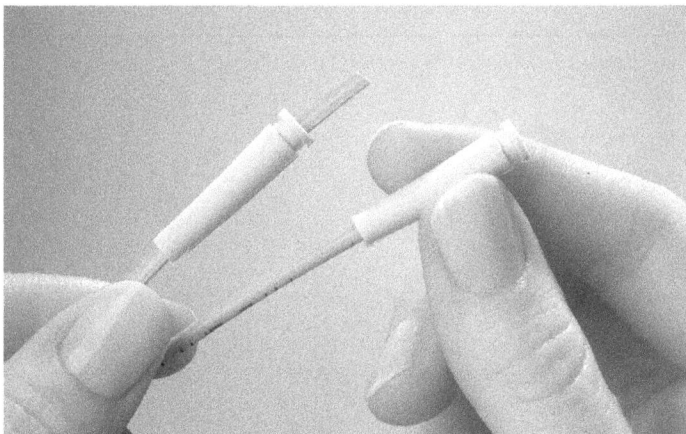

Figure 6.71A
Slide on boots.

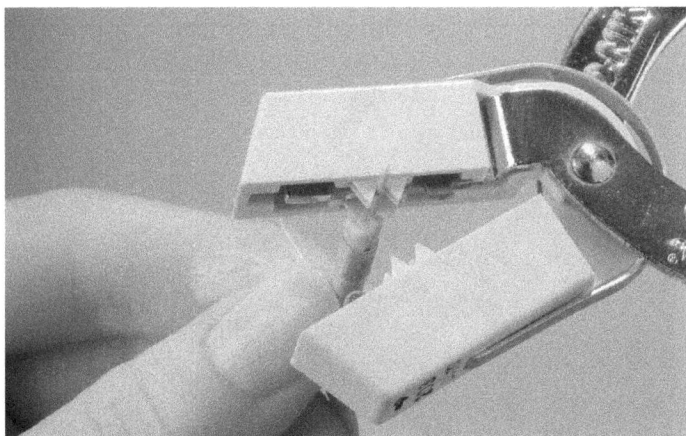

Figure 6.71B
Prepare cable jacket, aramid yarn, crimp sleeves and buffer.

Figure 6.71C
Dispense adhesive until a small bead forms on front of ferrule.

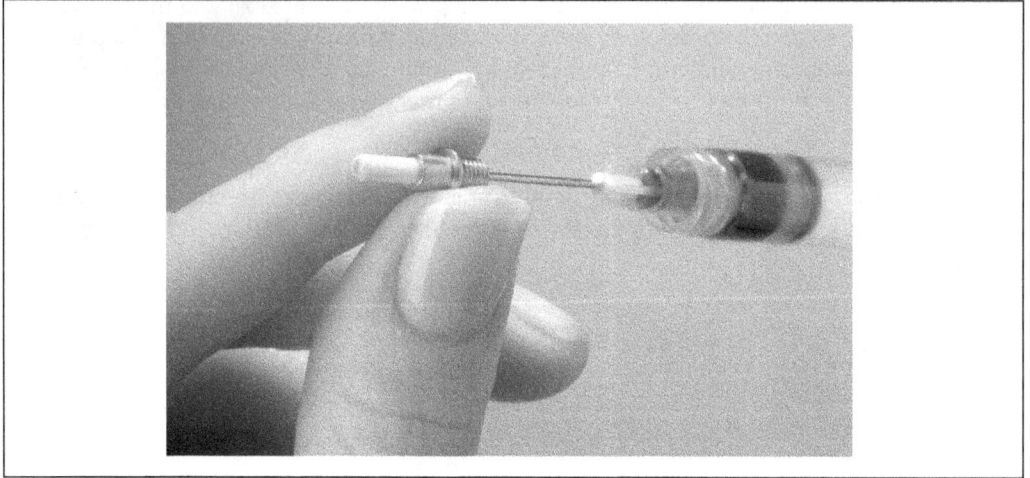

Figure 6.71D
Dip bare fiber into primer and install into rear of ferrule assembly.

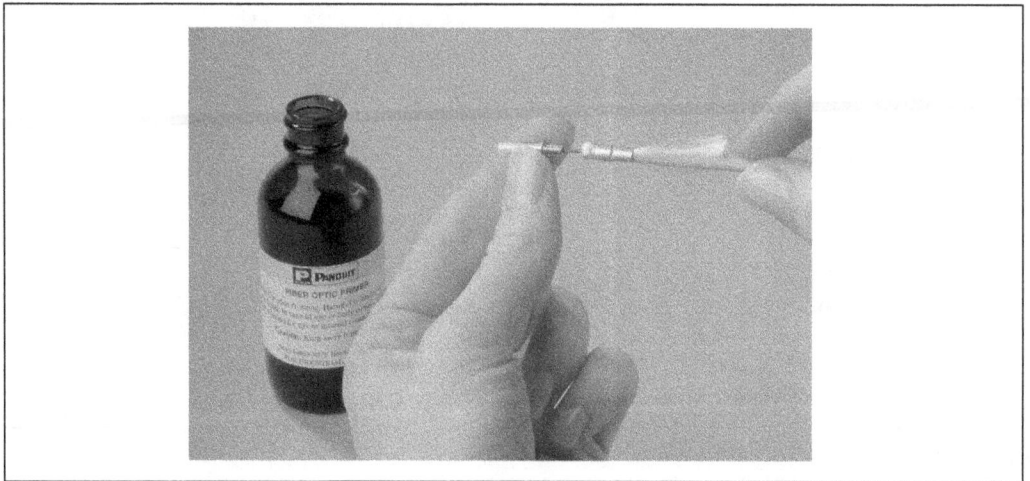

Figure 6.71E
Crimp both sleeve ends.

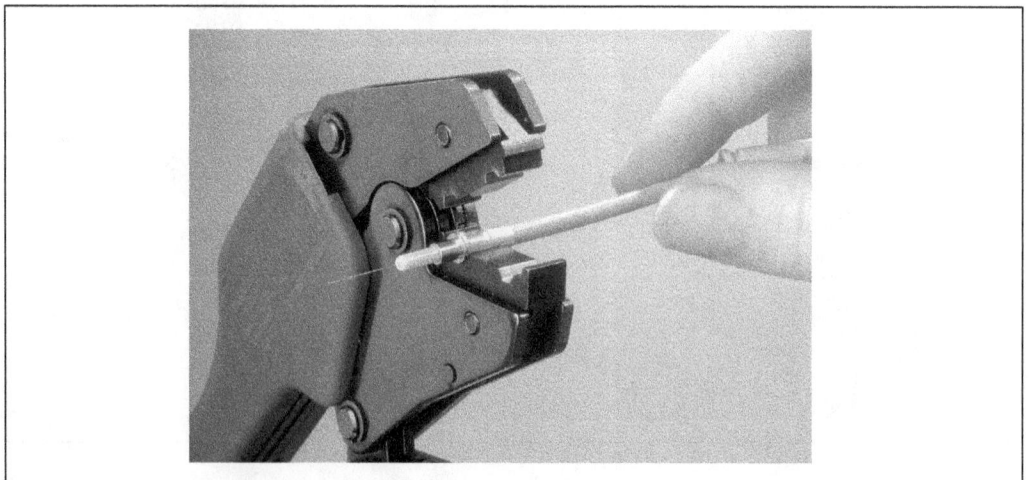

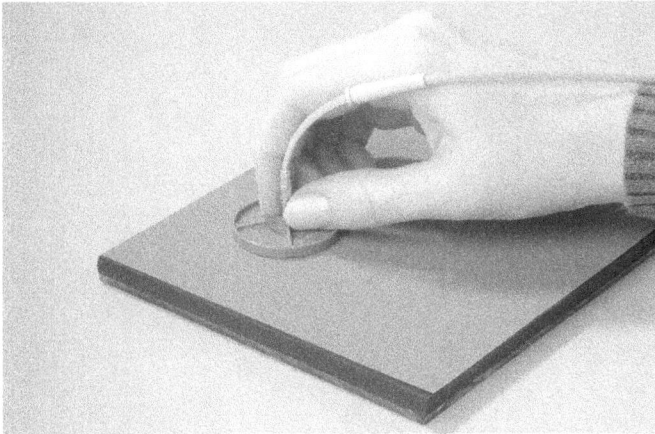

Figure 6.71F
After cleaving the fiber, polish.

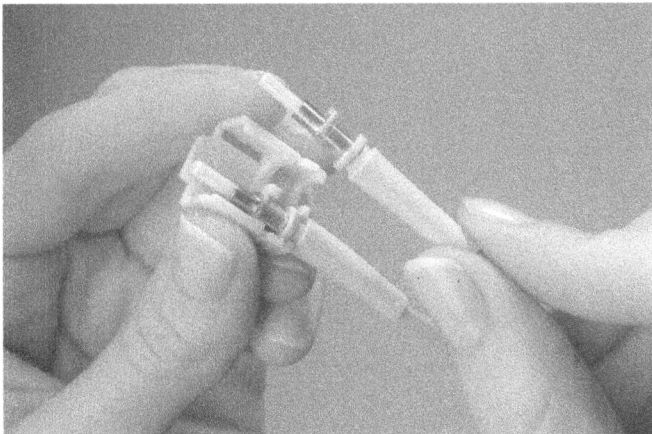

Figure 6.71G
Slide boots over crimp sleeves and insert ferrule assemblies into cap.

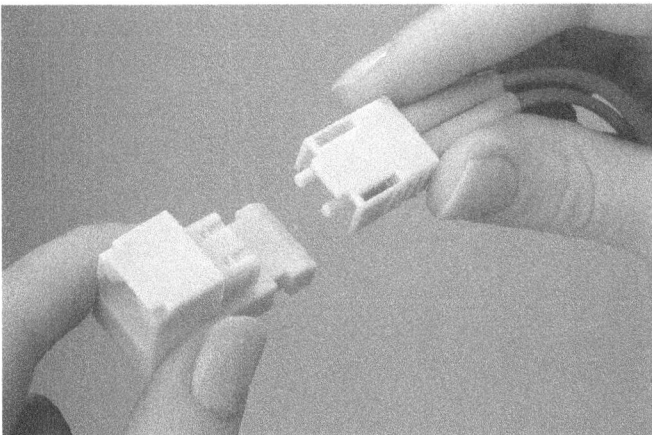

Figure 6.71H
Align cover over housing, slide and click into place.

TEST MODULE 6

1. Removing a stinger consists of using a can of air to blow it away.

   True   False

2. The most popular adhesive is epoxy, which produces the highest quality termina-tions.

   True   False

3. The epoxyless style connector is harder to install than the type that uses adhesives.

   True   False

4. A primary concern when working with adhesives is safety.

   True   False

5 Room temperature or heat-cured epoxy should be injected into the connector until a small bead appears on the outside of the ferrule.

   True   False

6. What are the most finicky of all adhesives?
   a) Hot melt
   b) Cyanoacrylates (CA)
   c) UV (ultraviolet)
   d) Anaerobic

7. What safety precautions should be taken when working with any fiber optic cable installation?
   a) Always wear safety glasses
   b) Take a bath after you are finished working
   c) Have alcohol available
   d) Wear a hard hat

8. What is the special tool made by manufacturers that makes connectorization easier?
   a) Fiber tool
   b) Crimping tool
   c) Jig device
   d) Snips

9. What tool can break glass by using a score-and-snap method of cutting the fiber?
   a) Crimping tool
   b) Optical fiber cleaver
   c) Clamp
   d) Fiber tool

10.What connector is very cost-effective but requires more time and a little more skill
to install?
a) Polish connector
b) UV-cured connector
c) Glass insert connector
d) Heat-cured (oven) connector

LAB EXERCISE

Demonstrate the use of required tools to strip cable, prepare a cable for termination including epoxy and epoxyless methods as well as inspect the connector for proper cleaving, cleaning and alignment.

Section 7.0
Splicing Fiber

Objective

Fuse two fibers together using either a mechanical splice or a fusion splice.

Outline

- *Types of Splicing*
- *Mechanical Splicing*
- *Fusion Splicing*

Learning Activity

Assessment: *Test Module 7*

Lab Exercise: *Demonstrate how to use tools required to accomplish the splicing task, then make two (2) types of splices including using a microscope to inspect work.*

Fiber Optic Splice

A splice is one of the most important elements of fiber optic transmission systems. Without splicing, single fibers could not be joined together to form longer fibers. In commercial buildings and campus environments, designers or installers often avoid splicing by installing a continuous length of cable, which of course is more economical and convenient. But in most installations, the existing cable plant, lengths of the runs, the condition of the raceways and conduit congestion, make splices unavoidable.

Types of Splicing

There are two types of splicing methods: fusion and mechanical. There are advantages and disadvantages to both. The method used will depend on several factors including application, preference of user, volume of splices, type of equipment the service provider has and the level of training of the installer.

The debate over which splicing method is better has gone on for a number of years. Fusion splicing still offers superior results for tensile strength, splice and return loss and environmental stability, but this method requires a very large initial capital investment and is very costly for training and for labor, but it offers low consumable costs. Consumables are all the disposable razors, knives, etc. As a result, fusion splicing has more long-term appeal, especially in the telephony market where cable sections are spliced together and buried in a vault. By contrast, mechanical splicing requires little investment both in capital equipment and labor, but has a higher consumable cost in each splice that is made. Mechanical splicing has become a popular alternative to fusion splicing in the premise environment where cables are more accessible and must be returned to service quickly.

Mechanical Splicing

Capillary Method of Splicing

This is one of the simplest splices to make and consists of nothing more than inserting the ends of the fibers into a thin round capillary tube. (See Figure 7.1.)

Once the fiber ends are prepared, they are inserted and pushed until the ends butt up against each other. This splice is usually preloaded with a matching gel and the fibers are held together by either a mechanical compression or friction by adjusting tunable coiling lines. The final splice we will make is the ULTRAsplice (Figure 7.2) by advanced custom applications.

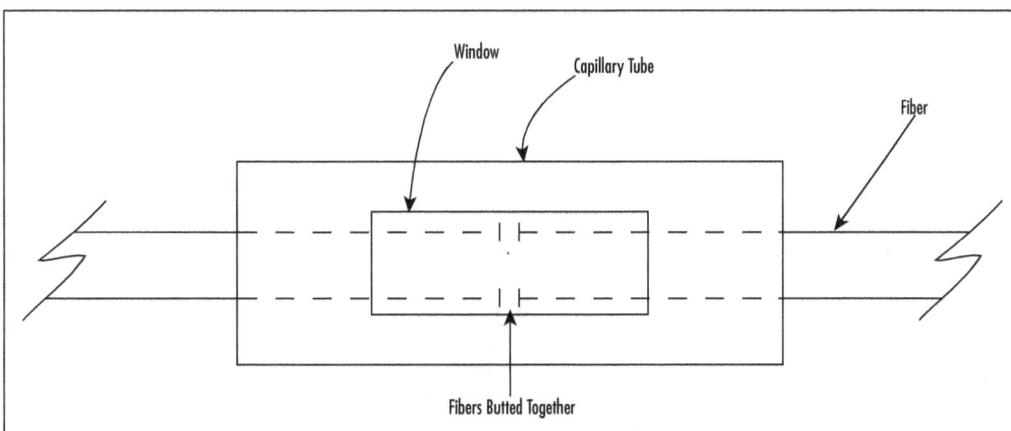

Figure 7.1

Window
Capillary Tube
Fiber
Fibers Butted Together

7.3

Figure 7.2

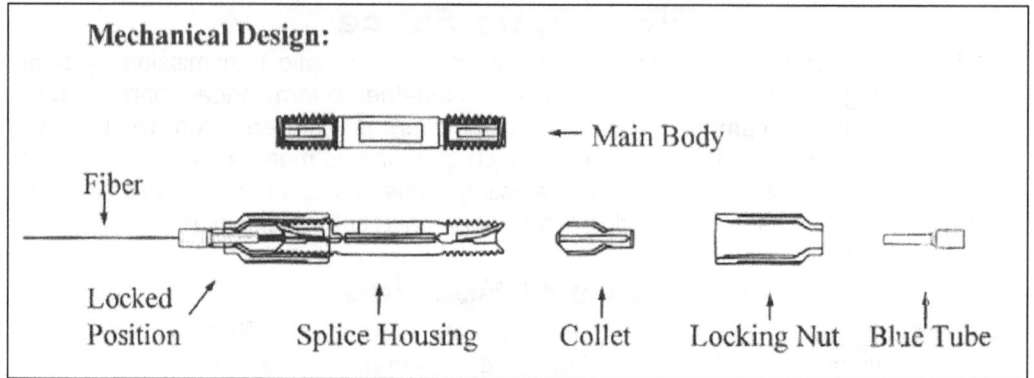

Mechanical Design:

Fiber

Main Body

Locked Position

Splice Housing   Collet   Locking Nut   Blue Tube

ULTRAsplice is a high-performance, low-cost, easy to install and fully mechanical fiber optic splice. Employing a visible glass capillary alignment member that is pre-loaded with index matching gel, the user can inspect the fiber location during installation. The splice is tunable and reusable for most any emergency (or permanent) installation requirement(s).

Prepare the fiber as shown in Figure 7.3. Prepare, clean and cleave the fiber ends. The cleave length should be 7-9 mm.

The ULTRAsplice comes pre-adjusted for 250 micron buffers, but any other buffer sizes WILL WORK after properly setting the gray locking nut position. Please note that the directions apply to either end of the ULTRAsplice, since they are equivalent in the mechanics.
- If working with 250 micron buffers, DO NOT TWIST THE GRAY LOCKING NUT AT ALL. The ULTRAsplice has been set to accommodate the buffer size during factory assembly.
- If working with OVER 500 micron buffers, proceed as follows:
  1. Remove the blue tube.
  2. Open the gray locking nut, in the counterclockwise direction until the bone-colored insert slightly moves back and forth. This action assures that the collet (bone-colored insert) is completely open and is ready to accept the larger diameter buffer.

The gray locking nut and collet have the same mechanical workings as an ordinary hand drill. When the gray locking nut is twisted in the counterclockwise direction (left), the internal collet "opens" in order for larger diameter buffer sizes to be accommodated. When twisting the gray locking nut toward the clockwise direction (right), it "closes" the internal collet onto the buffer. This action "grabs" the buffer (NOT THE FIBER) and has an average of 2.2 pounds of fiber retention. Twist the gray locking nut hand tight, until it cannot be easily twisted any further.

Hold the ULTRAsplice with the visible glass capillary (in the middle of the splice) facing toward you. Retrieve the previously cleaved fiber, and begin to insert it into the splice by placing the fiber into the blue tube (If UNDER 500 micron buffer is used) or the bone-colored collet (if OVER 500 micron buffer is used). Gently insert the fiber through the hole of the collet. You should be able to feel the buffer being grabbed.

ULTRAsplice has a molded in "spring arm" that grabs the buffer upon insertion. This arm only touches the buffer (not the fiber). At any given time, you can retract the fiber from the splice using a limited amount of force (unless the collet is closed) and the arm will continue to grab any other buffer size you insert, allowing for multiple quantity of use per splice.

Figure 7.3

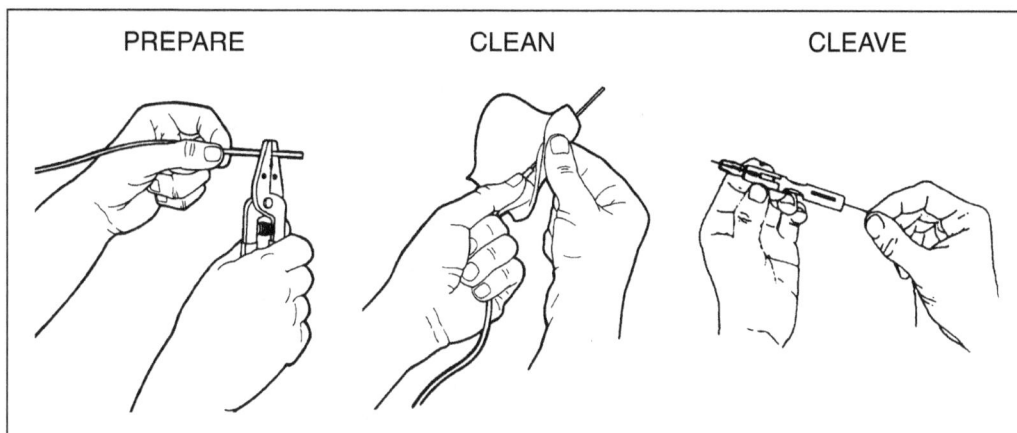

PREPARE   CLEAN   CLEAVE

Once you feel the buffer being grabbed, the fiber end is close to the entrance of the glass capillary. You should be able to see the fiber end through the visible glass capillary.

If cleaved to the proper length of 7 millimeters, the buffer stops automatically. The fiber should be in the middle of the glass capillary, or close as possible to the center. Once the fiber is in the middle, twist the gray locking nut closed (clockwise direction) until the gray locking nut is hand tight. Proceed to the other end.

Start inserting the other fiber end in the same manner as the first end, however, ensure careful attention is given in the following:

(a) The pre-loaded index matching gel acts like a hydraulic piston when the other fiber is being inserted. Therefore, insert the second fiber carefully, giving the gel enough time to "mold" itself around the fiber.
(b) The fiber should stop automatically, however, depending on the accuracy of the cleave length, the fiber could be too long, and possibly break if pushed too hard. Watch the fiber butt up to the other fiber end, through the visible capillary.

Once you have inserted the second fiber to your satisfaction, twist the gray locking nut closed (clockwise direction) – and the ULTRAsplice installation should now be complete.

To lock the gray locking nut, twist counterclockwise until hand tight. (See Figure 7.4.)

Tuning
If you are unsatisfied with the loss you may be achieving, you can simply "tune" the ULTRAsplice as follows:

(a) Twist one end open (counterclockwise direction) - Not more than 1/2 turn.
(b) Gently pull back (slightly) and twist the buffer, possibly even pushing it closer to the other fiber, until an achieved loss is obtained. If you open both ends, and push the fibers together (back and forth), you can possibly damage your cleave. It is recommended that only one end be loosened while tuning.
(c) Twist the gray locking nut closed (clockwise direction until hand tight).

CORELINK Splice by AMP
The CORELINK splice (Figure 7.5) is an easy-to-use high-performance mechanical splice for 125 micron singlemode and multimode fibers. Its unique clamping mechanism is activated with a simple key; no other tooling is required. The transparent material allows the craftsperson to both see and feel the fiber position. The CORELINK splice is remateable – either fiber can be removed and replaced in the splice at any time.

Figure 7.4

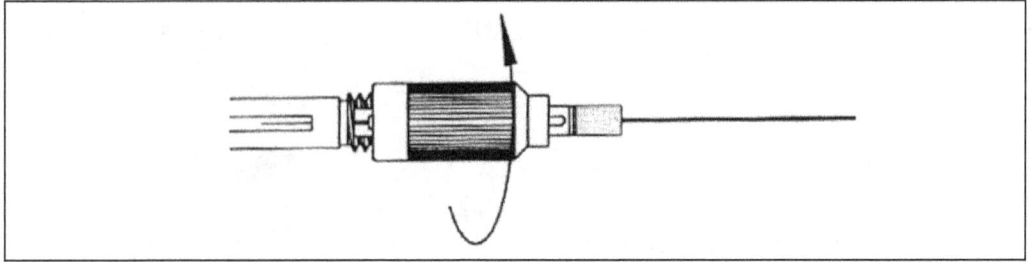

Figure 7.5

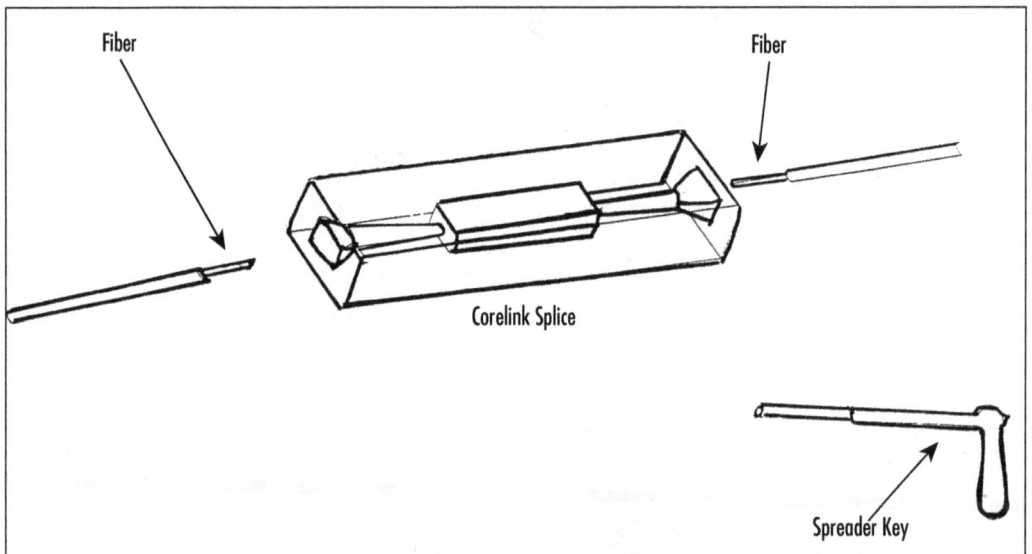

Fiber

Fiber

Corelink Splice

Spreader Key

Making the Splice

Begin by preparing the fiber as previously outlined and by following the manufacturer's recommendations.

- Remove the spreader keys from the key card supplied with the CORELINK splice. Holding the splice, insert the keys into the key entry ports (the holes closest to the edge of the splice). The key handle tabs should be parallel to the flat side and pointing away from the splice. Insert the keys all the way to their shoulders.
- Turn both key handle tabs 90 degrees downward. The splice is now open.
- Align each fiber with the entry port nearest the center of each end face. Insert the fiber slowly, making sure it travels smoothly through the channel into the center element. A 250-μm fiber's coating will stop at the edge of the aluminum element, whereas a 900-μm fiber's buffer will stop at the end of the wide part of the channel. Slide the fibers left or right and visually check that both gaps between the buffer and the aluminum element are equal. (See Figure 7.6.)
- While applying gentle, inward pressure with the thumb and forefinger, gently rotate the key handle turning it 90 degrees. Do NOT snap it closed. This locks the first fiber in place. Inspect the fiber to ensure it remained in the channel during locking.
- Maintain gentle, inward pressure on the second fiber while turning the key to ensure that the fibers butt against each other. Gently rotate the key when locking the second fiber to prevent the fiber tips from bouncing apart. Inspect the fiber to make sure it remained in the channel during locking. This completes the splice.
- To tune the splice, unlock one of the fibers and pull it back slightly, rotate it 90 degrees to tune it, then push it forward and re-lock it. Tuning is usually not necessary.

Figure 7.6

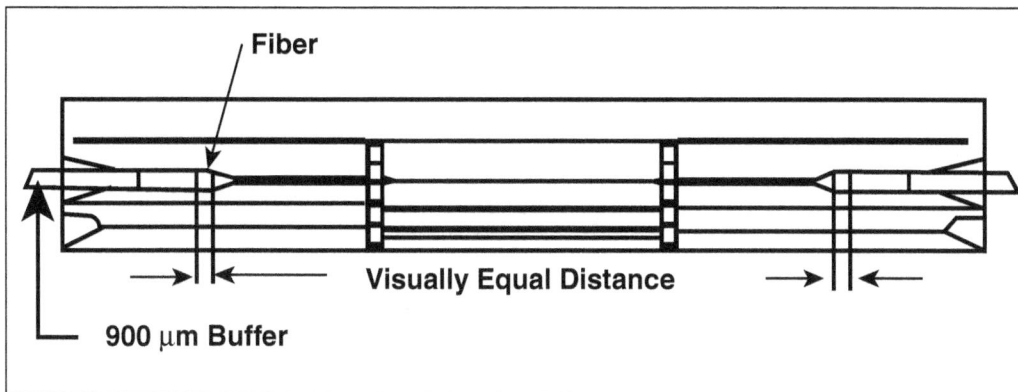

Figure 7.6

Fiberlok by 3M

This splice utilizes a metallic-splicing element held inside a molded plastic body and cap. (See Figure 7.7.)

This splice requires an assembly tool shown in Figure 7.8.

- Prepare fiber as previously outlined and by following manufacturer's recommendations.
- Check cleave length using the cleave length gauge on the Fiberlok Assembly Tool. Place the Fiberlok splice into the assembly. (See Figure 7.9.)
- Push the fiber down into the fiber retention pad on the proper side of the splice (Figure 7.10).
- Grasp the fiber about 1 inch (25 mm) from end and move it into the fiber alignment guide on the assembly tool (Figure 7.11).
- Continue pushing fiber straight through the alignment guide into the splice entrance port until resistance is felt. Total fiber travel should be about .75 inch (20 mm) before resistance is felt.
- Grasp first fiber near fiber retention pad and push fiber toward splice until about .1 inch (3 mm) bend develops (Figure 7.12).
- Continue pushing the second fiber straight through the alignment guide into the splice entrance port. As the second fiber passes the halfway point of total fiber travel, begin to watch for the bend in the first fiber to increase. This occurs when the end face of the second fiber contacts the first fiber and pushes the first fiber backward. Continue pushing the second fiber until the bend in the first fiber is approximately .3 to .4 inches (8 to 10 mm) (Figure 7.13).
- Push the first fiber back against the second fiber until there are equal bends in both fibers of approximately .2 to .3 inches (5 to 8 mm).
- Pivot the handle of the Fiberlok assembly tool down until it contacts the cap of the Fiberlok splice. Squeeze the handle of the assembly tool (Figure 7.14) in order to keep fibers and tool steady. When possible, secure the tool to a work surface for added support. A snap will be heard when the splice is actuated.
- Remove the Fiberlok splice from the assembly tool by first removing the fibers from the foam pads and then lifting the splice from the splice holding cradle (Figure 7.15).

Siecor CamSplice

Another type of mechanical splice is the CamSplice (Figure 7.16) by Corning Cable Systems. This splice requires no special tools, adhesives or curing time, and can be used for both loose tube and tight buffer fibers.

Figure 7.7

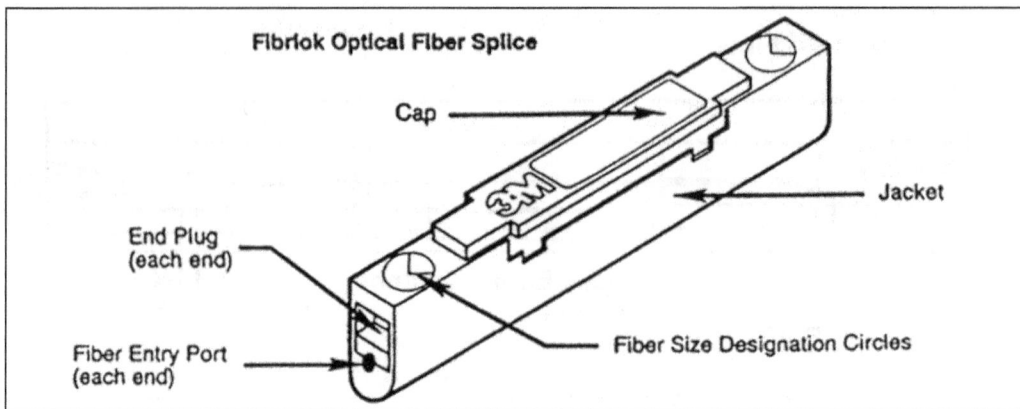

Fibrlok Optical Fiber Splice

Cap

End Plug
(each end)

Fiber Entry Port
(each end)

Jacket

Fiber Size Designation Circles

Figure 7.8

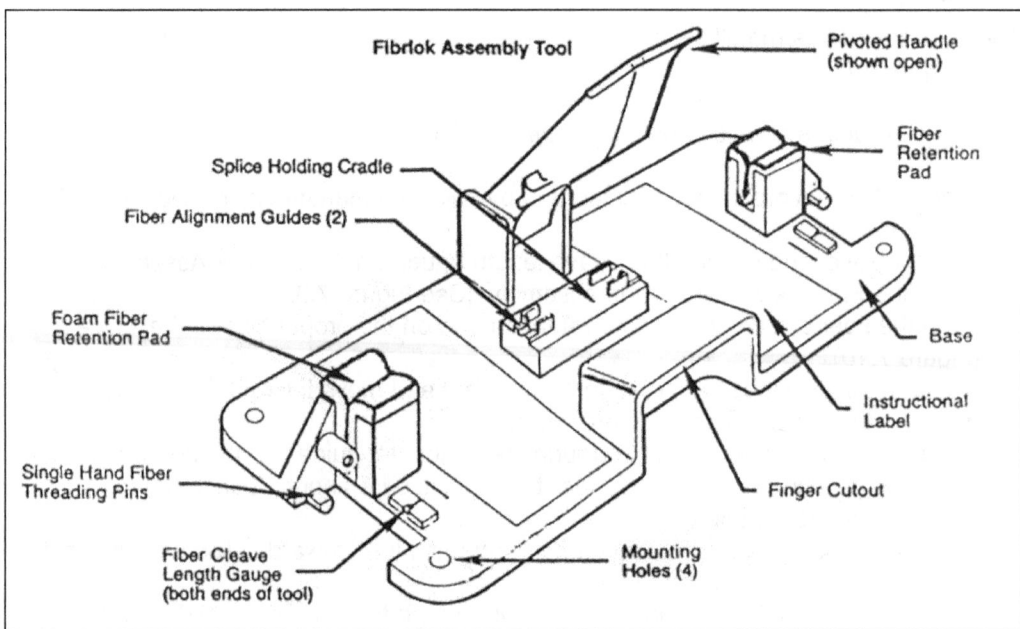

Fibrlok Assembly Tool

Pivoted Handle
(shown open)

Fiber
Retention
Pad

Splice Holding Cradle

Fiber Alignment Guides (2)

Foam Fiber
Retention Pad

Base

Single Hand Fiber
Threading Pins

Instructional
Label

Finger Cutout

Fiber Cleave
Length Gauge
(both ends of tool)

Mounting
Holes (4)

Figure 7.9

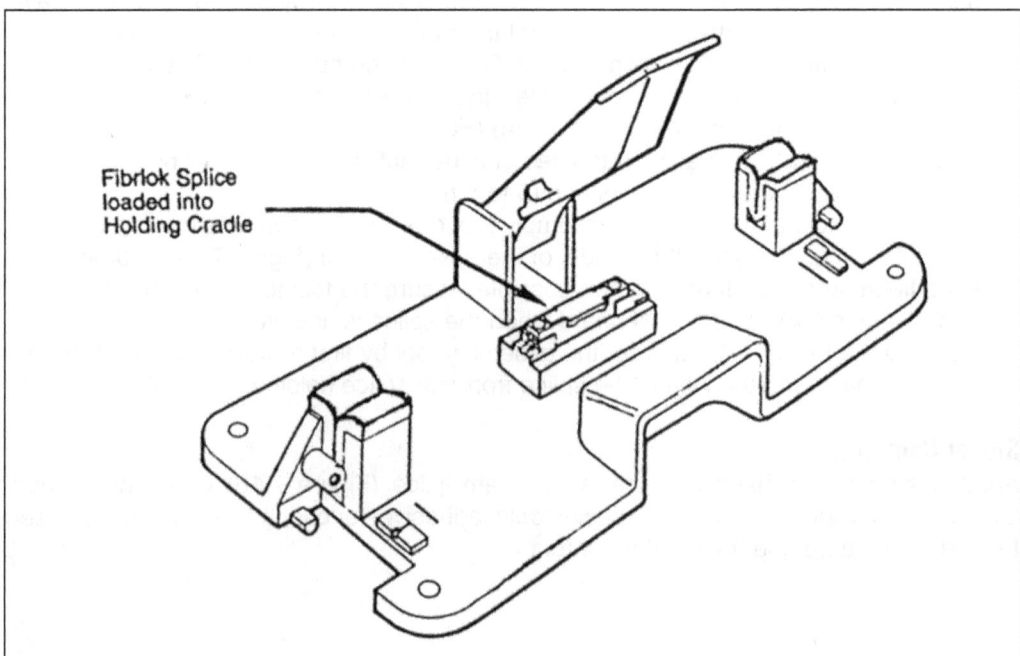

Fibrlok Splice
loaded into
Holding Cradle

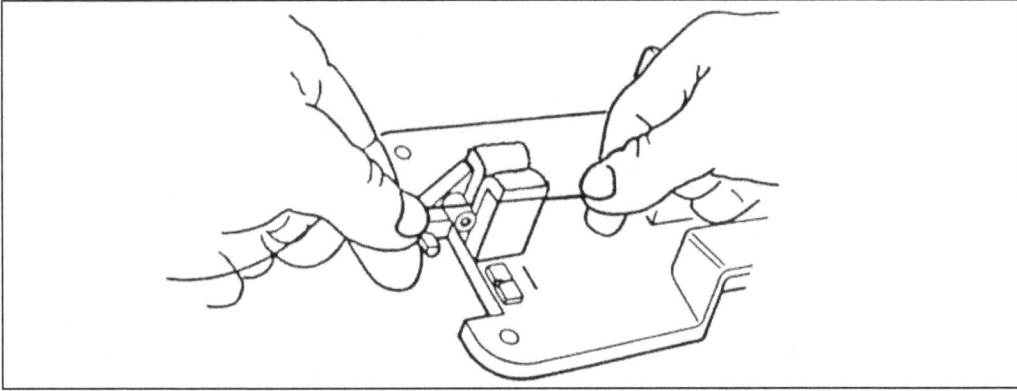

Figure 7.10

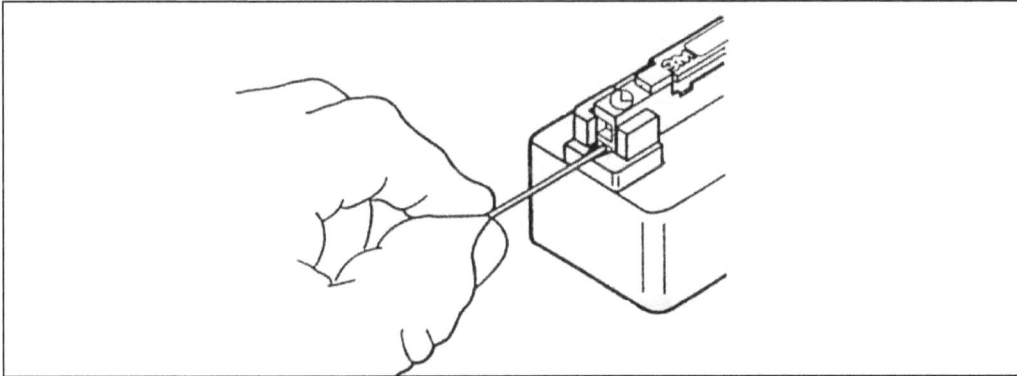

Figure 7.11

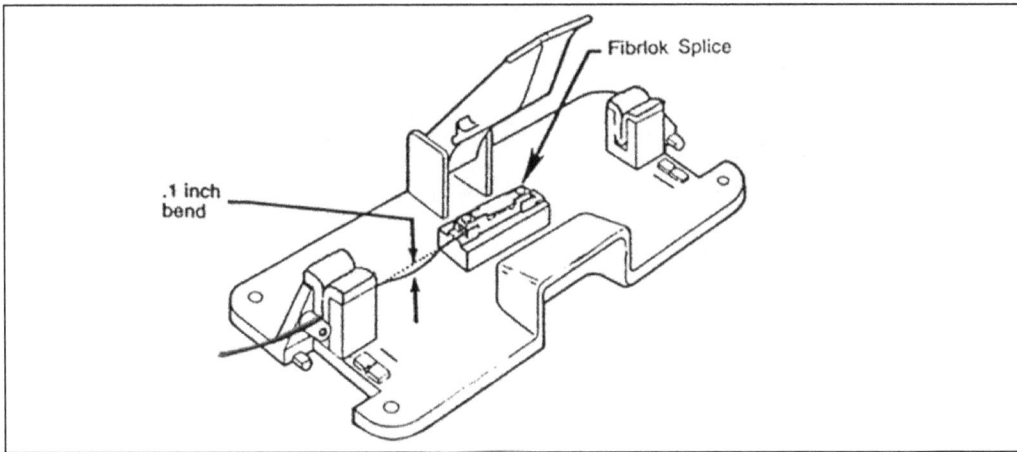

Figure 7.12

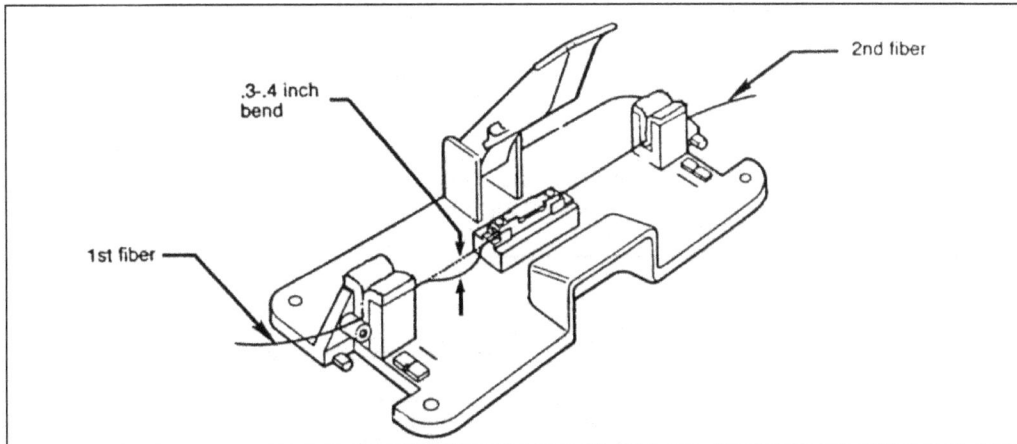

Figure 7.13

Figure 7.14

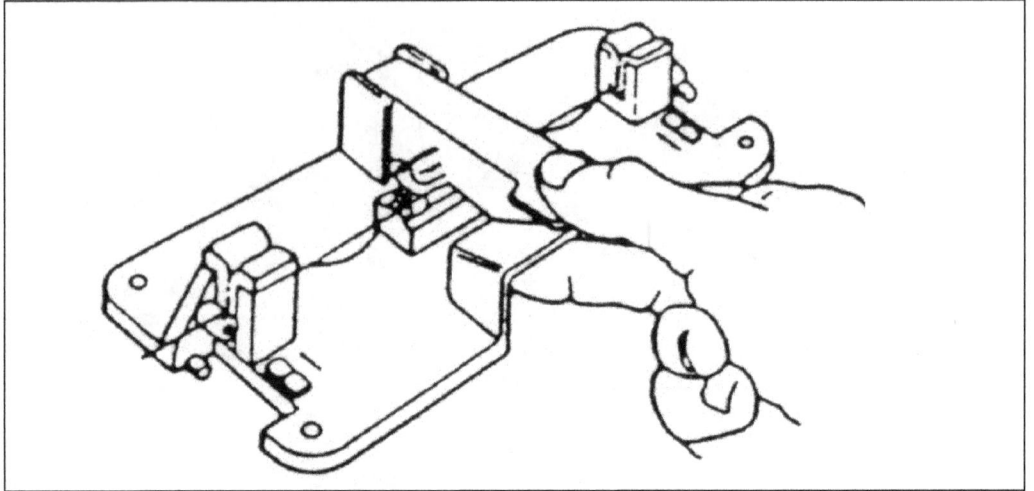

Figure 7.15

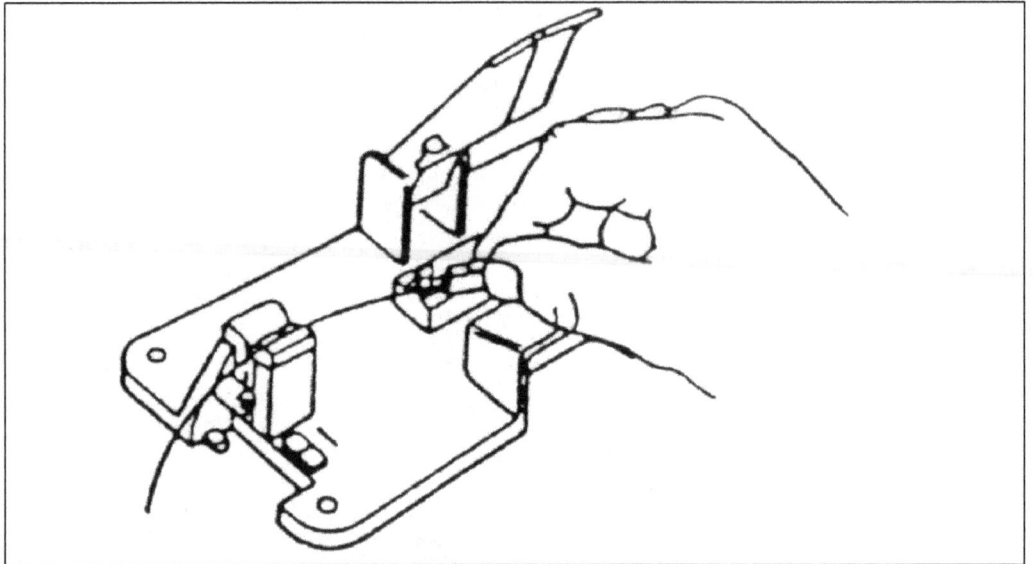

Figure 7.16

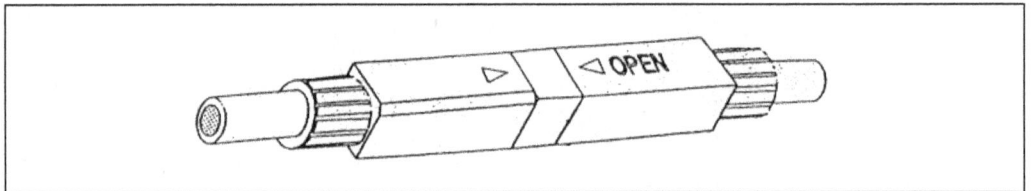

Notes

_____

_____

_____

_____

_____

_____

_____

_____

Follow these steps for using the CamSplice:

- Feed the fiber into the assembly tool, creating similar bends. (See Figure 7.17.)
- Once the fibers are in place, with the appropriate bends showing, rotate both levers to actuate the splice. (See Figure 7.18.)

To remove a completed CamSplice from the assembly tool:

- Gently release the fibers from both foam clamps.
- Working from the right side of the assembly tool, carefully lift the CamSplice out of the tool. (See Figure 7.19.)

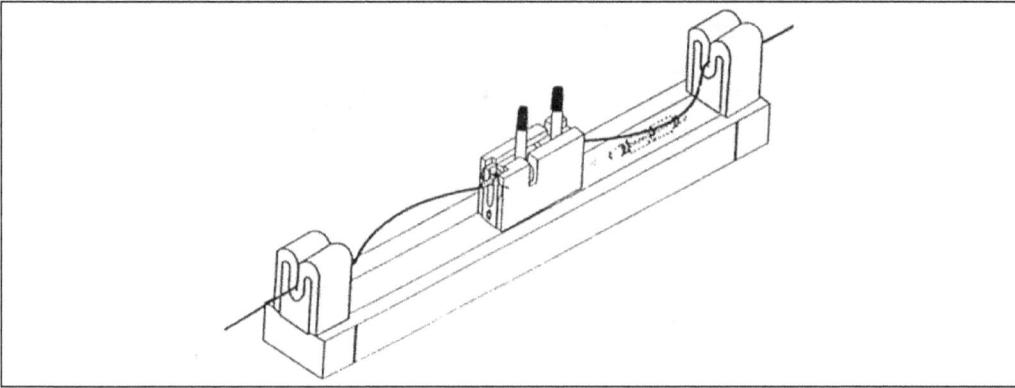

Figure 7.17

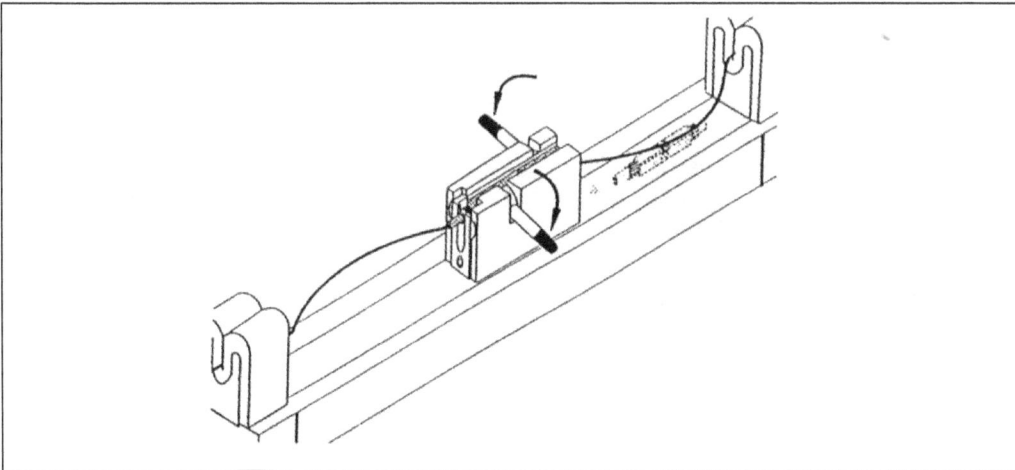

Figure 7.18

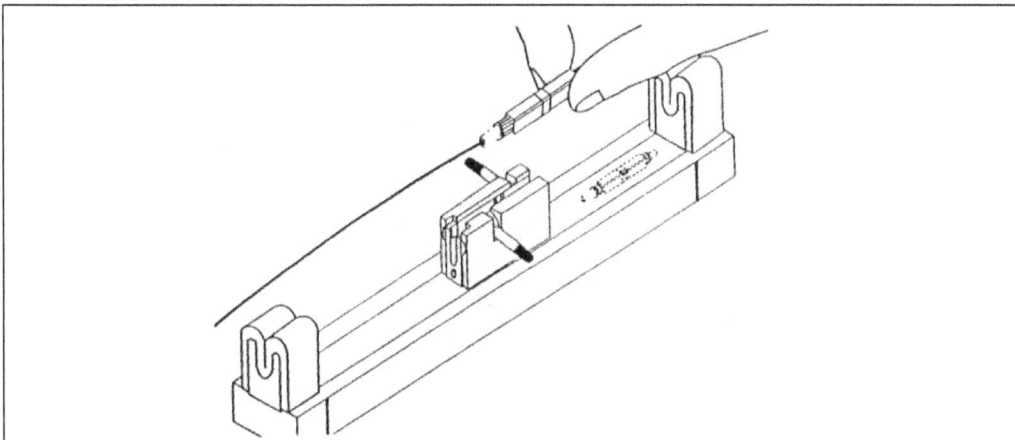

Figure 7.19

Secure the CamSplice in its splice tray or hardware. Follow the instructions provided with the splice tray or hardware to ensure that the splice and its fibers are properly strain-relieved.

Tuning. The CamSplice can be retuned or readjusted using an OTDR or power meter. This is accomplished by first rotating both levers back to their vertical position to open the splice. Rotate the left lever 45 degrees to tighten its fiber. This action provides a positive stop to butt the right fiber against while tuning. Pull back on the right fiber, slightly rotating it, and then rebut it against the left fiber. Check for loss improvement and repeat this step until you have achieved maximum performance. After optimizing the splice, rotate both levers down fully to actuate the splice. Check for splice loss again.

Fusion Splicing

Why splice?
Splicing is normally avoided in building or campus environments. However, long cable runs, crowded conduits and fire code restrictions sometimes make splicing necessary. (See Figure 7.20.) In many cases, it may be possible to simply add connectors to the fibers and join them together, a method that provides flexibility for future system reconfiguration and testing, especially if patch panels are used at the cable ends. Unfortunately, when this is done, signal loss, space requirements and - depending on the application - costs are increased. For nearly all outside joining points, splicing makes the most sense. This is why telephone and cable television companies do so much of it. Indoors, the decision to splice is made when the junction will be permanent and when it makes sense economically.

Why a Fusion Splice?
Mechanical splicing is often less expensive than fusion splicing, especially if you only make a couple of hundred splices a year. If you do a lot of fiber work, though, that $7 - $20 per mechanical splice starts to add up and afterward, all you have left to show for it are the empty boxes the materials came in. The fusion splicer, on the other hand, has a considerable initial cost, but practically nothing during its useful life. The fusion splice also has an advantage in that it reflects less light than the mechanical splice. For most local area networks (LANs) and digital applications, reflections are inconsequential, although high reflections can create serious problems for analog video signals. For this reason alone, cable television companies almost always use fusion splicing. For telephone companies and contractors, the choice between fusion and mechanical splicing remains primarily an economic one.

Figure 7.20

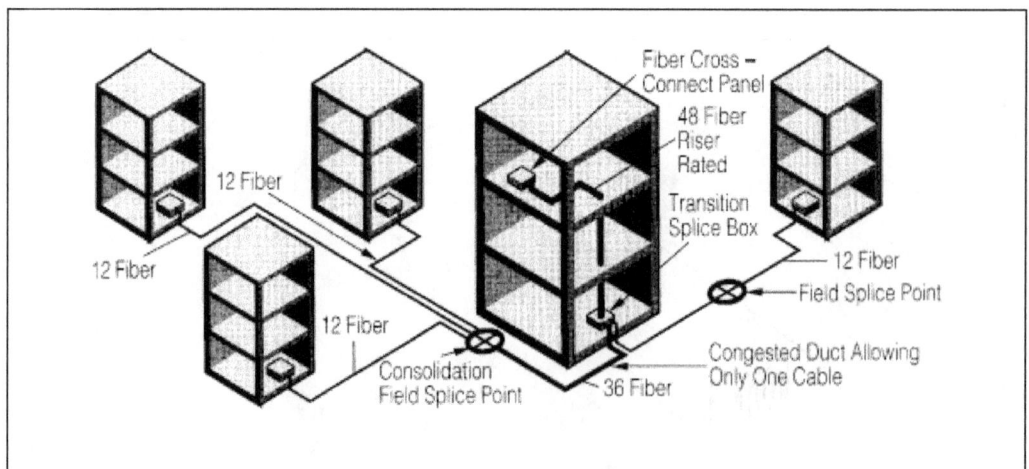

Fusion splicing works on the principle of an electric arc ionizing the space between the prepared fibers to eliminate the air and to heat the fibers to the proper temperature. There are two fusion splicing technologies; the local injection detection method and the profile alignment method. Both are excellent methods for fusing glass and both can produce an almost near to zero splice. Both methods have approximately the same procedure for fusing a splice.

Most fusion splicers (Figure 7.21) are menu driven and include many of the following features:

- Full programmability.
- Wide angle video monitors. The monitors are omni-directional, swivel type monitors that show two simultaneous views of the fiber.
- Built-in heaters. Heaters are used to heat shrink the tubing used for mechanical protection, which covers the new bare splice.
- Built-in cleavers.
- Fully automatic alignment. After the fibers are cleaved, cleaned and placed in the alignment groves, they will be aligned automatically.
- RS-232 port. Used for downloading information to a printer.
- Splice data memory compatible with all manufacturers of fiber cable. This report is needed to calibrate the type of fiber being installed.
- Test results from the fusion splice.

Each fusion splice will have several other options ranging from work lights to data output reports, which documents the dB loss of a finished splice.

Making a Fusion Splice
After carefully reading the manufacturer's operation specifications, begin by setting the adjustments for the size of fiber, coating and heat parameters recommended by the manufacturer.

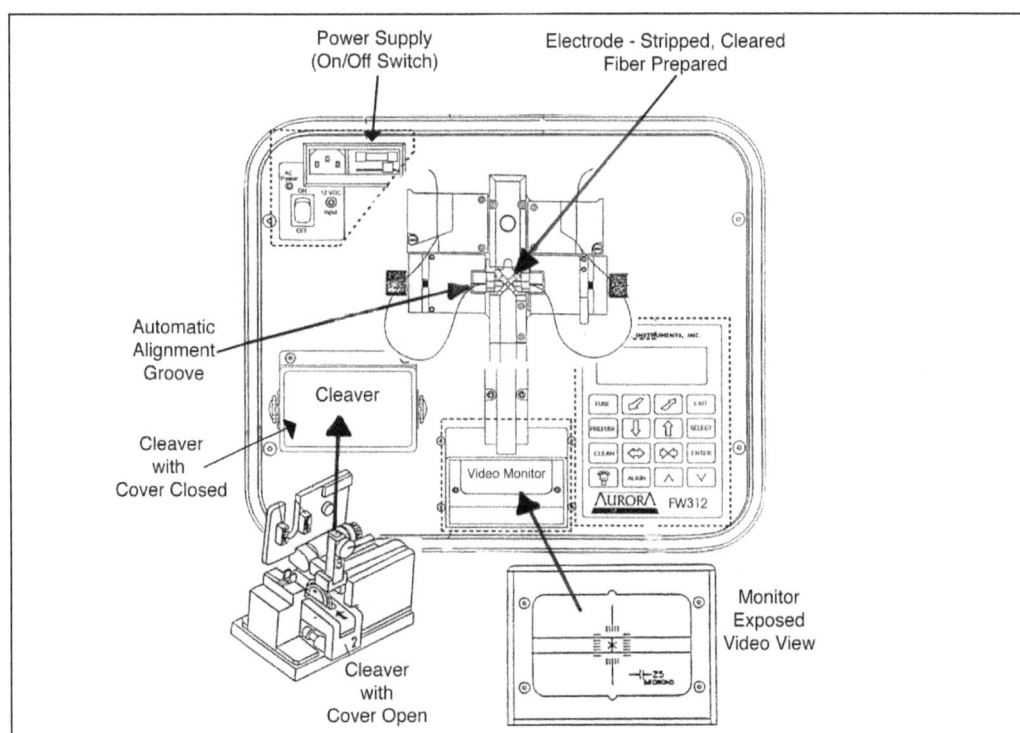

Figure 7.21

1. Determine the amount of coating to be stripped according to the manufacturer. (See Figure 7.22.)
2. Remove the buffer with a No-Nik tool. (See Figure 7.23.)
3. Clean the bare fiber and wipe with an alcohol soaked, lint-free tissue.
4. Open the flap of the cleaver and place the fiber in the cleaver so the end of the coating is at the 10-millimeter mark. Close the flap of the cleaver and then press downwards carefully. Open the flap before you remove the cut fiber in order to avoid damage to the end face of the fiber. (See Figure 7.24.)
5. Discard the broken fiber stub properly.
6. If using heat-shrink tubing, place a piece of tubing over either end of the fiber sliding far enough back not to interfere with the fusion process.
7. Open the electrode and fiber holding flaps. Grasp the fiber by the coating next to the bare fiber; place the fiber into the fusion splicer. Typically, the end of the fiber is placed so it stops between, but not past, the upper and bottom electrodes. (See Figure 7.25.) Repeat the same procedure for the other fiber.
8. Close the flaps.
9. Push the appropriate button, which will cause the unit to clean the ends of the fiber, to determine if the cleaving is acceptable, and to align the fibers.
10. Push the fusion button. The splice is now fused. A small arc of light will be visible around the cover.

If the splice was fused correctly, the splice loss will be displayed in decibels (dB) such as ‡ 0.1 dB on the monitor.

The splice can be viewed on the video monitor. Improperly joined fibers, caused by poor cleaving or cleaning, will not facilitate a proper splice. Examples of a good splice and a bad splice are illustrated in Figure 7.26.

Figure 7.22

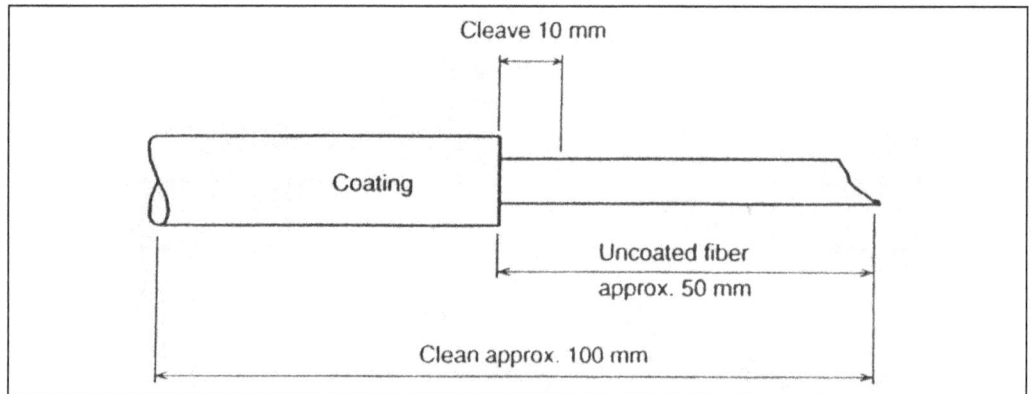

Figure 7.23

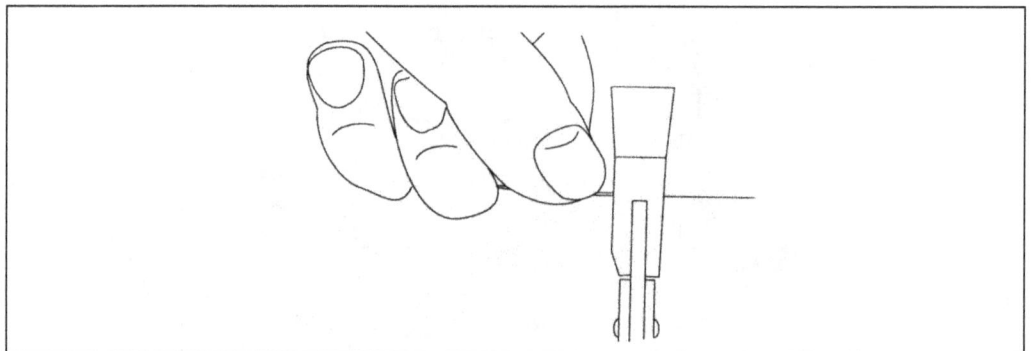

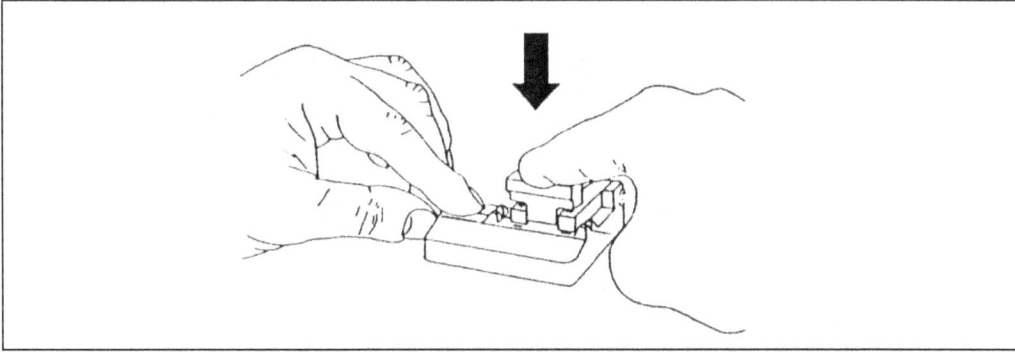

Figure 7.24

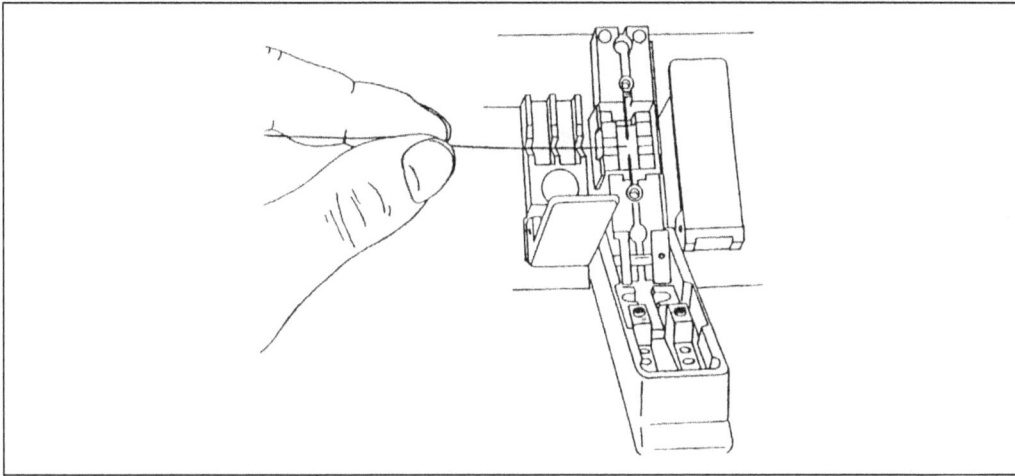

Figure 7.25

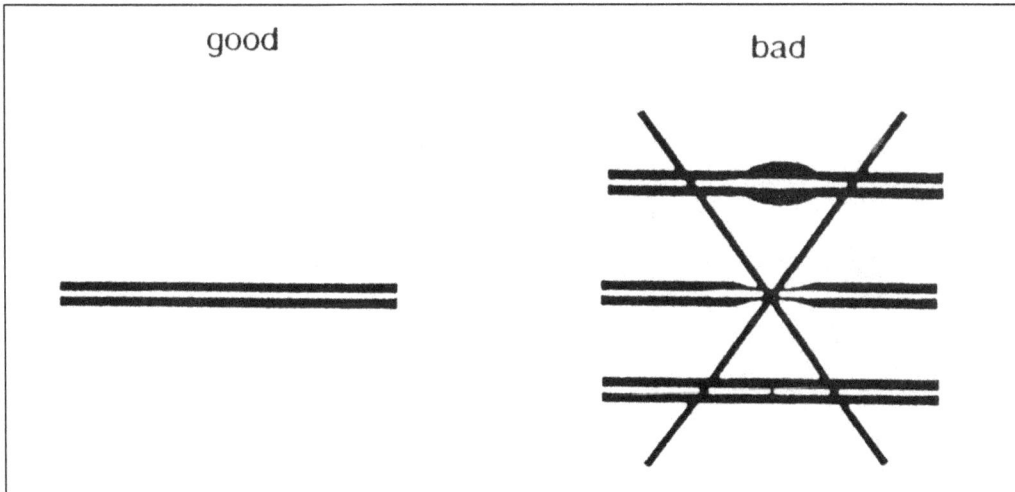

Figure 7.26

good

bad

Open the flaps over the electrodes and the fiber-holding V-grooves; remove the completed splice from the fusion splicer.

If using heat-shrink tubing protection, slide the device so it is centered over the splice, and then place the tubing in a heater. Heating is initiated by closing the cover (red LED lights) and is automatically ended after approximately two minutes. After the two minute time period, the shrinking is completed and the protected splice can be removed.

Remove the finished splice and place it in a splice tray. (See Figure 7.27.)

7.15

Figure 7.27

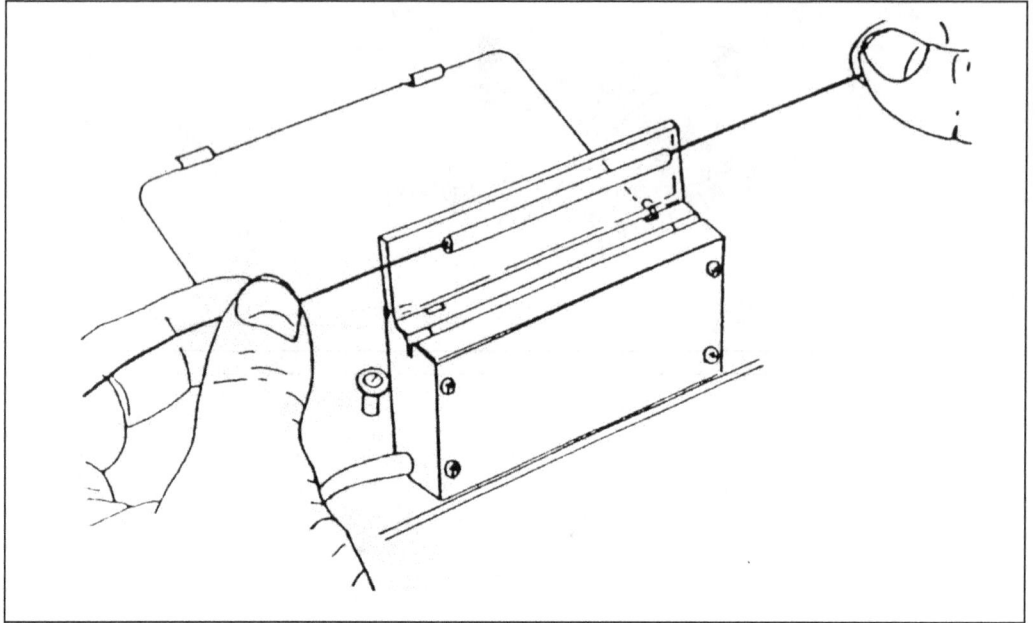

Before purchasing a fusion splicer, research the features you require and how much you wish to spend. There are several manufacturers of fusion splicers, with units ranging in price from $8,700 to as much as $30,000. After purchasing a fusion splicer, carefully read the directions, especially on start-up and troubleshooting.

TEST MODULE 7

1. Fusion splicing is often less expensive than mechanical splicing.

   True   False

2. Mechanical splicing is one of the simplest to make and consists of nothing more than inserting the ends of the fibers into a thin round capillary tube.

   True   False

3. Article 770 in the NEC defines the installation of fiber optic cables and raceways.

   True   False

4. One of the most important elements of fiber optic transmission systems is splice.

   True   False

5. The two types of splicing methods is fusion and mechanical.

   True   False

6. The CORELINK splice is an easy-to-use high-performance mechanical splice for 125 μm singlemode fiber.

   True   False

7. The Fiberlok by 3M utilizes a metallic-splicing element held inside a molded plastic body and cap.

   True   False

8. Splicing is always used in building or campus environments.

   True   False

9. Fusion splicing works on the principle of an electric arc ionizing the space between the prepared fibers to eliminate the air and to heat the fibers to the proper temperature.

   True   False

10. ULTRAsplice is a high-performance, low-cost fully mechanical fiber optic splice but hard to install.

    True   False

LAB EXERCISE

Demonstrate how to use tools required to accomplish the splicing task, then make two (2) types of splices including using a microscope to inspect work.

Section 8.0
Couplers

Objective

Examine the function of the coupler, including how the coupler distributes optical signals much like copper distributes electrical signals.

Outline

- *Basics of Couplers*
- *Types of Couplers*
- *Splitters*

Learning Activity

Assessment: Test Module 8
Lab Exercise: Familiarization of products on display.

Couplers

Most couplers manufactured today are passive. Passive optical devices are the functions performed by the propagation media through which the waves pass without the use of input energy other than that contained in the waves themselves. In fiber optics, a coupler is a device that enables a transfer of optical energy from one optical waveguide to another. Of course this is true of passive couplers if the total output power cannot be more than the total input power.

Fiber optic couplers serve the same function as "T" taps in copper-based networks. An optical coupler distributes optical signals just as a copper "T" would distribute the electrical signal. A coupler is a passive device that either divides or combines the optical signals.

Basics

Couplers are multiport devices. A port is an input or output point for light.

- Throughput port - when the splitting ratio between two output ports is unequal, the throughput port outputs the greater amount of power.
- Throughput loss - ratio of power at throughput port to the power at the input port.
- Tap port - when the splitting ratio between two output ports is unequal, the tap port outputs the lesser amount of power.
- Tap loss - the ratio of power at the tap port to the power at the input port.
- Directionality - ratio between unwanted power at input port four, and input power at input port one on a four-port coupler.

Splitters

Couplers usually serve two functions. Sometimes they divide one input into two or more output signals to drive multiple devices. Another function is the couplers combine two or more inputs from separate functional devices in an effort to provide input to another device. These functions are directional and if signals are transmitted in both directions through the system in the same manner as the bidirectional coupler, they are called splitters or combiners.

Types of Couplers

There are three basic types of couplers: the T, the tree and the star.

T Type Coupler
The T coupler has one input and two outputs. (See Figure 8.1.)

Tree Coupler
The tree coupler has a signal input and more than two outputs. (See Figure 8.2.)

Star Coupler
The star coupler has multiple inputs and outputs. (See Figure 8.3.)

Active Couplers
The active coupler is similar to the passive coupler except the active coupler can either generate or amplify light.

Figure 8.1

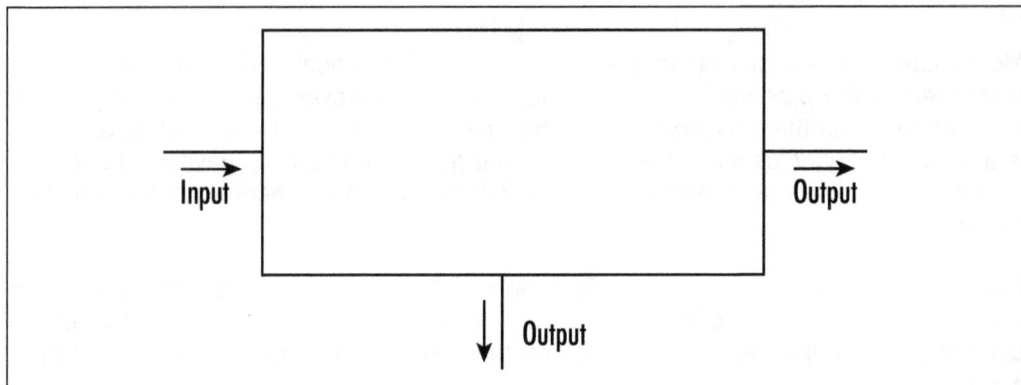

Input → Output

↓ Output

Figure 8.2

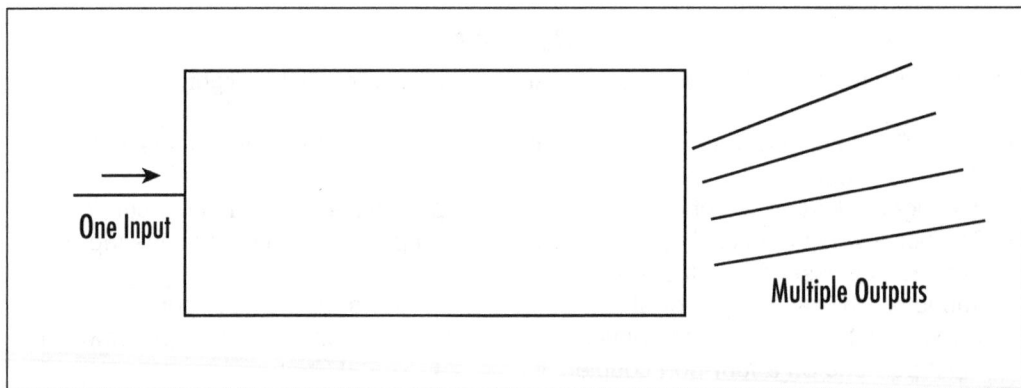

One Input →

Multiple Outputs

Figure 8.3

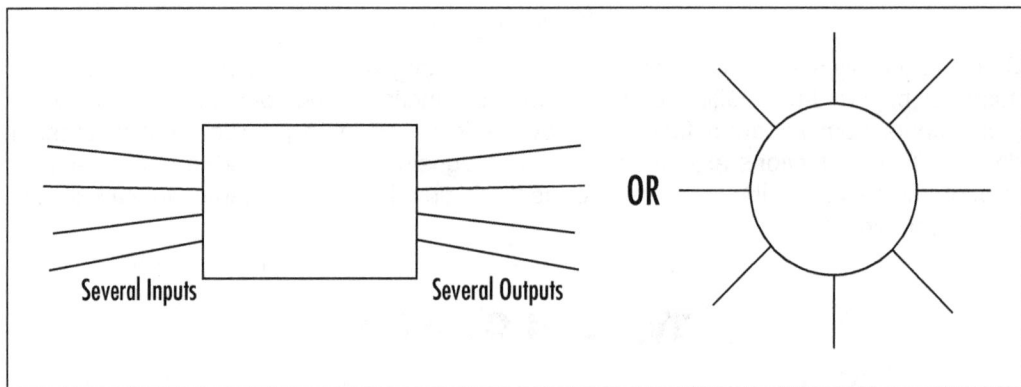

Several Inputs   Several Outputs   OR

TEST MODULE 8

1. Fiber optic couplers serve the same function as "T" taps in copper-based networks.

   True   False

2. There are three basic types of couplers, the T, the tree and the star.

   True   False

3. The T coupler has a signal input and more than two outputs.

   True   False

4. The tree coupler has one input and two outputs.

   True   False

5. The star coupler has multiple inputs and outputs.

   True   False

LAB EXERCISE

Familiarization of products on display.

Section 9.0
Light Sources

Objective

Understand how electronic signals are converted to lightwaves for fiber optic telecommunications.

Outline

- *Introduction to Light Sources*
- *Types of Light Sources*

Learning Activity

Assessment: Test Module 9
Lab Exercise: Familiarization with products on display.

Introduction

There are some important factors when selecting a light source for a fiber optic system. The light must be transmitted effectively at a wavelength by the optical fiber, commonly 780-850, 1300 or 1550 nm windows for glass fibers or 650 nm window for plastic fibers. The range of wavelengths is a factor, because the larger the range, the larger the potential chromatic dispersion. The light source must generate adequate power sending the signal through the fiber, but not enough that it causes nonlinear effects or distortion in the fiber or receiver. The output light must be modulated so it carries the signal and the light source must transfer its output effectively into the fiber.

Light Source

The two developments that made lightwave communication practical were the laser and glass fiber. (See Figure 9.1.)

Laser produces an intense beam of highly stimulated light, which travels in a straight path. The glass fiber used is of such purity that only a minute portion of a light signal emitted into the fiber is attenuated. With a laser source that is switched on and off at high speeds, the zeros and ones of a digital communications channel can be transmitted to a detector, such as a photodiode. The received signals are converted from light back to electrical pulses and passed on to equipment such as a multiplexer. (See Figure 9.2.)

In a fiber optic transmitter, the light source can be modulated by a digital or analog signal. For analog modulation, the input interface matches impedances and limits the input signal amplitude. For digital modulation, the original source may already be in digital form or, if in analog form, it must be converted to a digital pulse stream. For the latter case, an analog-to-digital converter must be included in the interface. (See Figure 9.3.)

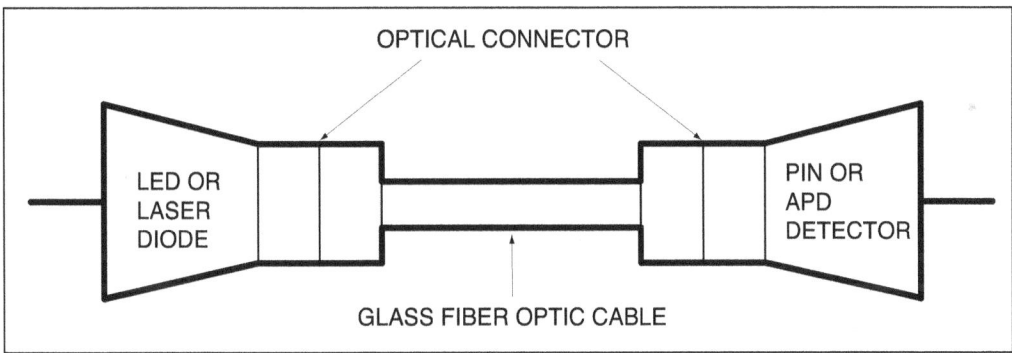

Figure 9.1

OPTICAL CONNECTOR

LED OR LASER DIODE

PIN OR APD DETECTOR

GLASS FIBER OPTIC CABLE

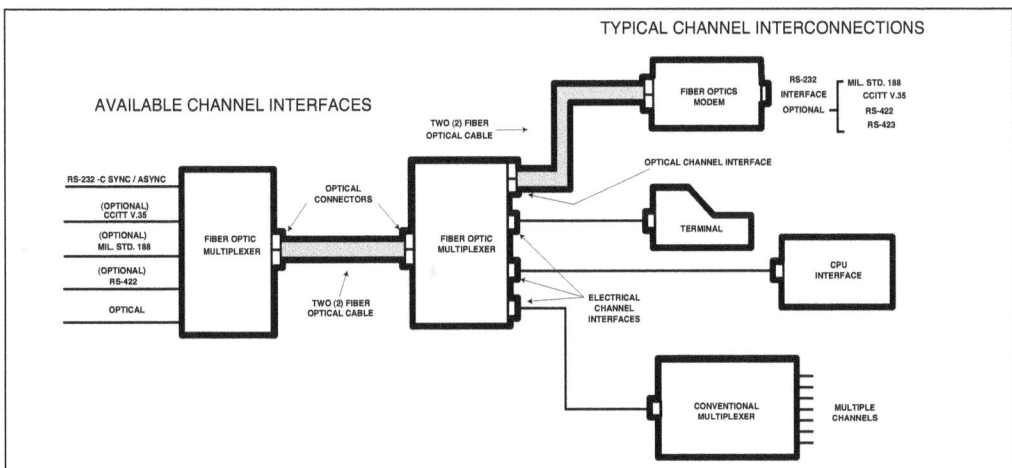

Figure 9.2

TYPICAL CHANNEL INTERCONNECTIONS

AVAILABLE CHANNEL INTERFACES

9.3

Figure 9.3

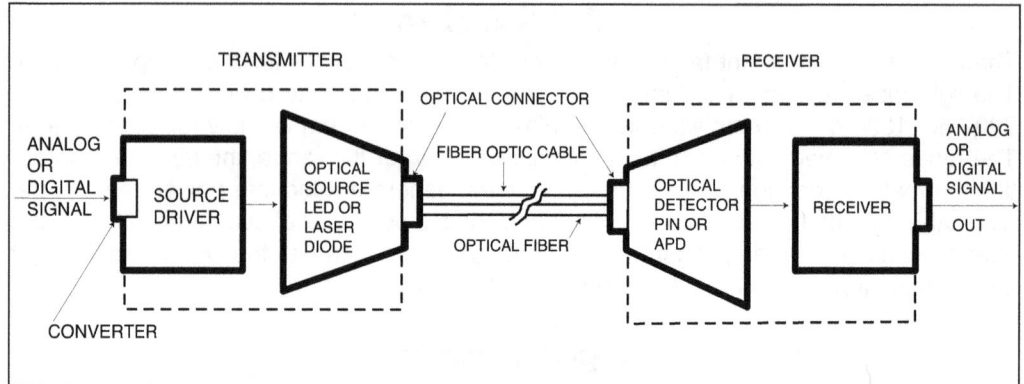

Semiconductors used as light emitters for fiber optic transmission fall into two categories, light emitting diodes (LEDs) and semiconductor lasers (often called diode lasers). Semiconductor devices fit well with standard electronic circuitry used in communication systems while fiber lasers and other compact solid-state lasers are used in some systems. In fiber optics, the lasers are usually referred to as injector laser diodes (ILDs). Two types of ILDs are commonly available, the single heterostructure and the double heterostructure. Although the single heterostructure ILD is high in output power, it has a limited duty cycle and requires an unusually high driving current.

The double heterostructure ILD, like an LED, is not limited by its duty cycle. Its optical output capability can be as high as the 20mW level. However, ratings in the range of three to 10mW are most common. Light emitted from ILDs is nearly monochromatic. That is, its spectral linewidth is typically two to five nanometers, (Nanometer (nm) = one billionth of a meter, or 10-9 meters.) The nanometer is a convenient unit for designating the wavelength of visible electromagnetic radiation (namely, the wavelength of light). The wavelengths extend from 750 nm at the highest infrared energy level (near-infrared) down to 390 nm lowest energy level (ultraviolet). The rated wavelength of an ILD is usually near 900 nm and its rise time to full output is in the range of 0.1 to 2 nanoseconds.

Given their high speed, higher power and narrow linewidth, laser sources lend themselves well to fiber optic transmission applications involving long distance and wide bandwidth. However, they are not necessarily essential in cases where distances are more moderate. The development of lower-loss multimode fibers and super luminescent devices (LEDs with high output power and narrow linewidth) can provide lower cost fiber optic data links in many environments.

Trade-offs on Cost and Performance

Students of engineering expect light sources with the most desirable characteristics to cost the most. The cheapest light sources are LEDs with slow rise times, large emitting areas and relatively low output power. The most expensive are diode lasers with narrow bandwidths emitting the 1300- and 1550- nm wavelengths where optical fibers have the lowest losses. The higher-power and narrower-line emission of lasers are at a premium, with the narrowest-line lasers costing more. LEDs usually have a longer lifetime than some lasers.

Functional Differences

Lasers are more powerful than LEDs because they convert electrical input power to light more efficiently and have higher drive currents. Lasers have a concentration of stimulated emission leading to a narrower beam. Laser lifetimes are somewhat shorter than LEDs because of the higher currents and optical power, but is not an important factor except where temperatures of GaA1As lasers cannot be controlled.

Attenuators

While transmitters are made to produce standard power levels, sometimes those levels are too high. This may occur in networks where the legs are different lengths or that have more fiber junctions than others.

Attenuators can be used to reduce transmitter output to safe levels because receivers can only handle limited input powers. On rare occasions, they may be used at the receiver.

Light Emitting Diodes

For links of moderate distance, our primary concern here, an LED source is generally appropriate. There are currently three popular types of LEDs: (See Figure 9.4.)
- Surface emitters
- Edge emitters
- Burrus-type

In the Burrus-type LEDs, an etched well defines the limit of the emitting surface area. The structure of the other two types is self-explanatory. All three often have a focusing lens bonded integrally to the semiconductor to improve the optical efficiency, further increasing their suitability for fiber optic applications.

Commercially available LEDs vary in wavelength from 565 nm (green) to 1,300 nm (infrared), with 660 nm (red) being quite common for such applications as panel lighting and other sorts of readouts. However, when used as a fiber optic transmitter, an LED should produce light that has the lowest attenuation for the fiber being used. In this case, between 800 and 900 nm or above 1,000 nm. The actual line-width of the LED typically ranges from 25 to 40 nm.

The drive requirements for an LED source typically falls in the 50 to 150 mA range; rise time will usually be somewhere between three and 20 nanoseconds. Typical power output ratings are around one milliwat, although there are some devices rated at more than twice this level.

LED emission profiles are very different from those of ILDs. There is a comparatively wide beam-width, as well as divergence, in LEDs, making the NA (numerical aperture) specification a critical piece of information. The actual amount of rated LED power launched into a fiber depends on the relationship between the two components' NAs. In any case, given their response (rise time) and spectral range, LEDs are generally adequate where data

Figure 9.4

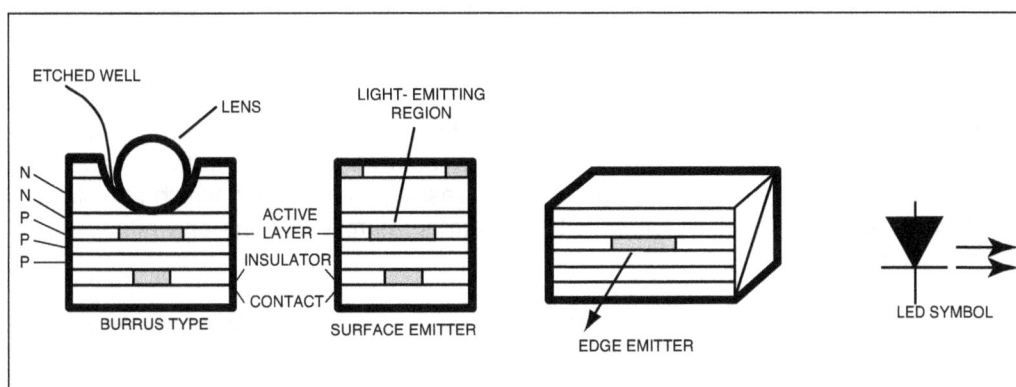

ETCHED WELL
LENS
LIGHT- EMITTING REGION
N
N
P
P
P
ACTIVE LAYER
INSULATOR
CONTACT
BURRUS TYPE
SURFACE EMITTER
EDGE EMITTER
LED SYMBOL

rates are under 10 MHz, 20 Mbps and distances are under one kilometer. In other words, they meet the needs of most data links.

Light Sensors for Data Links - A Number of Alternatives

At the receiving end of a fiber data link, a photodetector senses light and converts it to electricity for subsequent processing. In a very basic system, optical power input is all that is modulated in transmission. Therefore, light is all that must be sensed. Advances in multiplexing have brought about the use of other parameters, including frequency, in some data links. For the most part, direct (coherent) detection of light is all that is required. Since a detector's photoactive region is generally large in relation to a typical fiber's exit NA, fiber-to-detector coupling presents fewer problems with respect to power loss.

There are several types of detectors that are commonly used in fiber optic data links:
- Photodiodes
- PIN photodiodes
- Phototransistors
- Avalanche photodiodes (APDs)

The essential differences that render a particular type more or less appropriate in a given application are evident in the following brief descriptions.

Photodiodes
Standard photodiodes, operating in the photocurrent mode, provide good linearity, speed and stability as fiber optic receivers, but produce no gain.

PIN Photodiodes
These are usually chosen over the simpler photodiodes when data transmission is involved. The PIN photodiode's performance is improved with an undoped intrinsic layer that lowers device capacitance and allows very good frequency response (typically one GHz). The five to 10 volt biases normally required can be provided by fairly straightforward circuitry. Although it provides no intrinsic gain, the PIN photodiode is relatively inexpensive and easy to use. Ideally, a fairly strong light input to a PIN photodiode is provided to compensate for its low sensitivity and signal-to-noise ratio.

Phototransistor
These devices have relatively good responsivity and gain, but have high device capacitance and thus slow (2 nanoseconds) rise time. The gain-bandwidth is typically around 100 MHz or less and responsivity around 35 uA/uW. One two-stage variation, the photodarlington, provides even higher gain, but only at a cost of speed.

Avalanche Photodiodes (APDs)
APDs offer better gain and sensitivity than PIN photodiodes. Although somewhat more expensive, they provide inherent front-end signal gain of 50 to 500 without sacrificing speed. On the minus side, APDs require a high reserve bias voltage which complicates the drive circuitry. Also, they are sensitive to temperature variations and often require compensating systems.

A good example of a light source is the fiber optic modem used to connect two PCs together using fiber optic cable.

Using an RS-232 serial port to each PC, fiber optic modems contain all the transmitting and receiving elements necessary to carry out transmission asynchronously at data rates of 19.2 Kbps. This type of link usually has an operating wavelength of 820 nm and a range of 1 1/2 miles using a 62.5/125 micron fiber cable.

These units also can be configured for synchronous transmission up to 19.2 Kbps to a distance of approximately two miles. (See Figure 9.5.)

Put a couple of ST connectors on a pair of fiber optic cables and you have two PCs communicating.

Figure 9.5

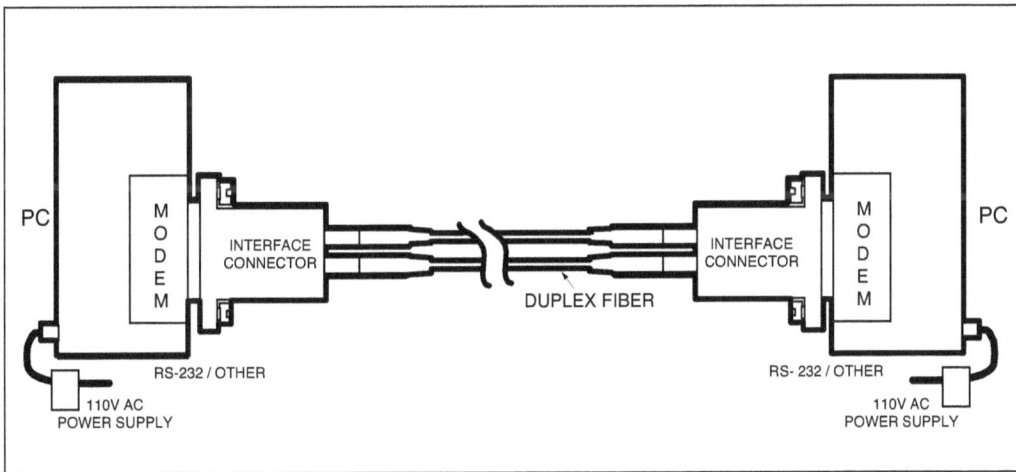

TEST MODULE 9

1. What are the two developments that made lightwave communication practical?

2. List the three major light emitting diodes.

   •

   •

   •

3. Semiconductors used as light emitters for fiber optic transmission fall into what two categories?

   •

   •

4. Are lasers or LEDs more powerful and why?

5. Name four types of detectors commonly used in fiber optic data links.

   •

   •

   •

   •

LAB EXERCISE

Familiarization with products on display.

Section 10.0
Cable Handling and Pulling

Objective

Describe the types of mechanical devices required to pull cables and examine the restrictions and limitations that should be considered when pulling and handling fiber optic cabling.

Outline

- Introduction to Cable Pulling
- Gripping Techniques
- Bending Fiber
- Cable Slack

Learning Activity

Assessment: Test Module 10

Lab Exercise: Attach pulling devices to indoor cable that are required to pull cable inside the building and into the cabinets.

Introduction to Cable Pulling

Installing fiber optic cable is similar to pulling Category 5 cable, but with a few more restrictions. These include refraining from pulling directly on the fiber and avoiding tight loops, kinks, knots or tight bends. Experience shows that inside fiber optic cable, if properly handled, can be installed with greater ease than copper cables because it's lighter.

The first step in proper cable installation technique is to review the cable specification. The two most important specifications are tensile loading and bend radius. It is very important to adhere to these limits. To effectively utilize all of the available strength in the cable, the strength member must be used.

Gripping Techniques

For cables using aramid yarn alone as the strength member, the jacket can be removed exposing the aramid. The aramid should be tied in a knot with the pull rope so the jacket will not be inadvertently used for strength. Optionally, the jacket can be tied into a tight knot before pulling. After pulling, the knot should be cut off. (See Figure 10.1.)

For cables using aramid yarn and a central member, a pulling grip should be used. (See Figure 10.2.) The strength member(s) should be attached independently. This can be accomplished by weaving the strength member into the fingers of the grip and then taping it together. All strength members should be gripped equally to ensure a proper distribution of tension.

Bending Fiber too Tightly

A common problem is bending the fiber on too tight a radius. The bending radius is always important in a static condition. However, it becomes even more important under tensile loading, because the tensile stresses due to bending are added to those due to pulling. A minimum bending radius of 10 cable diameters must be maintained over long-term, static conditions.

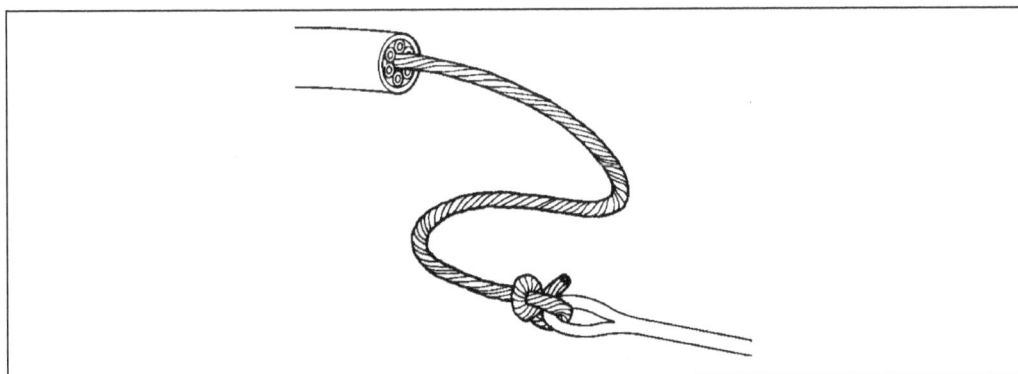

Figure 10.1

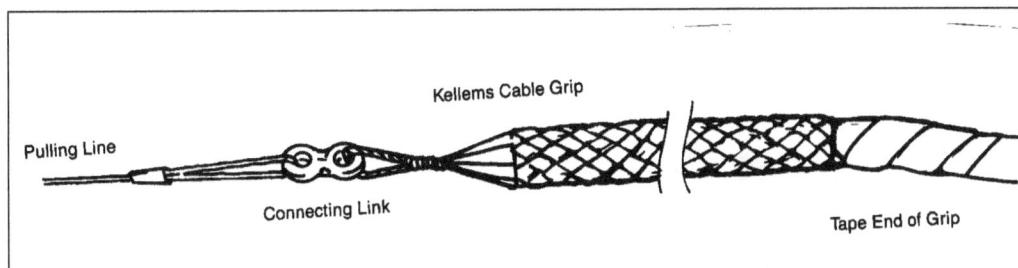

Figure 10.2

Pulling Line

Kellems Cable Grip

Connecting Link

Tape End of Grip

10.3

Figure 10.3

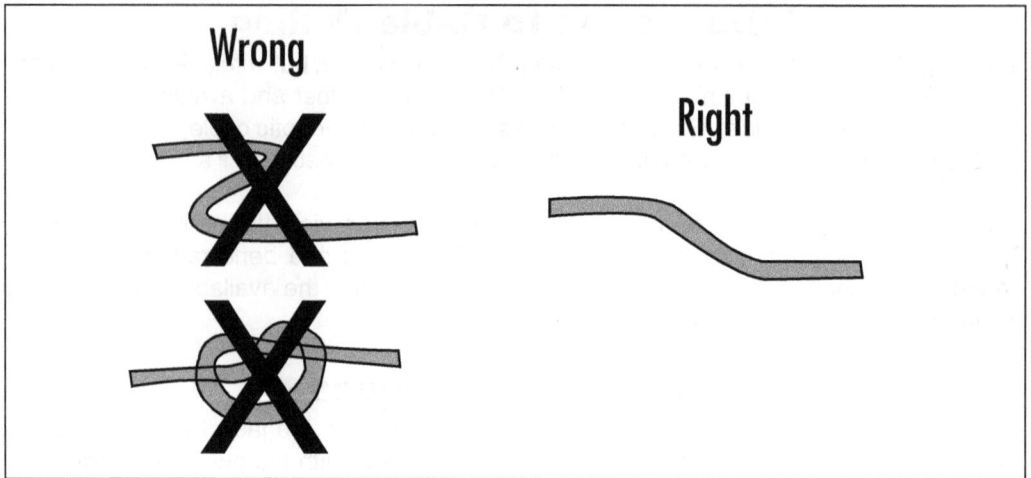

Wrong

Right

Figure 10.4

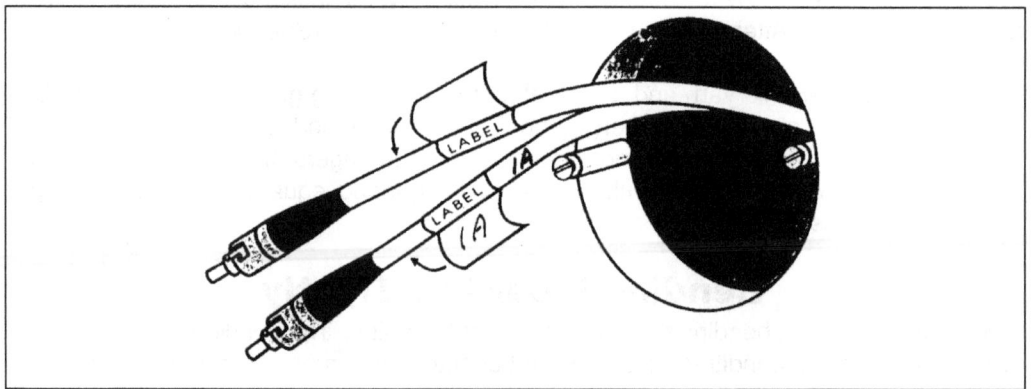

Figure 10.5

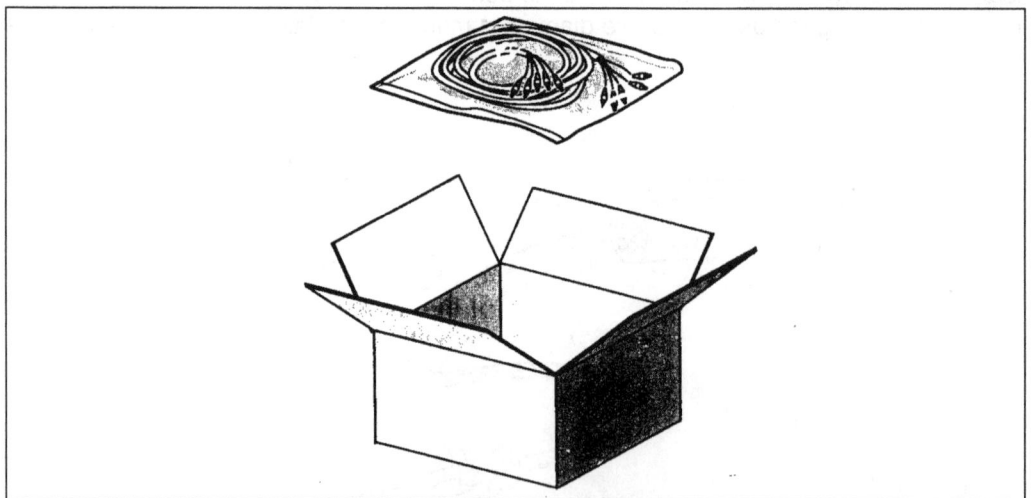

When cable is placed under a tensile load while being pulled, a minimum of 20 cable diameters is recommended. It should be noted that a design in which a cable is placed by hand into a tray allows a tighter radius than one where installation will be carried out by pulling the cable in. Always follow manufacturer guidelines for bend radii.

The Cable Reel

Most fiber optic cable is shipped on lightweight reels. When pulling cables from the reel, respect the minimum bending radius and maximum pulling forces.

Marking Cable

Before pulling, mark the cables. Use labels as the major line of marking and as a back-up. Remember to mark both ends the same.

Indoor preconnectorized fiber cables are shipped in boxes and the connectors are placed on top in a crown fashion.

Cables should be removed from the reel with the reel in a vertical position without the cable being pulled across the reel itself.

At level 50, it should be pulled to each closure position such as outlets, forms, racks and patch panels. (See Figure 10.7.)

Figure 10.6

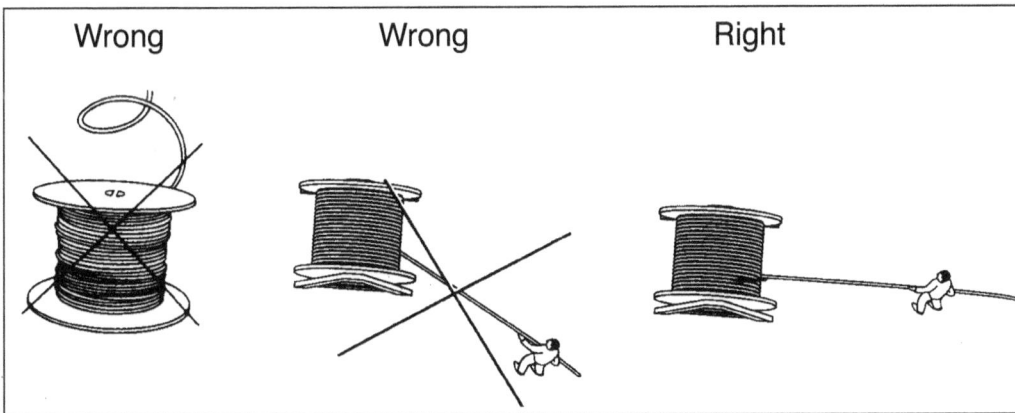

Wrong   Wrong   Right

Figure 10.7

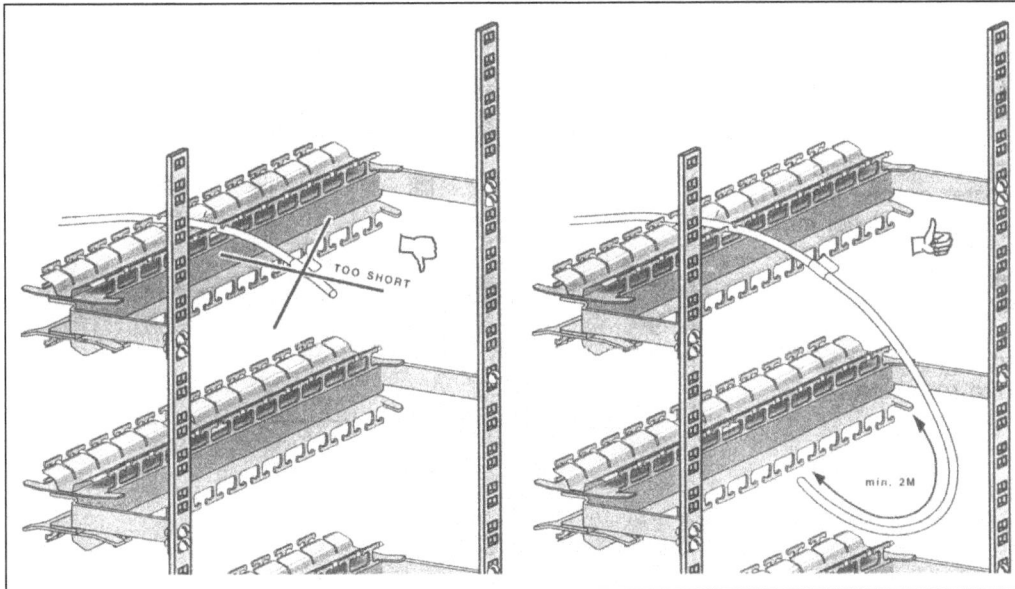

Figure 10.8

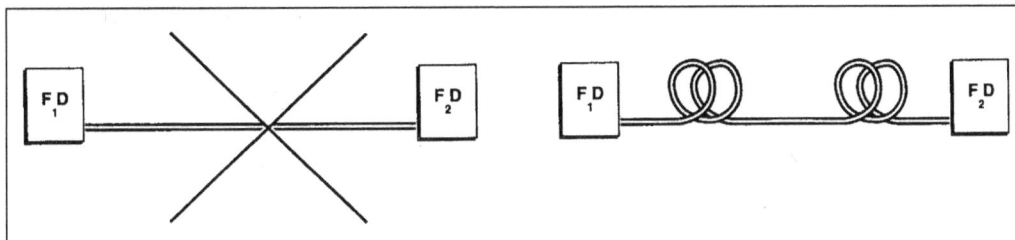

Leaving Cable Slack

Leave some extra slack, about five meters at different places on the link. This will make repairs on a broken cable easier and would allow movement of the fiber at a later date if necessary.

TEST MODULE 10

1. Installing fiber optic cable is similar to pulling Category 5 cable, but with a few more restrictions.

   True   False

2. The two most important specifications in cable installation are tensile loading and bend radius.

   True   False

3. A common problem is bending the fiber on too loose a radius.

   True   False

4. Most fiber optic cable is shipped on lightweight reels.

   True   False

5. The first step in proper cable installation technique is to review the cable specification.

   True   False

LAB EXERCISE

Attach pulling devices to indoor cable that are required to pull cable inside the building and into the cabinets.

Section 11.0
Indoor Hardware

Objective

Examine the hardware used in specific areas to house the cable and terminating points of the fiber cable, as well as the cabinets and the patch panels that make up the entire distribution system.

Outline

- *Definition of Hardware*
- *EIA/TIA 568 Commercial Building Wiring Standards*
- *Wiring Standards*
- *Patch Panel*
- *Splice Panel*

Learning Activity

Assessment: Test Module 11
Lab Exercise: Assemble a cabinet and fiber tray.

Definition of Hardware

To begin any project design, you must first know the fiber count (how many fibers) and the path (routing) of the cable. Keep in mind that the main purpose for hardware is to protect and organize splice and termination points. Hardware can be divided into two categories, indoor and outdoor. We will be dealing with indoor hardware as applicable to the commercial buildings and campus environments.

Fiber is not just thrown into an installation. You need to know the number of fibers to pull to each location and the purpose for each fiber. Then you can plan your hardware and type of cable to be used. Next, plan the conduit, duct and innerduct. The designer must decide in advance on the networks, support systems and topologies to be used, such as Ethernet, Token Ring, voice, data, video, imaging, control and industrial application. Once the application is determined and the fiber and connection/splice is known, the route must be determined. Then, decide on locations for the hardware cabinets, racks and associated panels.

EIA/TIA 568 Commercial Building Wiring Standards

There are rules for fiber distribution established by the EIA/TIA 568 Committee. (See Figure 11.1.) The EIA/TIA 568 Commercial Building Wiring standard is based on a hierarchical star for the backbone and a single star for horizontal distribution. Rules for backbone wiring include a 2,000 meter (6,560 feet) maximum distance between the main cross-connect and the telecommunications closet, with a maximum of one intermediate cross-connect between the main cross-connect and the telecommunications closet.

Main Cross-Connect

The main cross-connect (MC) (Figure 11.2), also referred to as the main distribution frame (MDF), should be in close proximity to or in the same location as the data center or PBX. This placement ensures a centralized management point for reconfiguration of the fiber optic network. Equipment in the MC should be capable of the following:

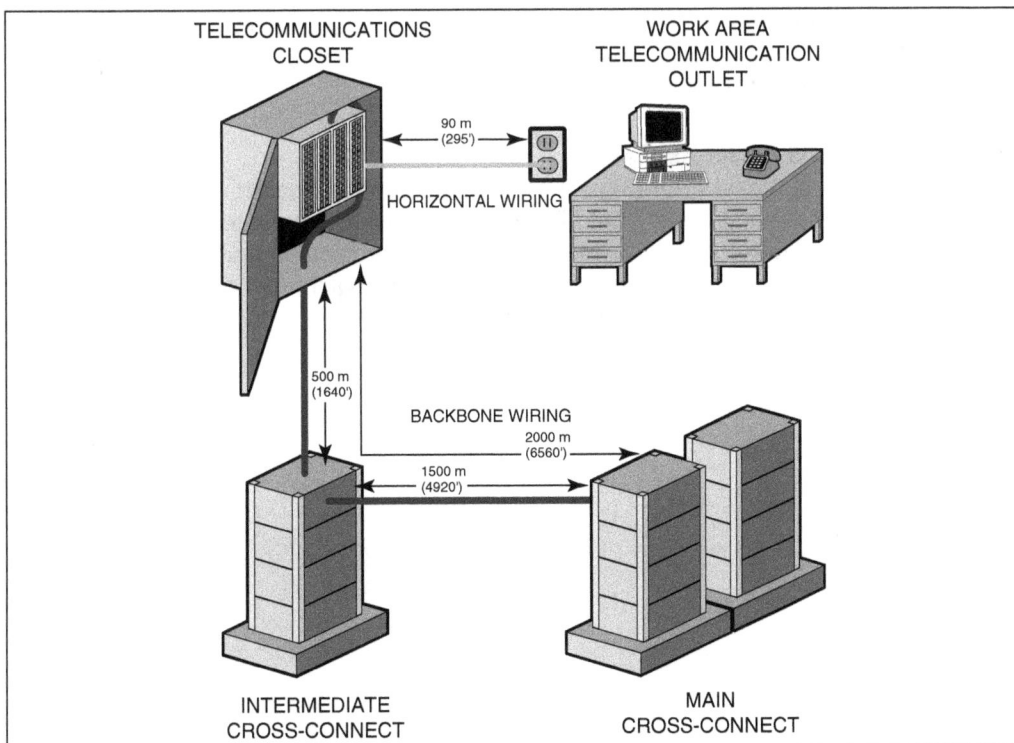

Figure 11.1

Figure 11.2

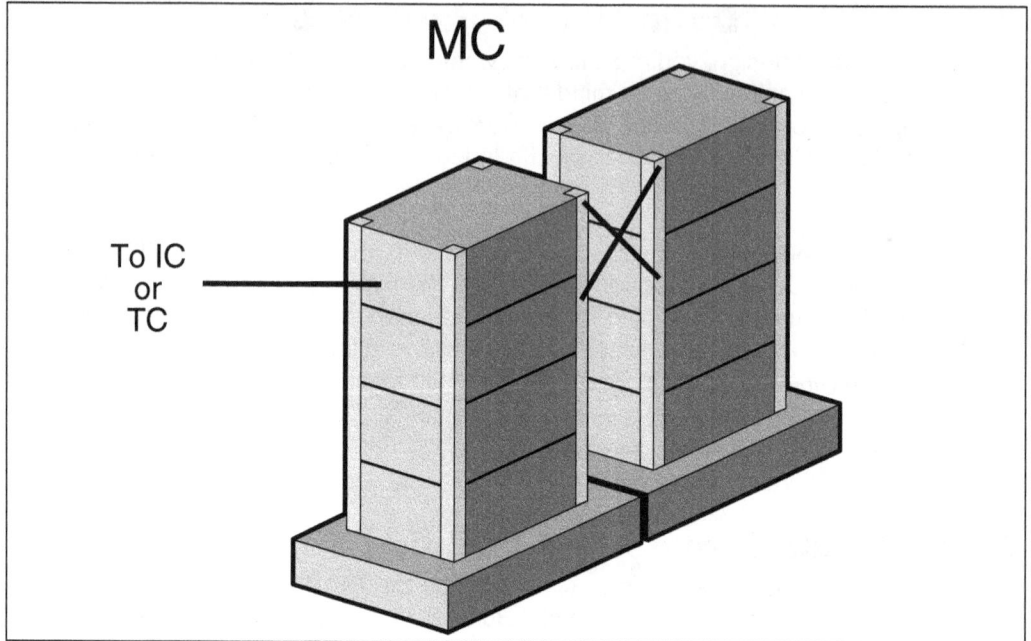

- Handling large fiber counts
- Accepting either direct termination of equipment or pigtail splices and pre-terminated assemblies for splicing at the frame
- Providing jumper storage and reconfiguration capabilities
- Allowing for growth

Typically, MC equipment can be installed in racks or wall cabinets.

Intermediate Cross-Connect

The intermediate cross-connect (IC) (Figure 11.3) typically connects the intrabuilding cable plant to the interbuilding cable plant. It is smaller in scope with lower fiber counts than the MC. Products in the IC may need to be wall-mounted. The size of the IC will determine which products will be used.

Telecommunications Closet Hardware

Telecommunications closet (TC) (Figure 11.4) equipment makes the transition from the backbone to the horizontal cable plant. Requirements for backbone terminations will typically involve lower fiber counts, jumper routings to electronics and products that can be either wall or rack mounted. The horizontal cable plant equipment must accommodate higher fiber counts and a variety of hybrid cable types.

TC

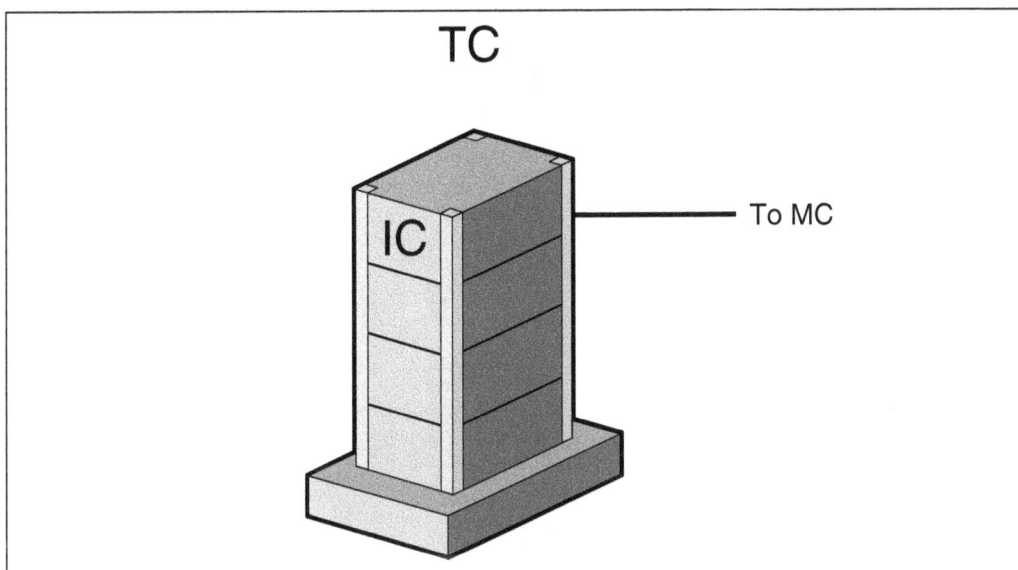

IC

— To MC

Figure 11.3

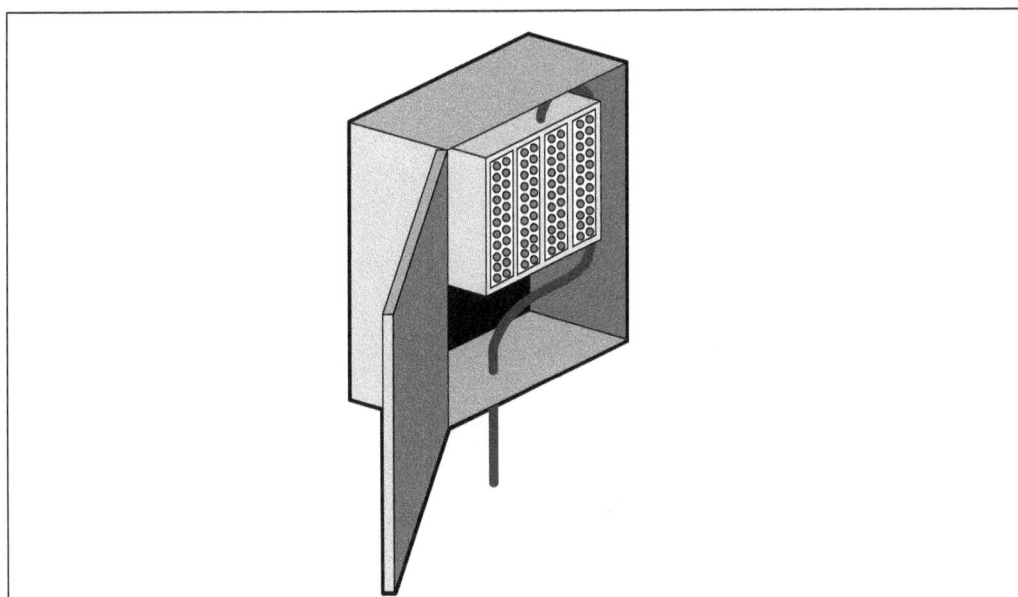

Figure 11.4

Work Area Telecommunications Outlet

The work area telecommunications outlet (Figure 11.5) is the end point of the horizontal wiring. The cables may be fiber only or a combination of fiber and copper.

In each of the above distribution points, one or more of the following hardware components can be installed.

Splice Panel

The splice panel changes outside cable to riser or plenum cable, or breaks out the outside cable to individual buffered fibers. After being spliced, the fibers are outed to the appropriate splice tray and positioned to prevent damage. The splice panel can accommodate fusion or mechanical splices, and can be mounted in a rack or wall cabinet. (See Figure 11.6.)

Figure 11.5

Figure 11.6

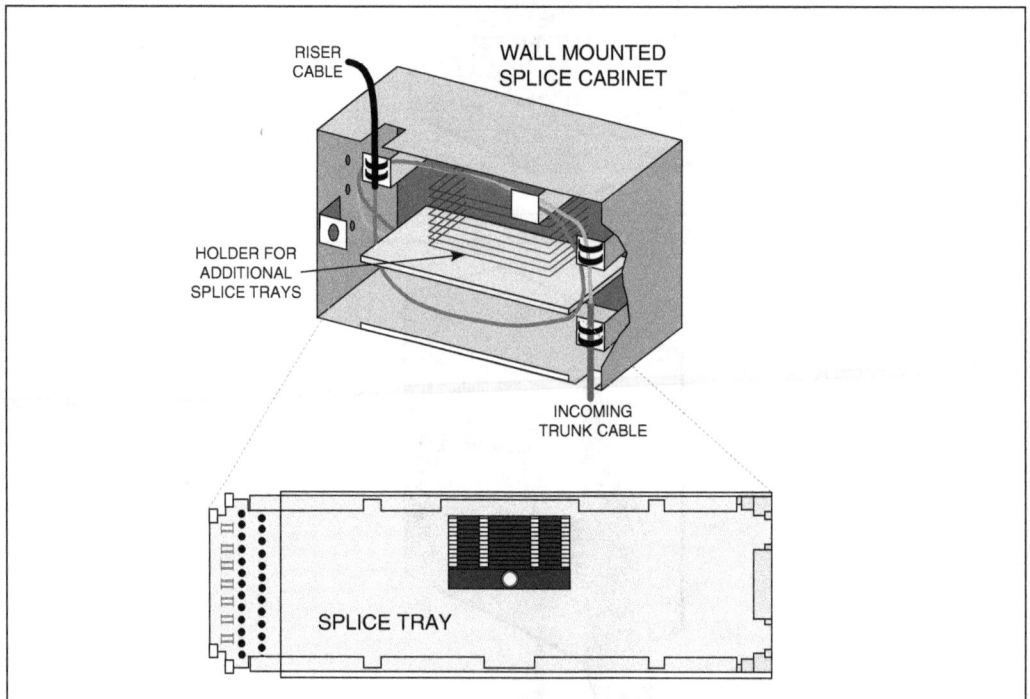

Patch Panel

A patch panel provides a centralized location for patching fibers, testing, monitoring and restoring riser or trunk cables. Figure 11.7 shows an incoming riser cable being terminated on one side of a coupling panel with a connector. On the other side of the coupling panel is another group of connectors terminating the local area network (LAN) cables. In small offices, the patch panel can be used as a main cross-connect or intermediate cross-connect.

To order this type of panel, you must know the type of connectors for the coupling panels, the number of connections needed, the type of cabinet, whether the panel will be wall or rack mounted and the size of the cabinet required for the number of panels to be installed.

Panels

When designing your cabinets, you can select the panel with the type of coupling to fit the connectors you have chosen. Figure 11.8 shows several panels that are currently available.

You can also order coupling panels without the couplers (Figure 11.9) and install these yourself. Sometimes the type of connectors will not be known until the cut-over date.

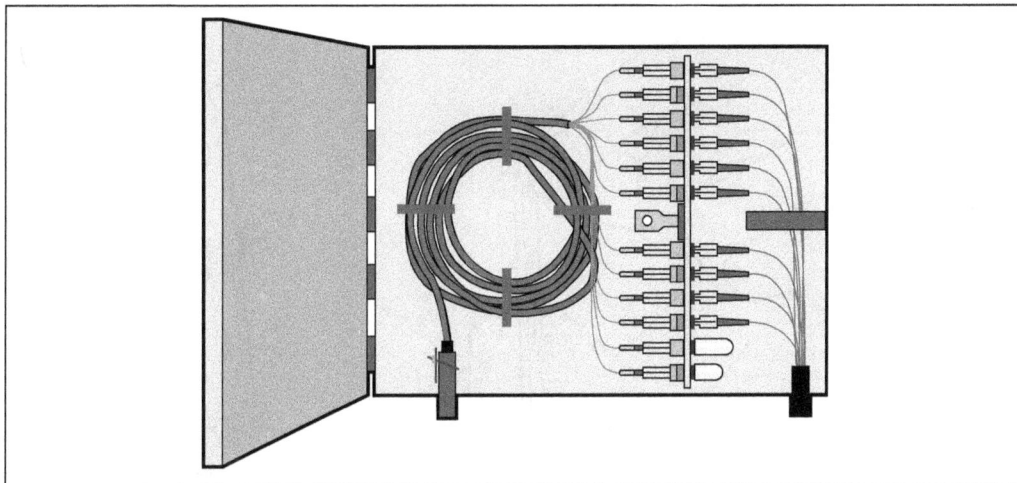

Figure 11.7

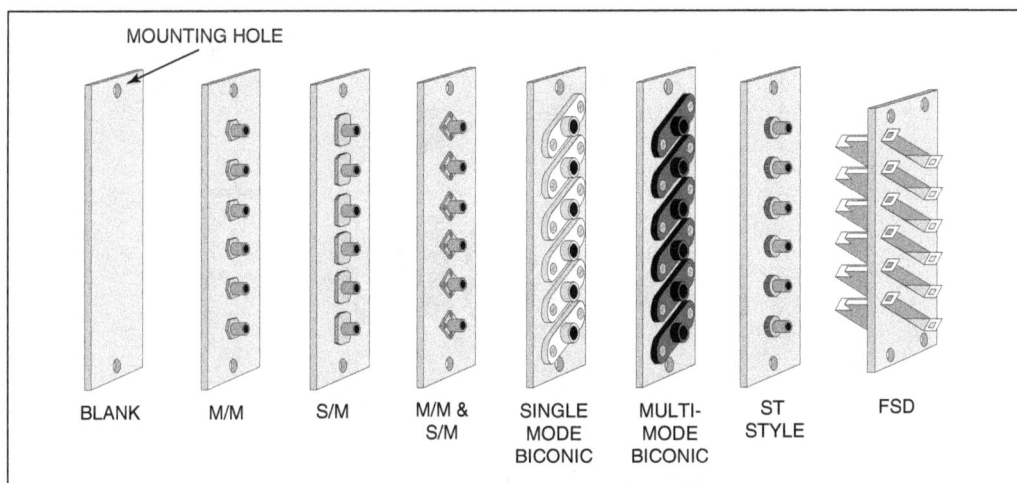

MOUNTING HOLE

BLANK   M/M   S/M   M/M & S/M   SINGLE MODE BICONIC   MULTI-MODE BICONIC   ST STYLE   FSD

Figure 11.8

Each manufacturer has a unique design for stacking equipment panels. One such method involves stacking the panels on a relay rack. (See Figure 11.10.)

There are two methods for arranging and placing fiber hardware. The first method attaches the panels on top of each other in a relay rack arrangement. The second method places the panels in a cabinet and each cabinet is installed on the wall above each other or side by side. (See Figure 11.11.)

When ordering any cabinet, make sure it is equipped with cable terminating equipment, such as ground clamps, a place to secure the dielectric central member, Kevlar or other fill material and cable ring or channel guides.

Pigtail Splicing

A fiber optic pigtail splice is a fiber cable that has been factory connectorized on one end with an optical connector. The other end remains unterminated. (See Figure 11.12.)
The unterminated end of the pigtail is spliced to the fiber requiring termination. Customized cabinets for coupling panels and splice trays with pigtails can be ordered (Figure 11.13) and used to mount the splice cabinet in a relay rack.

Many options and variations are available to choose from when designing your system. Working with a good manufacturer can help meet the specific needs of your design.

11.7

Figure 11.9

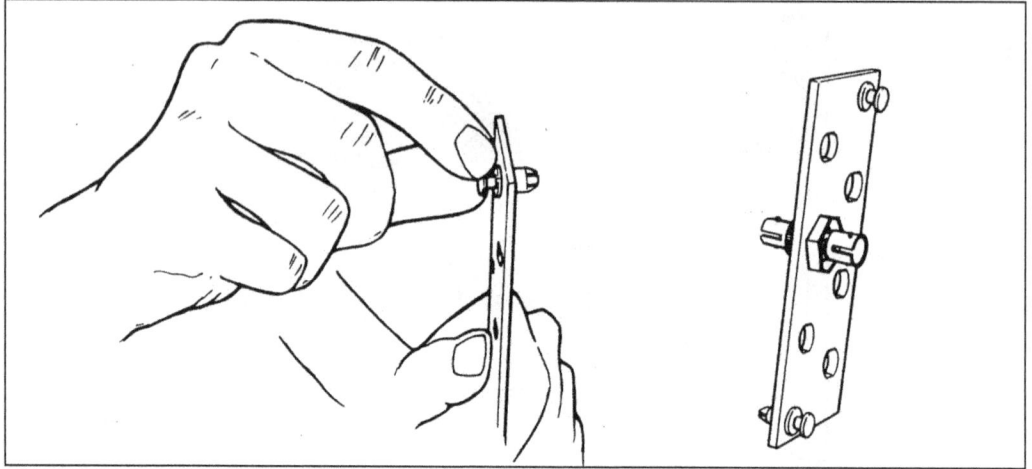

Figure 11.10

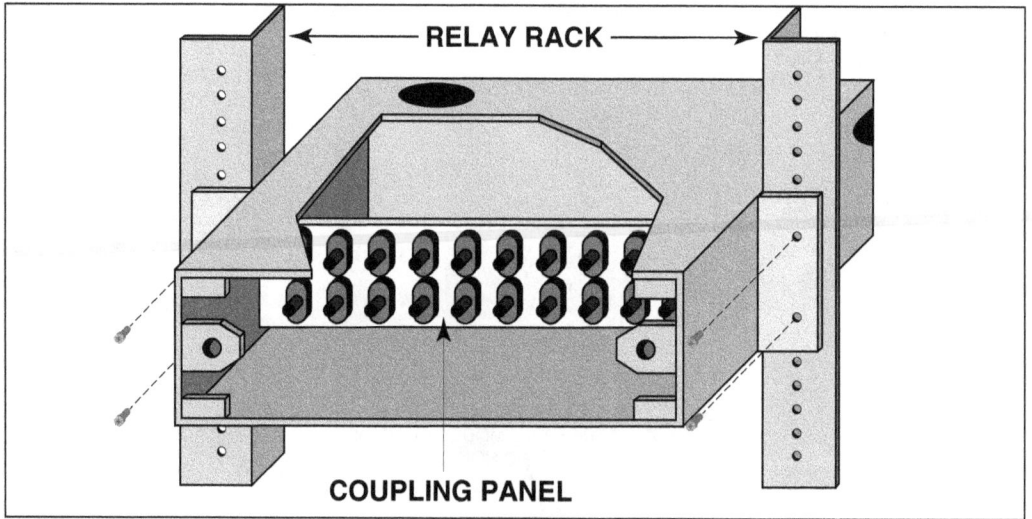

RELAY RACK

COUPLING PANEL

Figure 11.11

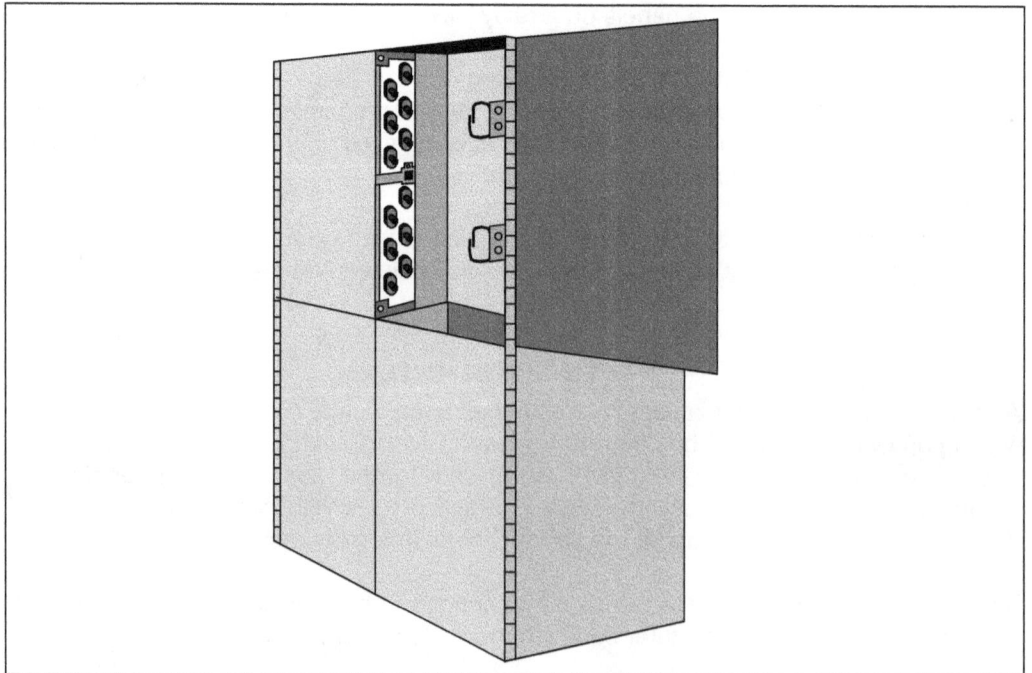

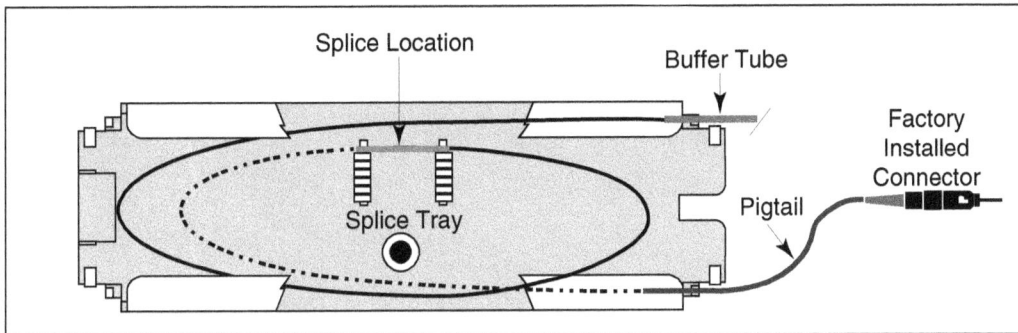

Figure 11.12

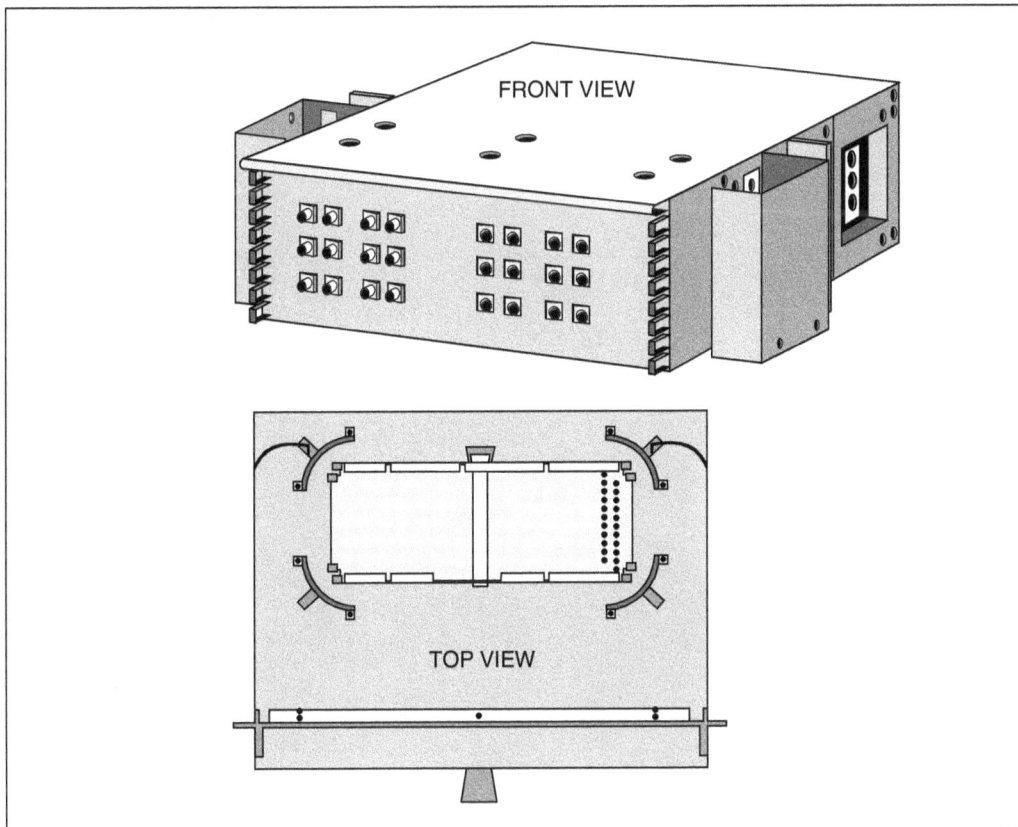

Figure 11.13

TEST MODULE 11

1. A fiber optic pigtail splice is a fiber cable that has been factory connectorized on one end with an optical connector.

   True   False

2. The main-cross-connect (MC) typically connects the intrabuilding cable plant to the interbuilding cable plant.

   True   False

3. Hardware can be divided into two categories, indoor and intrabuilding.

   True   False

4. The splice panel can be rack or wall mounted and accommodates both fusion and mechanical splices.

   True   False

5. In small offices, the patch panel can be used as a main cross-connect or intermediate cross-connect.

   True   False

LAB EXERCISE

Assemble a cabinet and fiber tray.

Section 12.0
Building Construction Simplified

Objective

Understand building construction from interior walls to outside brick work, and learn pertinent building codes such as; National Electrical Code including designations OFNP, OFCP, etc.

Outline

- *Knowledge of Building Construction*
- *Walls – Empty, Insulated and Conduit in the Walls*
- *Underfloor Duct*
- *Computer Flooring*
- *Ceiling Space*
- *Article 770 - Optical Fiber Cables and Raceways*
- *Types of Cables Specified by the National Electrical Code*

Learning Activity

Assessment: Test Module 12
Lab Exercise: Examine cables with NEC listings.

Knowledge of Building Construction

To be effective, installers should be knowledgeable of building construction. This is especially true of wall, ceiling and floor construction because the knowledgeable installer will have a better idea of what is likely to be the clearest, most unobstructed path to run the cable. Fortunately, there are certain kinds of construction now common in most newer buildings that make the job of running and concealing cable relatively easy. Newer types of construction have allowed for ready-made openings, slots and cavities in floors and ceilings. Problems generally occur in older buildings where the paths do not already exist.

Let's look at the different types of wall, ceiling and floor construction - both new and old - and some techniques for running cable through each type.

Walls

Years ago, the standard for constructing walls was wood framing, over which wet plaster was troweled after a lath (a metal mesh) was secured to the framing. In older buildings, and to some degree in new, smaller buildings, this type of wall construction still exists. This type of wall is labor intensive for the cable installer.

About 95 percent of new construction uses gypsum board, which is nailed or screwed to either wood or metal studs (metal is more common). The wood studs are held in place by a top plate and sole plate (bottom track). (See Figure 12.1.)

Studs are secured to the top and sole plate, which are spaced 16 or 24 inches on center. Both wood and metal/steel studs are installed typically on 16-inch centers (Figure 12.2); however, some installations can use 24-inch centers. Figure 12.3 illustrates a cut-away look at both wood and metal studs.

Once the walls are up, the wall material is then attached to the metal framing members, studs and plates, with sheet metal screws. The wall material itself is almost always gypsum board, also known as plasterboard or Sheetrock. It normally comes in panels that are four-foot wide, eight- and ten-foot long, and in thicknesses of 1/2-inch, 5/8-inch and one-inch. The 5/8-inch thickness is most often used in new office buildings.

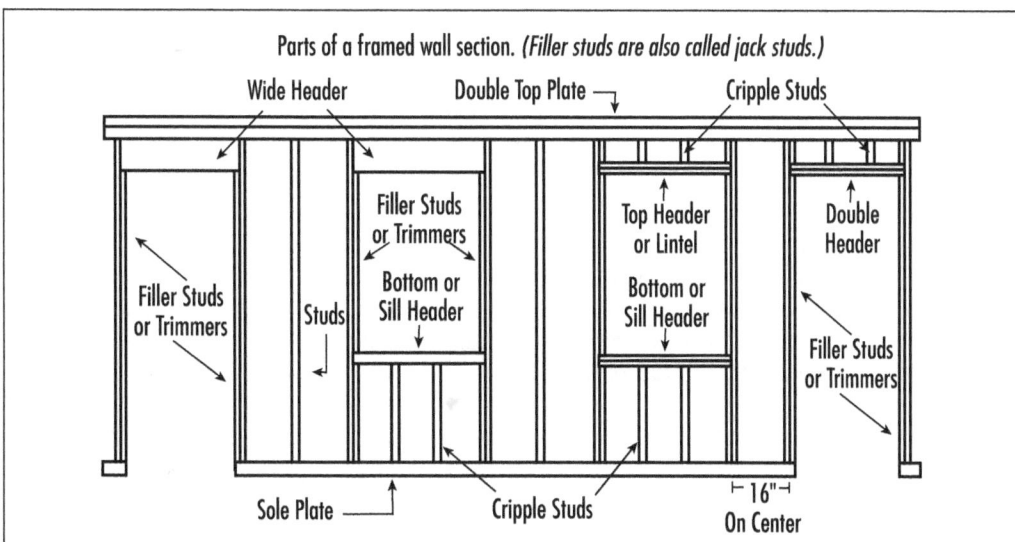

Figure 12.1

12.3

Figure 12.2

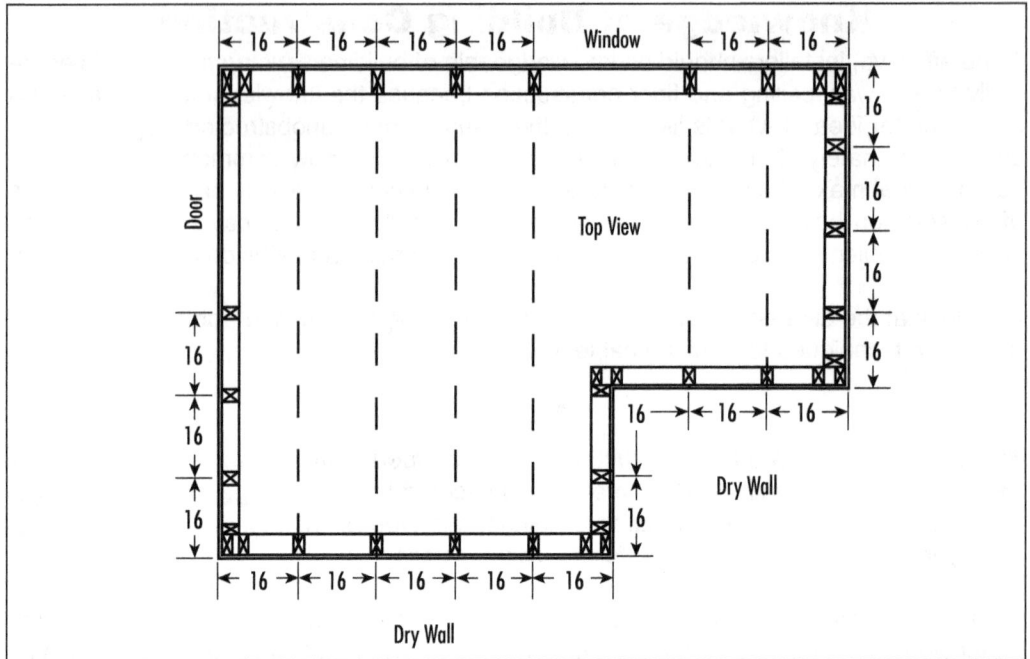

Figure 12.3

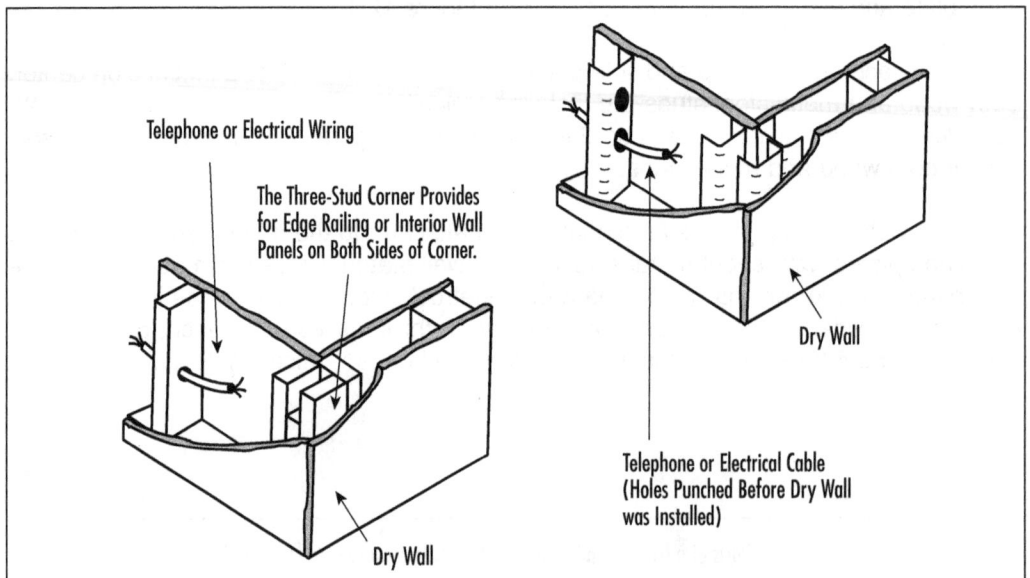

Empty Walls (No Insulation)

New buildings are commonly constructed of a steel framework with concrete slabs that serve as floors, and whose undersides serve as ceilings for the floor below. Studs commonly run all the way from slab floor to slab ceiling, but the plasterboard walls stop a foot or two short of the ceiling, leaving a clear, open space. Because of this, a suspended ceiling can be installed well beneath the slab, and form an open but hidden area to run cables, wire and pipe. Optical fiber cables can be run along the suspended ceiling structural supports and down through the empty walls to emerge from holes drilled into the walls at jack openings. (See Figure 12.4.)

Path for Wire - Through space above ceiling, down along inside wall, and out through jack hole.

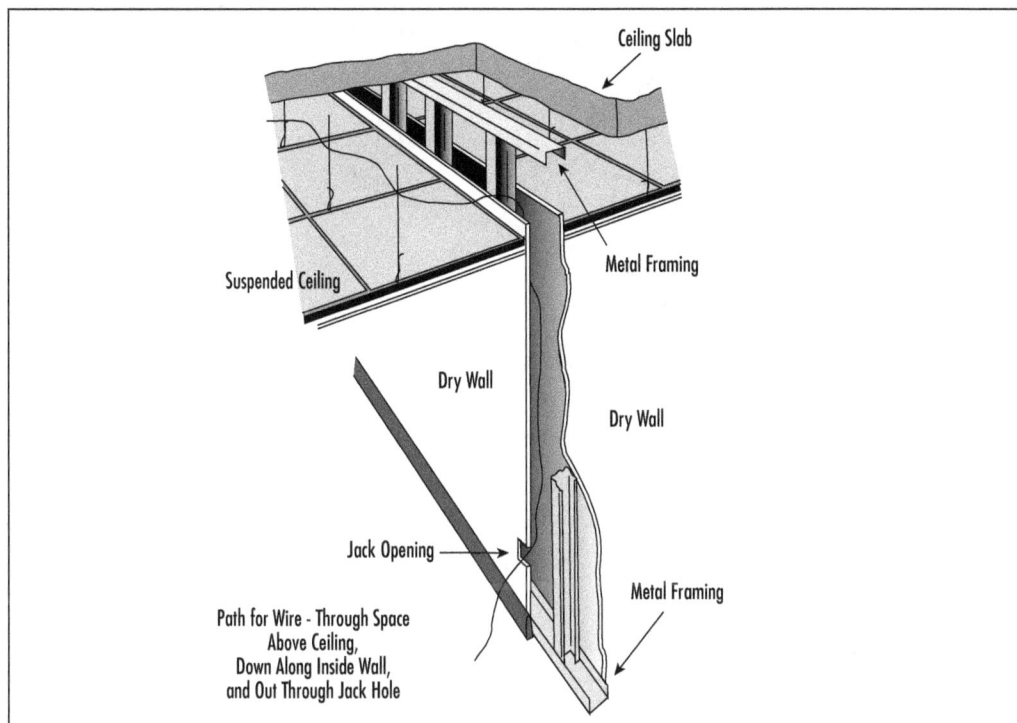

Figure 12.4

Figure 12.4 labels: Ceiling Slab, Metal Framing, Suspended Ceiling, Dry Wall, Dry Wall, Jack Opening, Metal Framing, Path for Wire - Through Space Above Ceiling, Down Along Inside Wall, and Out Through Jack Hole

Insulated Walls
Some walls contain insulating material between the studs. Walls between different offices are usually insulated for soundproofing. Core or perimeter walls, which separate office space from external halls, may be insulated for fireproofing. In both cases, insulated walls usually run all the way to the underside of slabs. Pulling cable through these walls is harder than through empty walls and required penetrations.

Walls with Conduit
In some cases, walls will also contain conduit designed for running cable to outlets. Unless they are already full, these conduits make it much easier to run cable in the wall. There is no way to tell immediately whether a wall contains conduit; but, once an installer gains access to the space above, he will be able to ascertain this fact. (See Figure 12.5.)

Floors
Figure 12.6 has been provided as an aid in developing a better understanding of floors. Although the various methods of flooring are quite numerous, this figure will provide a general insight into floor construction.

Underfloor Duct
This is another popular kind of cable housing. Although it differs by manufacturer, it is essentially a series of hollow ducts for running telephone cable, electrical wire and optical fiber cable. (See Figure 12.7.)

Computer Flooring
One of the most popular floors is computer or raised flooring (See Figure 12.8.) This consists of a series of tile or carpet-covered metal panels mounted on aluminum locking pedestals. Beneath the panels is open space. You can remove the panels or use a special plunger device to pick them up.

Figure 12.5

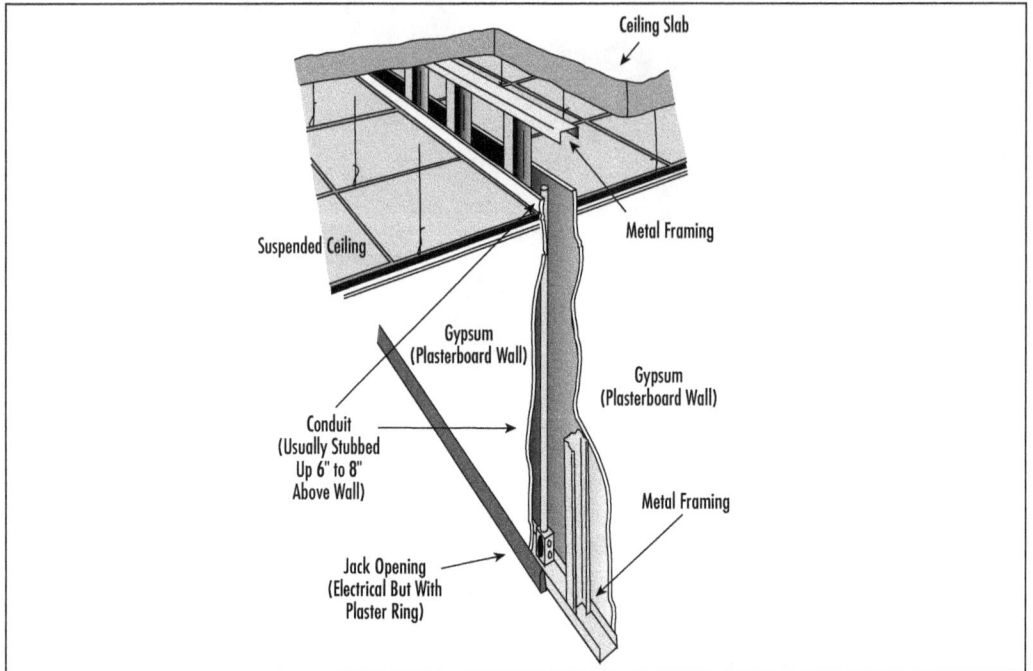

Ceiling Slab

Metal Framing

Suspended Ceiling

Gypsum
(Plasterboard Wall)

Gypsum
(Plasterboard Wall)

Conduit
(Usually Stubbed
Up 6" to 8"
Above Wall)

Metal Framing

Jack Opening
(Electrical But With
Plaster Ring)

Figure 12.6

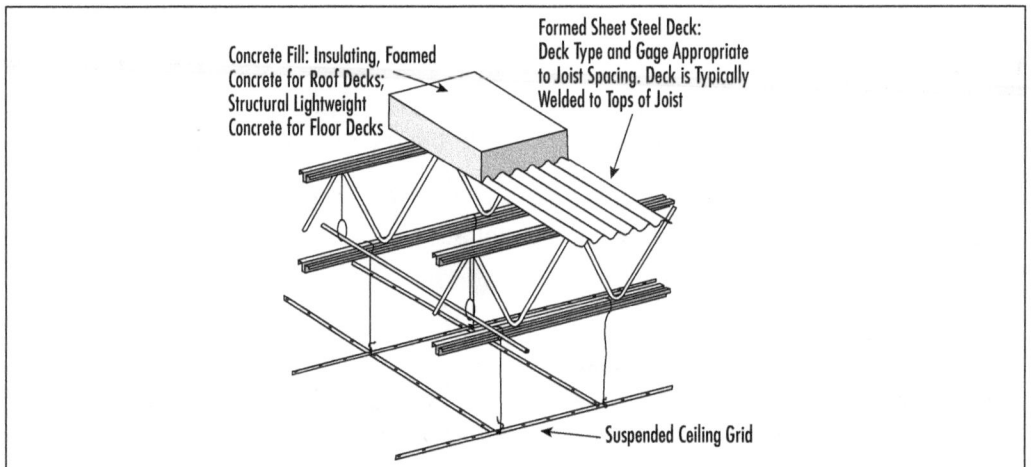

Concrete Fill: Insulating, Foamed
Concrete for Roof Decks;
Structural Lightweight
Concrete for Floor Decks

Formed Sheet Steel Deck:
Deck Type and Gage Appropriate
to Joist Spacing. Deck is Typically
Welded to Tops of Joist

Suspended Ceiling Grid

Figure 12.7

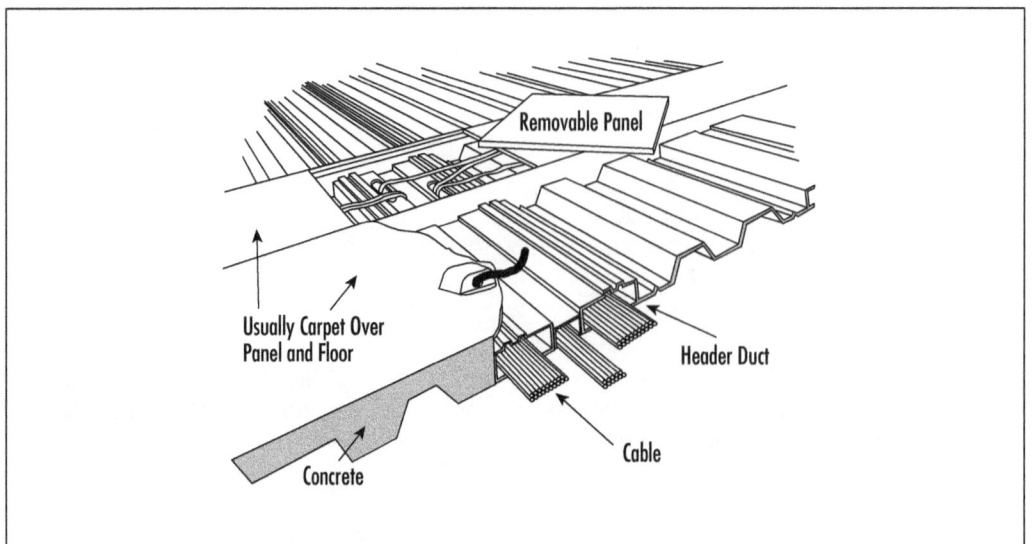

Removable Panel

Usually Carpet Over
Panel and Floor

Header Duct

Cable

Concrete

Figure 12.8

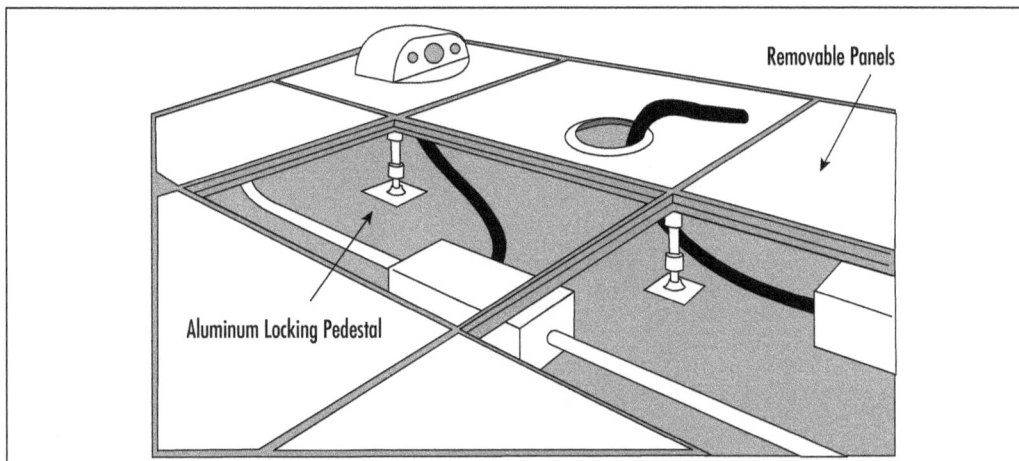

Removable Panels

Aluminum Locking Pedestal

Ceilings

Ceiling distribution systems are common in old and new building constructions because of the advantage of common access to many different areas of a structure.

Suspended Ceilings

The ceilings in new office buildings are commonly the suspended type. A metal framework or grid consisting of cross tees and runners hangs from the ceiling with various kinds of wires or metal struts. Cross tees are long, metal-lipped pieces and Ts are metal sections shaped like the letter T that intersect with the runners. The ceiling panels are laid in the runners and cross tees, which are lipped. (See Figure 12.9.) Such panels are commonly two feet by four feet and are available in a wide variety of textures and colors. They are most commonly fissured and have some sound-absorbing ability.

Suspended ceilings are easy to access. Just push up a panel and move inside. Above the ceiling are J-hooks, traps, raceways or other kinds of support hardware for cables.

Open Office Concept

The design of office space without walls is the newest method of construction. It originated with the use of the portable column (Figure 12.10), which is still prevalent. Partitions, or movable walls, are another type of wall used in an open office environment. The walls are made of metal and are easy to disassemble. They have slots along the top within the vertical posts and along the baseboards. The sections covering the slots can be pried off with a screwdriver and the cable easily run through the slots.

Demountable walls usually have glass panels mounted on the top that run to the ceiling. Cable enters these types of walls either from a post that runs from the ceiling, down into the side panels, or up through the floor, then into the post and through the rest of the panel. (See Figure 12.11.)

Furniture Flooring Concept

New office furniture is being designed to serve multiple purposes. Its functions include serving as walls/partitions and columns/poles. The furniture/walls are equipped in most cases with cable management channels which allow for various methods of installation according to Category 5 radius bend requirements.

There are many ways cable can be routed from the communications closet to the furniture system. They can be broadly classified into four main categories, ceiling, floor, perimeter wall and interior structural column. (See Figure 12.12.) Each method usually

Figure 12.9

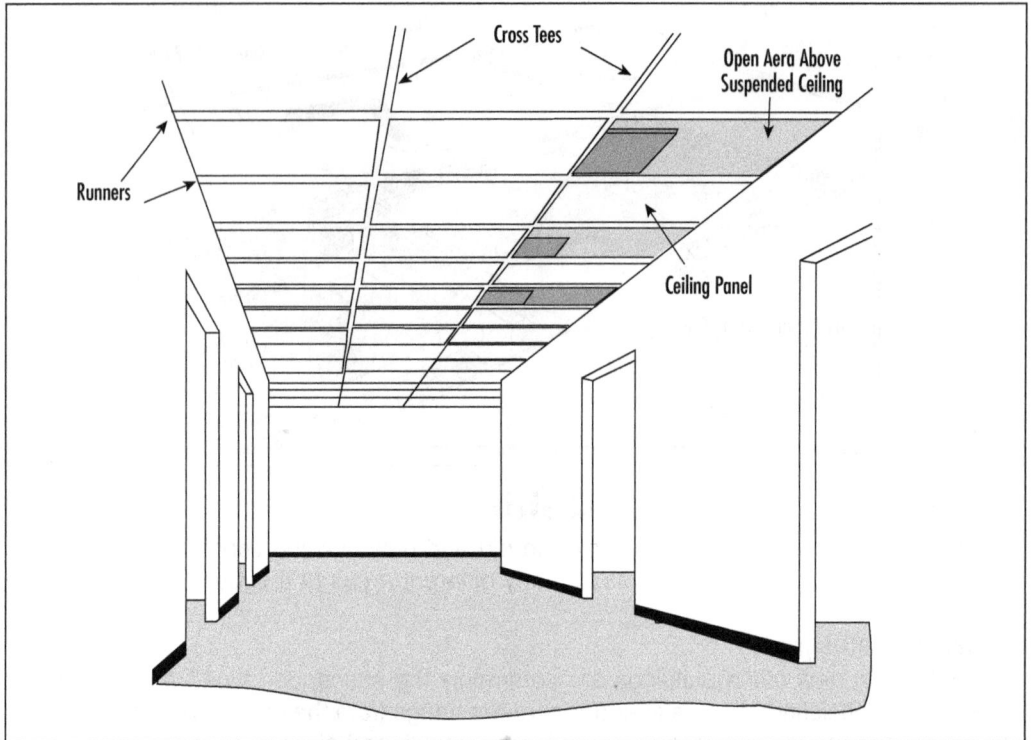

Cross Tees

Open Aera Above
Suspended Ceiling

Runners

Ceiling Panel

Figure 12.10

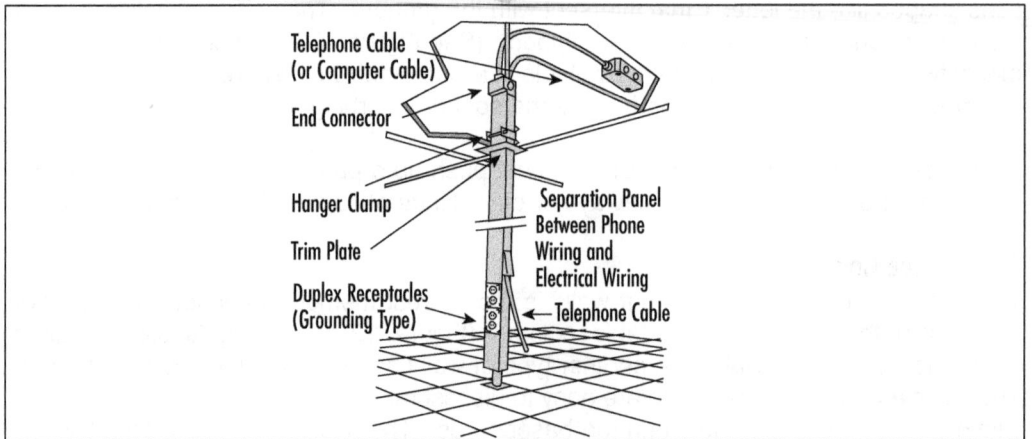

Telephone Cable
(or Computer Cable)

End Connector

Hanger Clamp

Trim Plate

Separation Panel
Between Phone
Wiring and
Electrical Wiring

Duplex Receptacles
(Grounding Type)

Telephone Cable

Figure 12.11

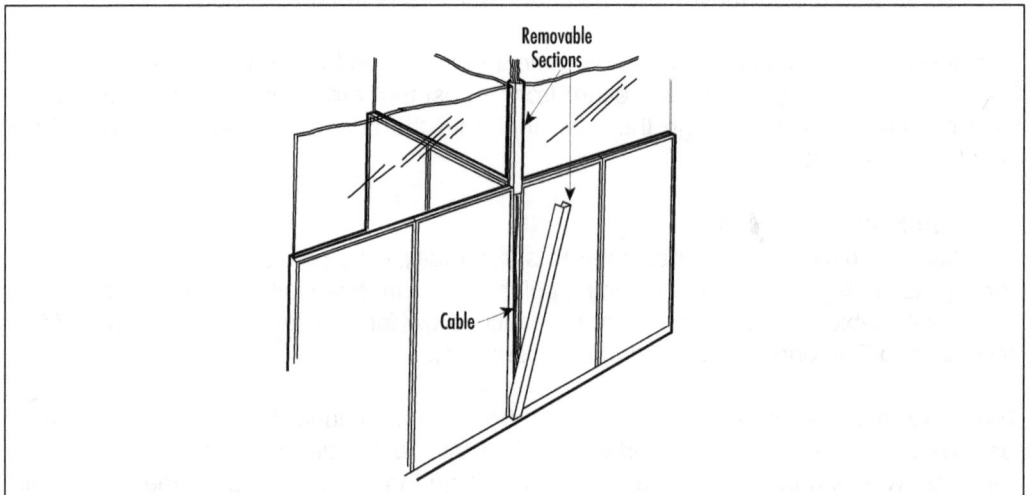

Removable
Sections

Cable

Figure 12.12

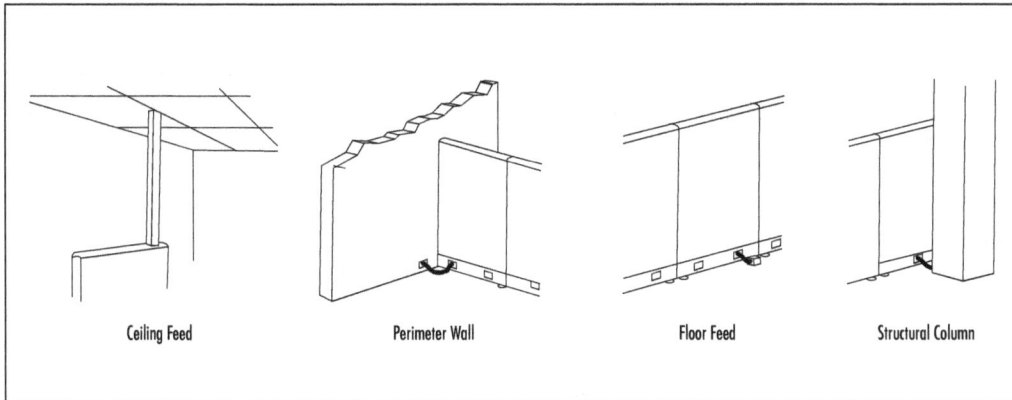

Ceiling Feed   Perimeter Wall   Floor Feed   Structural Column

requires a different entry point into the furniture and each method presents some unique challenges.
Remember that building construction varies. Always be familiar with the working environment. The payoff will be a much easier installation project.

Article 770 - Optical Fiber Cables and Raceways

A. General
770-1. Scope. The provisions of this article apply to the installation of optical fiber cables and raceways. This article does not cover the construction of optical fiber cables and raceways.

770-2. Locations and Other Articles. Circuits and equipment shall comply with (a) and (b) below.

(a) Spread of Fire or Products of Combustion. See Section 300-21.
(b) Ducts, Plenums, and Other Air-Handling Spaces. Section 300-22, where installed in ducts or plenums or other space used for environmental air.
Exception to (b): As permitted in Section 770-53 (a).

770-3. Optical Fiber Cables. Optical fiber cables transmit light for control, signaling and communications through an optical fiber.

770-4. Types. Optical fiber cables can be grouped into three types.

(a) Nonconductive. These cables contain no metallic members and no other electrically conductive materials.
(b) Conductive. These cables contain noncurrent-carrying conductive members such as metallic strength members and metallic vapor barriers.
(c) Composite. These cables contain optical fibers and current-carrying electrical conductors, and shall be permitted to contain noncurrent-carrying conductive members such as metallic strength members and metallic vapor barriers. Composite optical fiber cables shall be classified as electrical cables in accordance with the type of electrical conductors.

770-5. Optical Fiber Raceway System. A raceway system designed for enclosing and routing only nonconductive optical fiber cables. Where optical fiber cables are installed in a raceway, the raceway shall be of a type permitted in Chapter 3 and installed in accordance with Chapter 3.

Exception: Listed optical fiber raceway.

(FPN): Plastic innerduct commonly used for underground or outside plant construction may not have appropriate fire safety characteristics for use as an optical fiber optic cable wiring method within buildings.

770-6. Cable Trays. Optical fiber cables of the types listed in Table 770-50 shall be permitted to be installed in cable trays.

(FPN): It is not the intent to require that these optical fiber cables be listed specifically for use in cable trays.

770-7. Access to Electrical Equipment Behind Panels Designed to Allow Access. Access to equipment shall not be denied by an accumulation of wires and cables that prevents removal of panels, including suspended ceiling panels.

B. Protection
770-33. Grounding of Entrance Cables. Where exposed to contact with electric light or power conductors, the noncurrent-carrying metallic members of optical fiber cables entering buildings shall be grounded as close to the point of entrance as practicable or shall be interrupted as close to the point of entrance as practicable by an insulating joint or equivalent device.

For purposes of this section, the point of entrance shall be considered to be at the point of emergence through an exterior wall, a concrete floor slab, or from a rigid metal conduit or an intermediate metal conduit grounded in accordance with Article 250.

C. Cables Within Buildings
770-49. Fire Resistance of Optical Fiber Cables. Optical fiber cables installed as wiring within buildings shall be listed as being resistant to the spread of fire in accordance with Sections 770-50 and 770-51.

770-50. Listing, Marking and Installation of Optical Fiber Cables. Optical fiber cables in a building shall be listed as being suitable for the purpose, and cables shall be marked in accordance with Table 770-50.

Exception No. 1: Optical fiber cables shall not be required to be listed and marked where the length of cable within the building does not exceed 50 feet (15.2m) and the cable enters the building from the outside and is terminated in an enclosure.

(EPN): Splice cases or terminal boxes, both metallic and plastic types, are typically used as enclosures for splicing or terminating optical fiber cables.

Exception No. 2: Conductive optical fiber cable shall not be required to be listed and marked where the cable enters the building from the outside and is run in rigid metal conduit or intermediate metal conduit and such conduits are grounded to an electrode in accordance with Section 800-40(b).

Exception No. 3: Nonconductive optical fiber cables shall not be required to be listed and marked where the cable enters the building from the outside and is run in raceway installed in compliance with Chapter 3.

(FPN No. 1): Cable types are listed in descending order of fire resistance rating. Within each fire resistance rating, nonconductive cable is listed first, since it may substitute for the conductive cable.

Table 770-50. Cable Markings

Cable Marking	Type	Reference
OFNP	Nonconductive optical fiber plenum cable	Sections 770-51(a) and 770-53(a)
OFCP	Conductive optical fiber plenum cable	Sections 770-51(a) and 770-53(a)
OFNR	Nonconductive optical fiber riser cable	Sections 770-51(b) and 770-53(b)
OFCR	Conductive optical fiber riser	Sections 770-51(b) and 770-53(b)
OFNG	Nonconductive optical fiber general-purpose cable	Sections 770-51(c) and 770-53(c)
OFCG	Conductive optical fiber general-purpose cable	Sections 770-51(c) and 770-53(c)
OFN	Nonconductive optical fiber general-purpose cable	Sections 770-51(d) and 770-53(c)
OFC	Conductive optical fiber general-purpose cable	Sections 770-51(d) and 770-53(c)

(FPN No. 2): See the referenced sections for requirements and permitted uses.

770-51. Listing Requirements for Optical Cables and Raceways. Optical fiber cables shall be listed in accordance with (a) through (d) below, and optical fiber raceways shall be listed in accordance with (e) and (f) below.

(a) Types OFNP and OFCP. Types OFNP and OFCP nonconductive and conductive optical fiber plenum cables shall be listed as being suitable for use in ducts, plenums and other space used for environmental air and shall also be listed as having adequate fire-resistance and low-smoke-producing characteristics.

(FPN): One method of defining low-smoke-producing cables is by establishing an acceptable value of the smoke produced when tested in accordance with the Test for Fire and Smoke Characteristics of Wires and Cables, NFPA 262-1990 (ANSI) to a maximum peak optical density of 0.5 and a maximum average optical density of 0.15. Similarly, one method of defining fire-resistant cables is by defining maximum allowable flame travel distance of 5 feet (1.52 m) when tested in accordance with the same test.

(b) Types OFNR and OFCR. Types OFNR and OFCR nonconductive and conductive optical fiber riser cables shall be listed as being suitable for use in a vertical run in a shaft or from floor to floor and shall also be listed as having fire-resistant characteristics capable of preventing the carrying of fire from floor to floor.

(FPN): One method of defining fire-resistant characteristics capable of preventing the carrying of fire from floor to floor is that the cables pass the requirements of the Standard Test for Flame Ptropagation Height of Electrical and Optical Fiber Cable Installed Vertically in Shafts, ANSI/UL 1666-1986.

   (c) Types OFNG and OFCG. Types OFNG and OFCG nonconductive and conductive general-purpose optical fiber cables shall be listed as being suitable for general-purpose use, with the exception of risers and plenums, and shall also be listed as being resistant to the spread of fire.

(FPN): One method of defining resistance to the spread of fire is for the damage (char length) not to exceed 4 feet 11 inches (1.5 m) when performing the "Vertical Flame Test - Cables in Cable Trays," as described in test Methods for Electrical Wires and Cables, CSA C22.2 No. 0.3-M 1985.

   (d) Types OFN and OFC. Types OFN and OFC nonconductive and conductive optical fiber cables shall be listed as being suitable for general-purpose use, with the exception of risers, plenums and other space used for environmental air, and shall also be listed as being resistant to the spread of fire.

(FPN): One method of defining resistant to the spread of fire is that the cables do not spread fire to the top of the tray in the "Vertical – Tray Flame Test" in the Reference Standard for Electrical Wires, Cables and Flexible Cords, ANSI/UL 1581-1985.

Another method of defining resistant to the spread of fire is for the damage (char length) not to exceed 4 feet 11 inches (1.5 m) when performing the "Vertical Flame Test — Cables in Cable Trays," as described in Test Methods for Electrical Wires and Cables, CSA C22.2 No. 0.3-M 1985.

   (e) Plenum Optical Fiber Raceway. Plenum optical fiber raceways shall be listed as having adequate fire-resistant and low-smoke-producing characteristics.
   (f) Riser Optical Fiber Raceway. Riser optical fiber raceways shall be listed as having fire-resistant characteristics capable of preventing the carrying of fire from floor to floor.

Please order a complete copy of the National Electrical Code and review all of Article 770 rules.

TEST MODULE 12

1. What is the newest method of office space construction?

2. Why should a cable installer be knowledgeable of building construction?

3. New office furniture is being designed to serve what functions?

4. Give the meaning of the following cable designations.

   OFCG_____

   OFNP_____

5. List three types of fiber optic cables grouped into types by the NEC- Article 770.

   •

   •

   •

LAB EXERCISE

Examine cables with NEC listings.

Section 13.0
Thoughts of a Designer

Objective

Describe the thought process of a designer when considering both fiber and copper issues starting with putting together a budget/loss sheet, things to know or variables, type of cables, today's backbone topology and new technologies.

Outline

- *Introduction to Loss Budget*
- *Variables*
- *Power Budgeting*
- *Fiber Loss*
- *Other Losses*
- *Margin*
- *Choice of Fiber Type*
- *Topology*
- *Backbone Cabling Distances*
- *Today's New Technology*

Learning Activity

Assessment: Test Module 13
Lab Exercise: None

Introduction

Among the top technical considerations for communication systems are loss budget and transmission capacity, or bandwidth. Loss budget should be calculated to be certain enough signal reaches the receiver to ensure adequate performance. Furthermore, you must calculate pulse dispersion, or bandwidth, to ensure the system can handle signals at the speeds you want to transmit. There are some guidelines to giving rough assessments, but in the real world, consider cost-effectiveness. This involves making trade-offs among various approaches and seeking the one giving the best performance at the most moderate cost.

Variables

There are many design variables that enter into the equation, directly or indirectly.

- Spectral linewidth of the light source
- Coupling losses
- Type of fiber (single or multimode)
- Light source output power (into fiber)
- Splice and connector loss
- Response time of the light source and transmitter
- Bit error rate or signal-to-noise ratio
- Receiver sensitivity
- System configuration
- Fiber attenuation and dispersion
- Fiber core diameter
- Fiber NA
- Signal coding
- Optical amplifiers
- Direct versus indirect modulation of transmitter
- Costs
- Receiver bandwidth
- Operating wavelength
- Wavelength-division multiplexing
- Type of couplers
- Switching requirements
- Number of splices, couplers and connectors

Power Budgeting

Power budgeting is similar to ensuring there is enough money to pay the bills. Enough light is needed to cover all optical transmission losses and deliver adequate light to the receiver achieving the desired signal-to-noise ratio or bit error rate. The design should leave extra margin above the receiver's minimum requirements allowing for system aging, fluctuations and repairs, such as splicing a broken cable. However, the power should not overload the receiver.

Warning: Know what power is specified where. There will be a loss of 3 dB if the transmitter manufacturer specifies output as peak power but the receiver manufacturer specifies average power.

The power budget formula is power transmitter - total loss + amplification = margin + receiver sensitivity when arithmetic is done in decibels or related units such as dBm. The simplicity of these calculations is the principal objective for using decibel units.

Optical amplification can offset loss in the system budget. Optical amplifiers are expensive, but the high cost can be justified in some cases. You wouldn't buy a $3,000 optical amplifier to replace a $100 laser source with a $10 LED, but it would be justified to avoid spending $10,000 on an electro-optic regenerator.

All losses in the system must be considered including:
- Loss in transferring light from source into fiber
- Fiber-to-receiver coupling loss
- Splice loss
- Fiber loss
- Coupler loss
- Connector loss

Fiber Loss

Fiber loss nominally equals the attenuation (in decibels per kilometer) times the transmission distance:

Total loss = (dB/km) x length

This is only an approximation for multimode fibers. One consideration is that measurements of fiber attenuation in dB/km do not consider transient losses occurring near the start of a multimode fiber. An LED with high NA and a large emitting area excites high-order modes that leak out as they travel along the fiber. Normally, this transient loss is 1 to 1.5 dB, concentrated in the first few hundred meters of fiber following the transmitter. The loss becomes less significant after a kilometer or two, but graded-index multimode fibers are rarely used over these long distances. Always allow transient loss in any system margin.

Another situation that may occur with graded-index fiber is unpredictable and uneven coupling of modes between adjacent lengths of fiber. Although such systems are extremely rare, these concentration effects make loss of long lengths of spliced graded-index fiber difficult to calculate.

Singlemode fibers are better behaved because they carry a single mode, avoiding differential mode attenuation.

Other Losses

Couplers, splices and connectors contribute significant losses in a fiber optic system. Fortunately, those losses are usually easy to measure and calculate. Couplers, splices and connectors have characteristic losses that can be multiplied by the number in a system to estimate total loss. However, there are two possible complications.

One complication, especially for connectors, is the variability of loss. A given connector may be specified as having a maximum loss of 1.0 dB and a typical loss of 0.5 dB. The maximum is the specified upper limit for that type of connector; no higher losses should show up in a system (unless the connector was installed improperly or is dirty). An average is the typical value, meaning that average connector loss should be 0.5 dB, but individual connectors may be higher or lower.

Total loss can be calculated in two ways for a system with four connectors. The worst-case approach is to multiply the highest possible loss (1 dB) by the number of connectors to get 4 dB. But, if the average connector loss is 0.5 dB, the most likely total loss is four times that or 2.0 dB. The sensible approach for a small number of connectors is to take the worst-case value, but a more realistic approach is to take the average loss for systems with many connectors or splices. Because detector overload can cause problems, you can run into trouble by seriously overestimating loss as well as by underestimating it.

Transient losses following connectors can further complicate the picture for multimode fibers. Connectors near the transmitter may increase transient losses by stripping away high-order modes, which would otherwise leak out farther along the fiber. Once light has traveled far enough to reach an equilibrium mode distribution (about a kilometer), a connector can redistribute some light to higher-order modes. These tend to leak out of the fiber - a milder form of transient loss than experienced with light sources.

Margin

A quantity that invariably figures in the loss budget is system margin, a safety factor for system designers. This allows for uncertainties in calculating losses, for minor degradation and for minor repairs of system components. Uncertainties are unavoidable because component losses are specified within ranges and components change as they are used and age. Margin also allows for repairs because of cable damage, which typically adds to cable loss.

Considering the application, the cost, the performance requirements and the ease of repair, the loss margin added by designers may be 3 to 10 dB.

Choice of Fiber Type

The type of fiber choice will have an enormous impact on the cost and performance of your system. The basic choice is singlemode or multimode fiber, but on a more detailed level, consider the various types of each available.

Singlemode fiber would be the choice if going more than a couple of kilometers. When transmitting at low speeds, it is possible to go farther, but the decision to not transmit at a higher speed must be made before going with singlemode. The premium grade is preferred for new dense-WDM systems. It is nonzero-dispersion-shifted fiber, with zero dispersion shifted to wavelengths longer than the standard erbium-doped fiber amplifier range of 1580 nm or longer. Dispersion-shifted fiber with zero dispersion around 1550 nm is vulnerable to four-wave mixing in WDM systems. Step-index singlemode fiber requires dispersion compensation and/or expensive narrow-line lasers (typically externally modulated distributed feedback types), unless operating at 1300 nm.

Multimode fiber is needed for spanning only short distances. There are more choices for shorter distances. Graded-index fibers are usually better for 100 m and up; 50/125-(m fibers have slightly more bandwidth than 62.5/125-(m fibers, but the smaller cores could mean higher connector losses. Larger-core step-index fibers collect light very well but have limited range because of their high dispersion. Plastic fiber's optical performance is poor but they are easier to terminate than glass fibers. Plastic fibers can be considered when links are short and termination is likely to be important and look for graded-index fibers, which improve bandwidth.

Other Guidelines

Some trade-offs that can affect cost and performance include the relative costs of components - generally cheaper at shorter wavelengths - and the low marginal cost of adding extra fibers to multifiber cables. Many of these balances shift with the pricing of commercial equipment. Some other rough-and-ready guidelines are listed starting with a few commonsense rules. Remember this is not a complete list.

- Time is valuable. Don't spend an entire day trying to save $5 on hardware or the result will be a net loss.
- Consider hiring a fiber optic contractor to install the fiber optic system or buy connectorized cables. The cost of hiring an expert may be less than learning the hard way.
- Installation, assembly, operation and support are not free. Installation and maintenance of fiber optic systems sometimes cost more than the hardware. Sometimes paying extra for hardware that is easier to install and service is more cost efficient.
- Save money by using standard mass-produced components rather than designing special-purpose components optimized for a particular application.

Here are some basic cost trade-offs people often face in designing fiber optic systems.
- The marginal costs of adding extra fibers to a cable are modest and less expensive than installing a second parallel cable. If reliability is an important factor, consider the extra cost of a second cable on a different route a worthwhile insurance premium.
- LEDs are less expensive and require less environmental protection than lasers, but they are harder to couple to small-core fibers and produce less power. Their broad range of wavelengths and limited modulation speed limit system bandwidth.
- Balance the performance of low-loss fiber, high-sensitivity detectors and powerful transmitters against price advantages of lower-performance devices.
- Fiber attenuation contributes less to losses of short systems than losses in transferring light into and between fibers.
- Because of their differences in component requirements, topology of multiterminal networks may have a considerable impact on system requirements and cost. Coupler losses could seriously restrict options in some designs.
- Rather than buy and install new cable, it usually is more sensible to upgrade transmitters and receivers to work with step-index singlemode fiber already installed.
- Fiber and cable become a larger fraction of total cost, having more impact on performance, the longer the system.
- Balance the advantages of eliminating extra components with the higher costs of the components needed to eliminate them. For example, it's hard to justify two-way transmission through a single fiber over short distances unless wavelength-division-multiplexing couplers are inexpensive, large installation savings are possible or system requirements permit only a single fiber.
- Low-loss, high-bandwidth fibers usually accept less light than higher-loss, lower-bandwidth fibers. Save money and overall attenuation over short distances, by using a higher-loss, more costly cable that collects light more efficiently from lower-cost LEDs. (Large-core multimode fibers are more expensive than singlemode fibers because of the economics of production and material requirements.)
- The cost of light sources and detectors for 1300 and 1550 are more than those for the 650- or 800- to 900- nm windows, although fiber and cable for the longer wavelength may cost less.
- 1300-nm light sources are less expensive than 1550-nm sources. They must have extremely low bandwidth to be reasonable or be used with dispersion-shifted fiber.
- High-power laser transmitters or optical amplifiers make sense in systems distributing signals to several terminals.
- Lasers with broader spectral linewidth are less expensive than narrow-line distributed-feedback lasers.
- Compare costs of high-speed TDM on single channels or lower-speed TDM at multiple wavelengths.
- Always consider future upgrade possibilities. To paraphrase one of Parkinson's laws, communications requirements expand to fill the available bandwidth. If a small extra investment now can allow expansion too much greater capacity in the future, it's generally worthwhile unless cost constraints are severe or the route lifetime is limited. (You don't want to spend extra money on a building that will be torn down in

two years.)

- Dark fibers (extra fibers installed in the original cable that were never hooked up to light sources) are sometimes available in existing cables.
- Leave room for expansion by adding WDM.
- Installing extra dark fibers in a cable can be worthwhile, because it allows more room for expansion.
- Always allow for repair and expansion. The cost of design margin into the system is much less than adding it afterwards.
- Remember to account for the costs of nonfiber hardware. For example, splice enclosures and patch panels, and any extra losses they involve and to allow room for installing them.
- Human actions - not defective equipment - cause most fiber optic failures. Consider ring topologies that can survive a single break. The extra time and money it takes to make important systems less vulnerable to damage always pays off. Label and document the system carefully, don't leave cables where people can trip over them or contractors can damage them.

Develop your own guidelines to add to this list as you grow more familiar with fiber optics.

Topology

Structured Cabling Systems

Designing a cabling backbone that is futureproofed is a difficult task. A cabling infrastructure is expected to support telecommunications connectivity for approximately 15 years. With this in mind, it does not pay to cut corners or use old design practices and technology. Since the cabling backbone is the main artery of information flow within a company, one must plan the design thoroughly and invest in new technology.

Backbones are supposed to provide interconnections between telecommunications closets (TCs), equipment rooms (ERs) and entrance facilities (EFs). Components of the backbone include backbone cables, main cross-connects (MCs), intermediate cross-connects (ICs), mechanical terminations, patch cords and jumper cordage used for backbone-to-backbone cross-connections.

Because its occupants' needs are dynamic, a building may require varying backbone requirements for cabling infrastructure. Growth and changes in service requirements should be accommodated without having to add or change cable. Usually, a maximum number of connections for copper and fiber optic cable media should be planned.

The Commercial Building Telecommunications Cabling Standard (ANSI/TIA/EIA 568A) gives several recommendations for the planning for backbones. Its complementary standard, Pathways and Spaces (ANSI/TIA/EIA 569), clarifies the subject and specifies the separation of the backbone cabling from EMI (electromagnetic impulse) sources such as electric motors, transformers and fluorescent light ballasts.

Topology

Topology is the term used to describe the layout or physical connections of a network. It can be thought of as a map of the roads between all the devices attached to a network. Examples of topologies are bus, ring and star. (See Figure 13.1.)

Communications backbones should use the conventional hierarchical star topology. (See Figure 13.2.) This star wiring scheme requires that each horizontal cross-connect (HC) in a TC is cabled to an MC or to an IC then to an MC, unless there is a need to use bus or

Figure 13.1

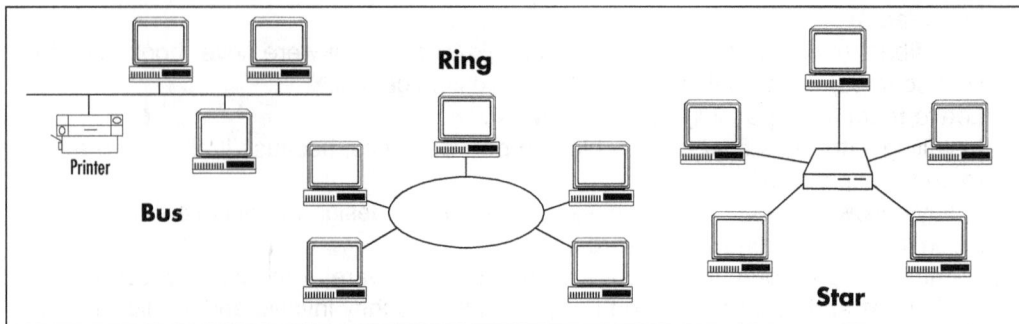

Ring

Bus

Printer

Star

Figure 13.2

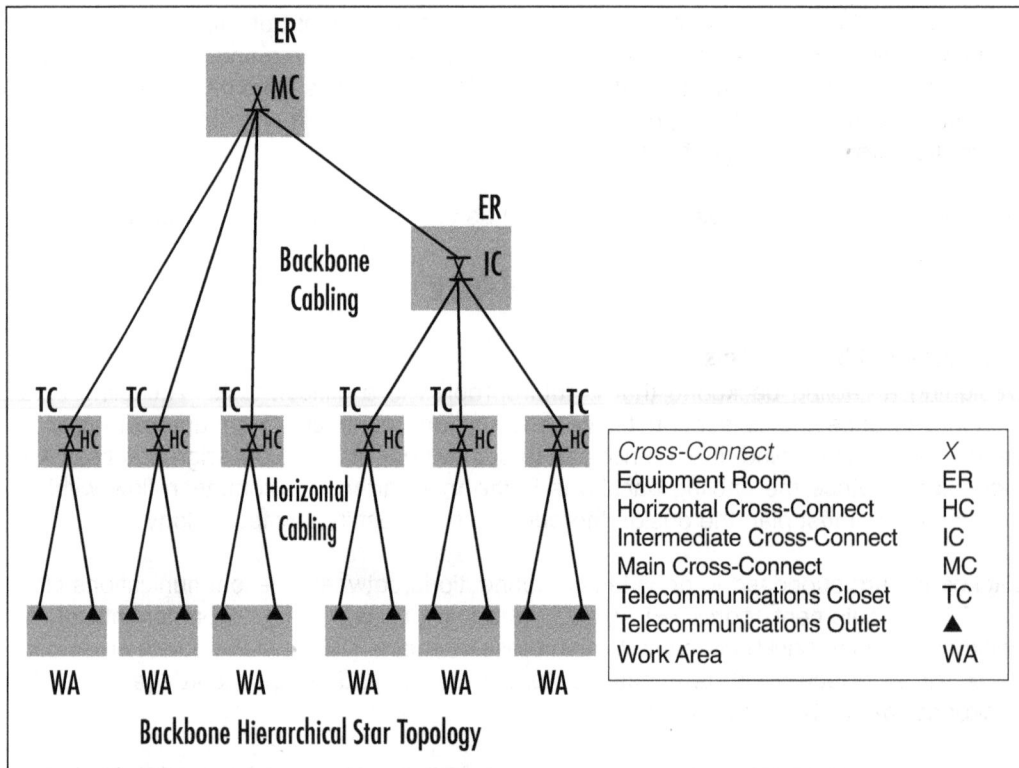

ER

MC

ER

Backbone
Cabling

IC

TC   TC   TC   TC   TC   TC

HC   HC   HC   HC   HC   HC

Horizontal
Cabling

Cross-Connect	X
Equipment Room	ER
Horizontal Cross-Connect	HC
Intermediate Cross-Connect	IC
Main Cross-Connect	MC
Telecommunications Closet	TC
Telecommunications Outlet	▲
Work Area	WA

WA   WA   WA   WA   WA   WA

Backbone Hierarchical Star Topology

ring configurations, which can be wired directly to the TC. Cabling of telecommunications equipment that connects directly to MCs or ICs should be less than or equal to 30 meters (98 feet) long.

Notice that this limits the wiring scheme to two levels or cross-connects. It reduces signal loss in circuitry and simplifies moves, adds and changes (MACs). This means the interconnection between any two HGs will pass between three or fewer cross-connects. In addition, only a single cross-connect will be passed to reach the MC.

A single backbone cabling cross-connect (the MC) may meet your cross-connect needs. Backbone cabling cross-connects may be located in TCs, ERs or EFs. Bridged taps may not be used for backbone cabling.

Non-star configurations (ring, bus or tree) can be accommodated by appropriate interconnections, electronics or adapters in the TC (Keep in mind the less than or equal to 30 meters rule.)

Recognized Cables

The sizes of sites and the varying range of services required dictates possible use of more than one choice of media. ANSI/TIA/EIA 568A recognizes four backbone transmission media types. They can be used individually or in combination.

1. 100 ohm UTP, four-pair, 24 AWG thermoplastic insulated solid conductors formed into four twisted pairs and enclosed by a thermoplastic jacket; or 100 ohm UTP multipair backbone cables in pair sizes greater than four pairs, assembled into binder groups of 25 pairs. The groups have distinctive colored binders and are assembled to form the core. A protective sheath consisting of an overall thermoplastic sheath covers the core. Note that four-pair and multipair screened twisted pair cables that meet transmission requirements for the above cables may also be used.
2. 150 ohm, STP-A, two-pair, 22 AWG, individually twisted pairs, with thermoplastic insulated solid conductors enclosed by a shield and an overall thermoplastic jacket.
3. 62.5/125-micron multimode fiber optic cable dominates the premise marketplace because it can be used with LED transmitters. Six to 12 fiber strands are preferred.
4. Singlemode fiber cable (higher bandwidth capabilities).

Media Selection

The different needs that services and applications may have for backbone media warrants consideration of multiple factors before selection of the backbone media is made. The size of the site and its population may dictate the choice of media. Other considerations

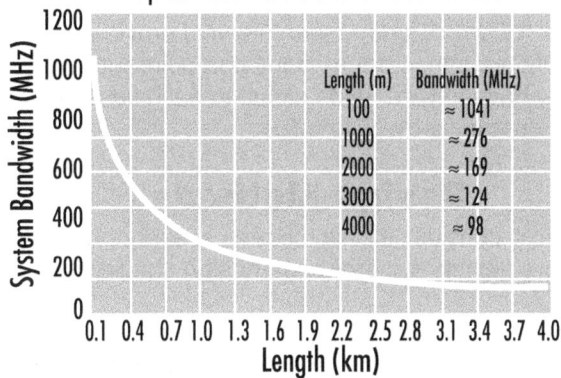

Figure 13.3

Typical System Bandwidth Utilizing ANSI/TIA/EIA 568A 62.5/125μm Optical Fiber Cable and a 1300 nm LED

Length (m)	Bandwidth (MHz)
100	≈ 1041
1000	≈ 276
2000	≈ 169
3000	≈ 124
4000	≈ 98

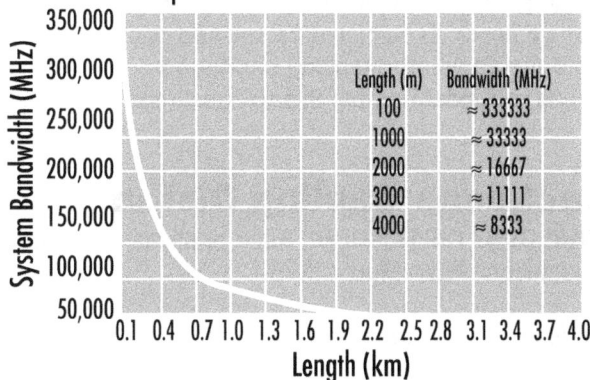

Figure 13.4

Typical System Bandwidth Utilizing ANSI/TIA/EIA 568A Singlemode Optical Fiber Cable and a 1310 nm Laser

Length (m)	Bandwidth (MHz)
100	≈ 333333
1000	≈ 33333
2000	≈ 16667
3000	≈ 11111
4000	≈ 8333

are a cable's flexibility with respect to supported services, cost and availability of connectors and termination practices. The media selected should also be based upon the benefits a cable can provide over the use of backbone cabling.

When one knows the quantity and type of occupants in a building, it is easier to choose a backbone media. A building that has occupants who come and go is different from a stable environment.

Without knowledge of the occupant's needs, one must plan on providing an array of services that supports everything that can be conceivably thought of. At this stage of design, it helps to group similar services together; voice, local area networks (LAN), display terminals and other digital services are typical categories. Individual types should be identified and the responding quantities projected within a group.

Voice circuits will have their own cross-connects (typically 66 or 110 blocks) connecting to the PBX. Keep in mind that each data drop location will probably require a voice connection and voice drops will be greater in number than data drops. Usually, 25 percent of the people in a company do not have PC access requirements because of the nature of their job, but they may require a phone. Voice backbones typically use Category 3 UTP and cross-connect cables punched down on both ends. Some designers are installing Category 5 UTP for voice, with the idea that those circuits may convert to data designation at a later date. Install Category 5 blocks and cordage in this instance.

Data circuits should use Category 5 UTP for up to 100 Mbps (megabits per second) operation and are installed with either blocks and/or patch panels with 110 or KRONE type terminations using patch cables for cross-connects.

Local exchange carriers (LECs) or independent service providers typically provide digital circuits. They are usually brought into a demarcation point with an RJ-21X block or jack for connection. You may have to extend the circuit to each subscriber within the building. This should be handled on a case-by-case basis.

Design Methods

A backbone design should consider the worst-case scenario; the higher the uncertainty for a given area, the more flexible the backbone design should be.

The best design methods promote multiple cable media selection and provide for redundant connectivity solutions. All cable media should share the same facility architecture, grouped within the same location for cross-connects, mechanical terminations, interbuilding and entrance facilities.

Backbone pathways cannot use elevator shafts. Cables are typically riser-rated or plenum-rated. For multiple floors, TCs should be stacked one above the other. Run conduit to protect the cable media where stacking cannot be accomplished. You must be creative when considering redundant paths; this will assure greater availability of service when a path is cut.

Backbone Cabling Distances

INTRA-AND INTER-BUILDING DISTANCES. Each application has its own limitation for maximum backbone distances. (See Figure 13.5.)

When the HC-to-IC distance is less than maximum, the IC-to-MC distance for optical fiber can be increased accordingly, but the total distance from the HC to the MC shall not

Figure 13.5

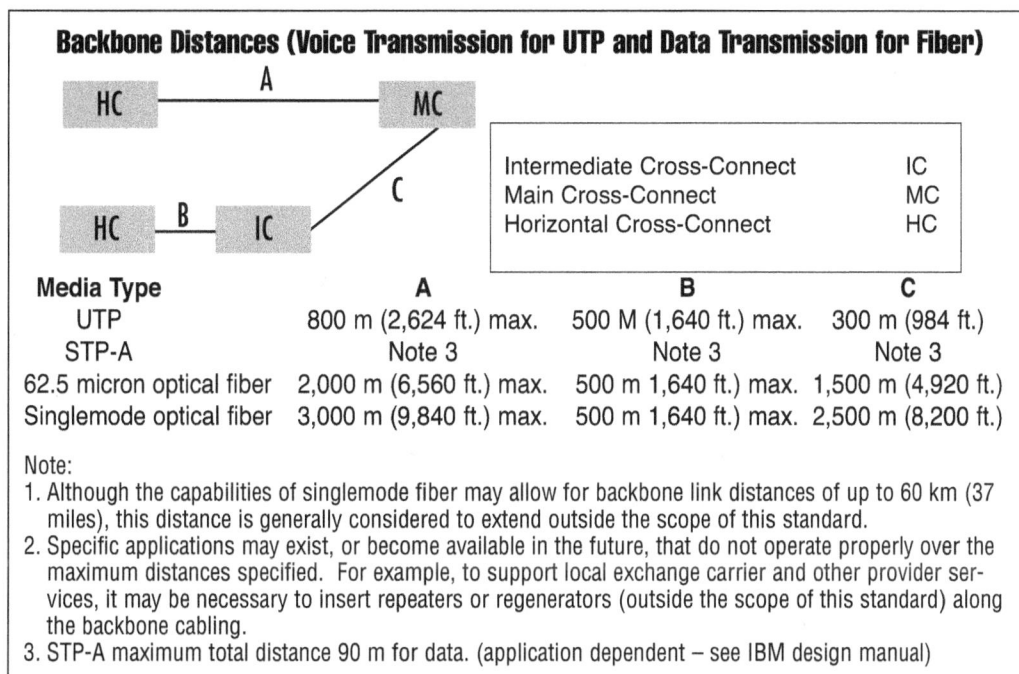

Backbone Distances (Voice Transmission for UTP and Data Transmission for Fiber)

Media Type	A	B	C
UTP	800 m (2,624 ft.) max.	500 M (1,640 ft.) max.	300 m (984 ft.)
STP-A	Note 3	Note 3	Note 3
62.5 micron optical fiber	2,000 m (6,560 ft.) max.	500 m 1,640 ft.) max.	1,500 m (4,920 ft.)
Singlemode optical fiber	3,000 m (9,840 ft.) max.	500 m 1,640 ft.) max.	2,500 m (8,200 ft.)

Intermediate Cross-Connect	IC
Main Cross-Connect	MC
Horizontal Cross-Connect	HC

Note:
1. Although the capabilities of singlemode fiber may allow for backbone link distances of up to 60 km (37 miles), this distance is generally considered to extend outside the scope of this standard.
2. Specific applications may exist, or become available in the future, that do not operate properly over the maximum distances specified. For example, to support local exchange carrier and other provider services, it may be necessary to insert repeaters or regenerators (outside the scope of this standard) along the backbone cabling.
3. STP-A maximum total distance 90 m for data. (application dependent – see IBM design manual)

exceed the maximum of 2,000 m (6,560 ft.) for 62.5 micron optical fiber cable or 3,000 m (9,840 ft.) for singlemode optical fiber cable.

When the HC-to-IC distance is less than maximum, the IC-to-MC distance for UTP cabling can be increased accordingly; but the total distance from the HC to the MC shall not exceed the maximum of 800 m (2,624 ft.).

When the HC-to-MC distance for UTP cabling can be increased accordingly; but, the total distance from the HC to the MC shall not exceed the maximum of 800 m (2,624 ft.).

It is standard practice to locate the main cross-connect near the center of a site. As mentioned earlier; TCs should be stacked above one another (like most electrical closets) to accommodate shorter backbone lengths. A computer room and its MC may be on a third floor, and it may support other floors via backbone cabling and TCs, ICs or HCs. The main idea is to have all communications collapsed back to the MC area that connects the computer utilities.

Notice that for any Category 5 UTP circuit, the limitation of 90 meters + 10 meters allowance for cross-connects and patch/line cords applies.

Some people make an error of interpretation of the distance requirements and think they can use 100 meters between the MC/IC and the HC. However, the length parameter will be exceeded.

A paradigm shift of the decentralization of network products, such as hubs and switches to the work group area, has occurred recently. Work group switches are being installed within the TCs to support users on that floor within the 100-meter distance. These switches can offer 10 to 100 Mbps dedicated domains for local users that are tied into the network via 100 Mbps uplinks using 100BASE-T or ATM technology. The same approach is being done for hubs (shared media) and this makes the pipelines larger for each work group. This is done by using a backbone run of Category 5 UTP (if it's within the 100 meter parameter) or multimode fiber, requiring a fiber port on the main core device and

Figure 13.6

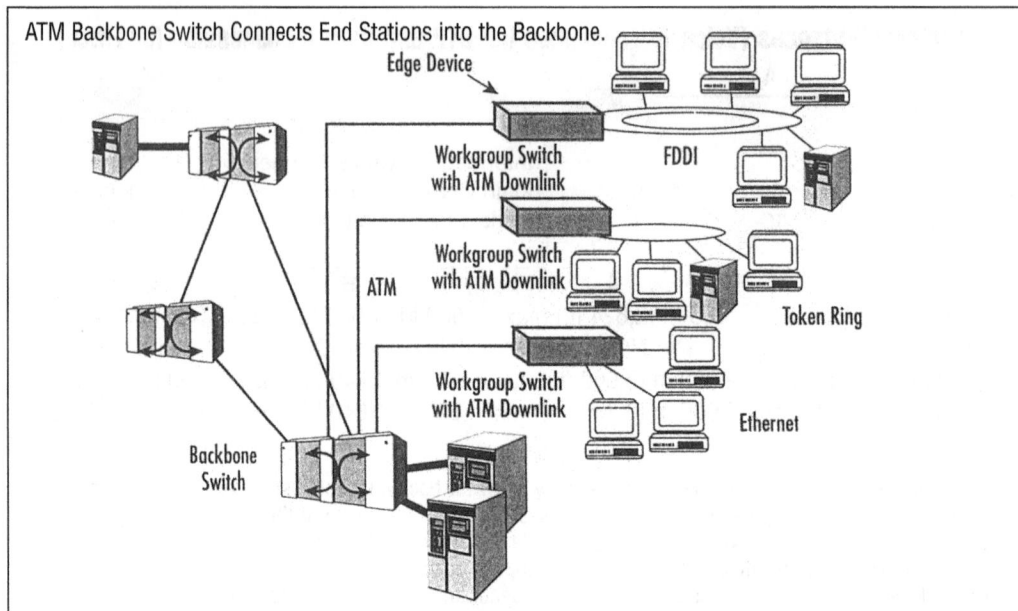

ATM Backbone Switch Connects End Stations into the Backbone.

also at the switch or hub in the closet. If one does not have a fiber port for the electron-ics and wants to go longer distances, there are 100 Mbps copper-to-fiber converters available at each end of a link with 100BASE-T ports.

It may be appropriate to install both copper (when length allows) and fiber backbone to be certain of future support. Fiber optic cabling and its termination products have really come close to closing the gap on cost with Category 5 UTP. It is possible to learn how to ter-minate and to polish fiber in a short period of time, such as a one-day class. In addition, fiber assemblies can be purchased pre-assembled if there are no fiber capabilities onsite (assuming the conduits/innerduct raceways are large enough to route the assembly with connectors attached). Remember to think proactively in these instances. Overkill regard-ing design will provide flexible future connectivity, whereas meeting the minimum requirements could force the re-cabling of an entire facility.

Use of Technology

It is one thing to identify recognized cable media and understand practices for installation, but it is another thing to evaluate proper technologies available that run on the media. The networking products used will dictate the respective protocols and their required support media.

At present, asynchronous transfer mode (ATM) is popular because it provides scaleable performance from 25 Mbps to 622 Mbps, depending on the media chosen. ATM was designed to carry data, voice and video, and its switched technology also avoids band-width contention. All other technologies offer ATM downlinks for connection to an ATM backbone.

Fiber distributed data interface (FDDI) has been around for quite some time and it has been effectively used as a backbone technology. It supports ring or star wiring schemes, operates at 100 Mbps, and supports distances of up to 2 km over fiber. FDDI was initial-ly designed with the intent of having redundant features whereby the network had self-healing capabilities in the event of a break in a fiber. To attain this, one must install the backbone with dual counter rotating rings and dual attached stations. Anyone who chooses FDDI as their backbone must live with the 100 Mbps speed and must also like redundancy. The only negative for FDDI is that the cost of the electronics and attachment devices are high when compared to other technologies.

Fast Ethernet (100BASE-T) provides 100 Mbps speeds and uses Category 5 UTP. It is used mostly for connection of server farms and for 100 Mbps dedicated port switching to high-bandwidth users. The use of Fast Ethernet for backbone technologies has some limiting distance factors relating to collision domains.

Collision domains of 100BASE-T are limited to 400 meters (point-to-point links), 300 meters (mixture of Category 5 UTP and fiber), and 205 meters for Category 5 UTP. Because of these limitations, designers should be careful when installing a 100BASE-T backbone.

TEST MODULE 13

1. Power budget formula is power transmitter - total loss + amplification = margin + receiver sensitivity when arithmetic is done in decibels or related units such as dBm.

   True   False

2. Fiber loss usually equals the attenuation (in decibels per kilometer) times the transmission distance.

   True   False

3. The type of fiber choice has no impact on the cost and performance of a system.

   True   False

4. It could be more cost-effective to hire a fiber optic contractor to install a fiber optic system rather than trying to learn on the job.

   True   False

5. Topology is the term used to describe the layout or physical connections of a network.

   True   False

6. Communications backbones should not use the conventional hierarchical star topology.

   True   False

7. Backbone pathways always use elevator shafts.

   True   False

8. The maximum backbone cable length distance is 2000 m from the HC to the MC for 62.5m.

   True   False

9. Asynchronous transfer mode (ATM) is currently popular because it provides scaleable performance from 50 Mbps to 900 Mbps, depending on the media chosen.

   True   False

10. Knowing the quantity and type of occupants in a building makes it easier to choose a backbone media.

    True   False

Section 14.0
Fiber Optic Cabling
for Data and Networking

Objective

Understand the basics of local area network concepts as they relate to fiber optic cabling and the types of applications for using fiber optics.

Outline

- Local Area Network
- Topologies
- Main Distribution Frame
- Fiber Optic Data Networks
- Ethernet and GIGI Ethernet
- Fiber Channel
- FDDI

Learning Activity

Assessment: Test Module 14
Lab Exercise: None

Local Area Network

Today's LANs
Today, LANs use multimode fiber, rather than the singlemode fiber common in long-haul telecommunications. Although singlemode fiber is inexpensive and very high in performance, it is not economical for LANs because it requires more expensive electronics and connectors than multimode fiber.

Some LAN topologies, such as buses, rely on tapping fiber for their very structure. In other words, they are based on the ability to tap the media with negligible loss of signal. However, the difficulty of tapping fiber also makes LAN bus topologies, for example, more difficult to implement than others.

Topologies

Point-to-Point
Point-to-point topology is very common today, requiring two nodes directly communicating, linked by normally a pair of fiber optics; one to transmit and one to receive. Applications for point-to-point include:
• Computer channels
• Terminal multiplexing
• Video transmission

The drawing shown in Figure 14.1 is of a cable assembly hook-up for a simple point-to-point fiber optic link.

Star Topology
Star LANs are arranged around a single hub that may act as a central controller for the network. (See Figure 14.2.) Transmission sent from one node or terminal must first pass through the hub. This hub can simply be a passive star computer or an active controller. Common star applications include:
• PBX
• Mainframe

Main Distribution Frame - Single Hub
The hub of the physical star is known as a main cross-connect (MC) and is commonly located in an equipment room or computer center. Use of an MC offers many advantages. (See Figure 14.3.)

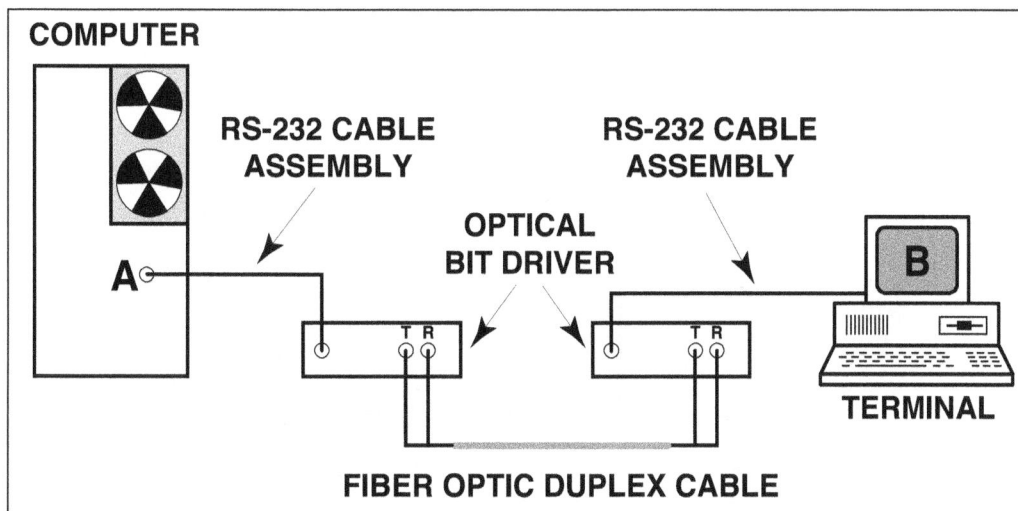

Figure 14.1

COMPUTER
RS-232 CABLE ASSEMBLY
RS-232 CABLE ASSEMBLY
OPTICAL BIT DRIVER
A
B
T R
T R
TERMINAL
FIBER OPTIC DUPLEX CABLE

Figure 14.2

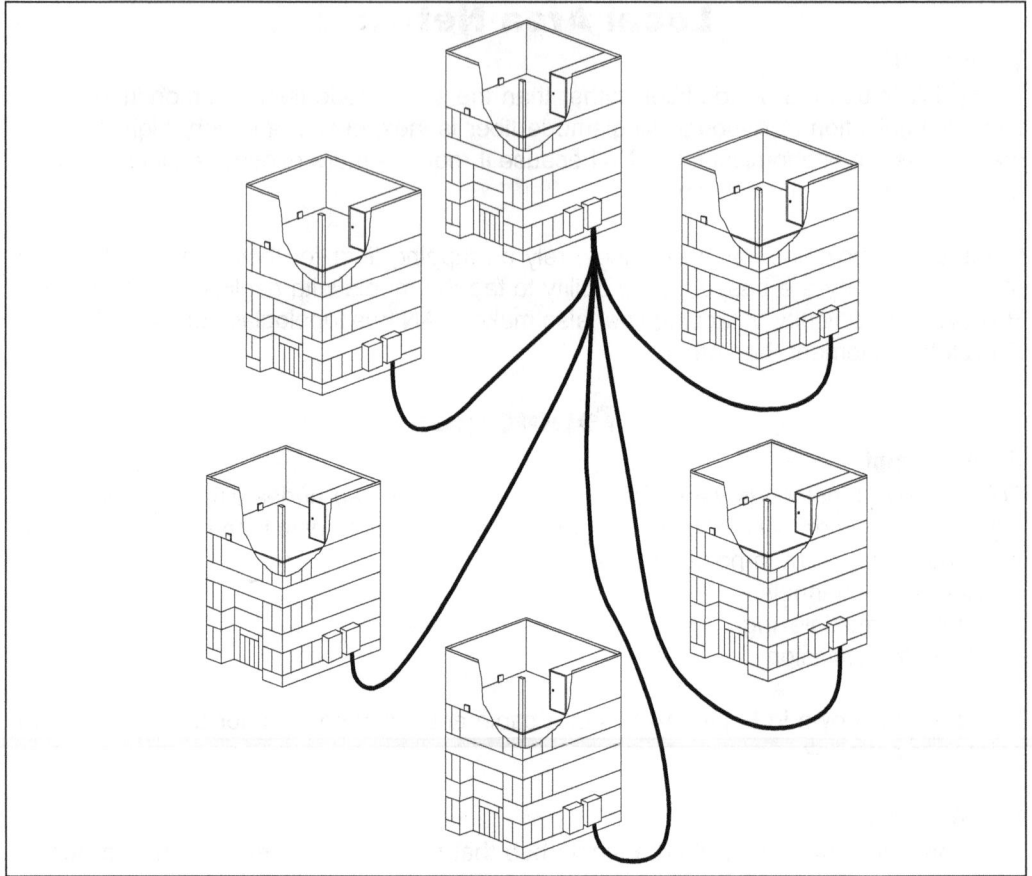

Figure 14.3

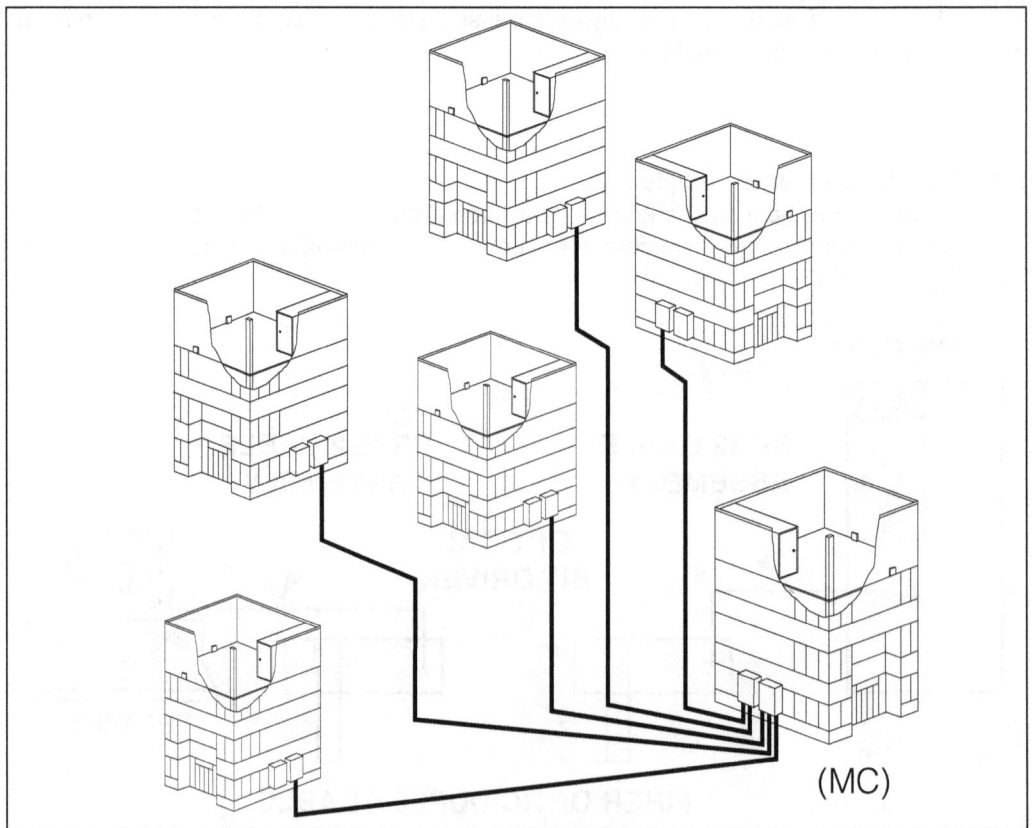

(MC)

14.4

- Provides a single point of control for system administration.
- System testing and reconfiguration of the system's topology can be accomplished from the hub.
- Security from unauthorized access can be maintained.
- Electro-optic components can be located in the hub today, while future components (such as active stars and optical switches) can be located at the hub as technology advances.

Main Distribution Frame - Stars Linked

In a campus application, multiple physical stars linked together may be appropriate when convenient hubs are separated by long distances. (See Figure 14.4.) In a building, each wiring closet will serve as an intermediate cross-connect (IC). An MC for building distribution should be planned. Planning the location of these hubs will avoid future physical rearrangement as fiber continues to penetrate the office environment.

Ring Topology

A ring consists of point-to-point completion with the various stations on the ring connected only to their adjacent stations and to no other stations. (See Figure 14.5.)

In a ring type network, all terminals are linked in a point-to-point series. If one part fails, the system is down unless bypass components are used. To avoid conflicting data

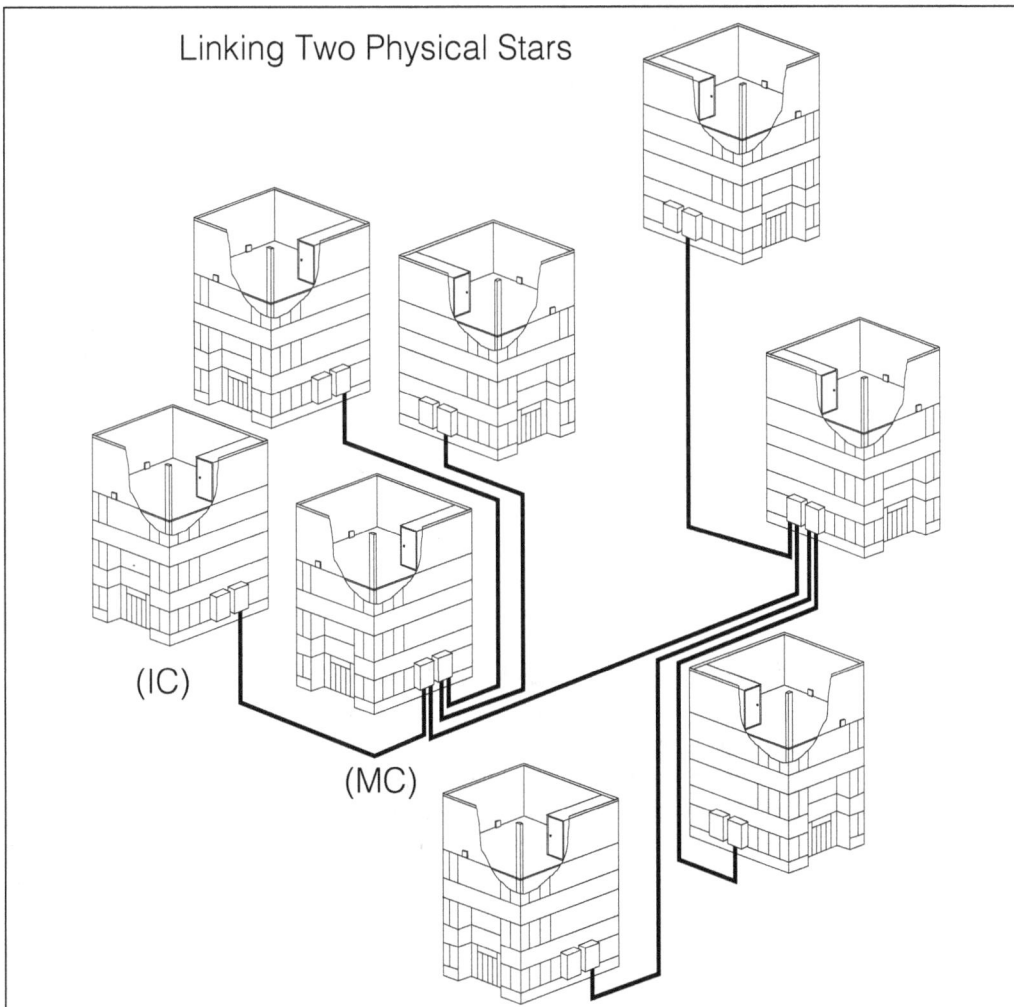

Figure 14.4

Linking Two Physical Stars

(IC)

(MC)

Figure 14.5

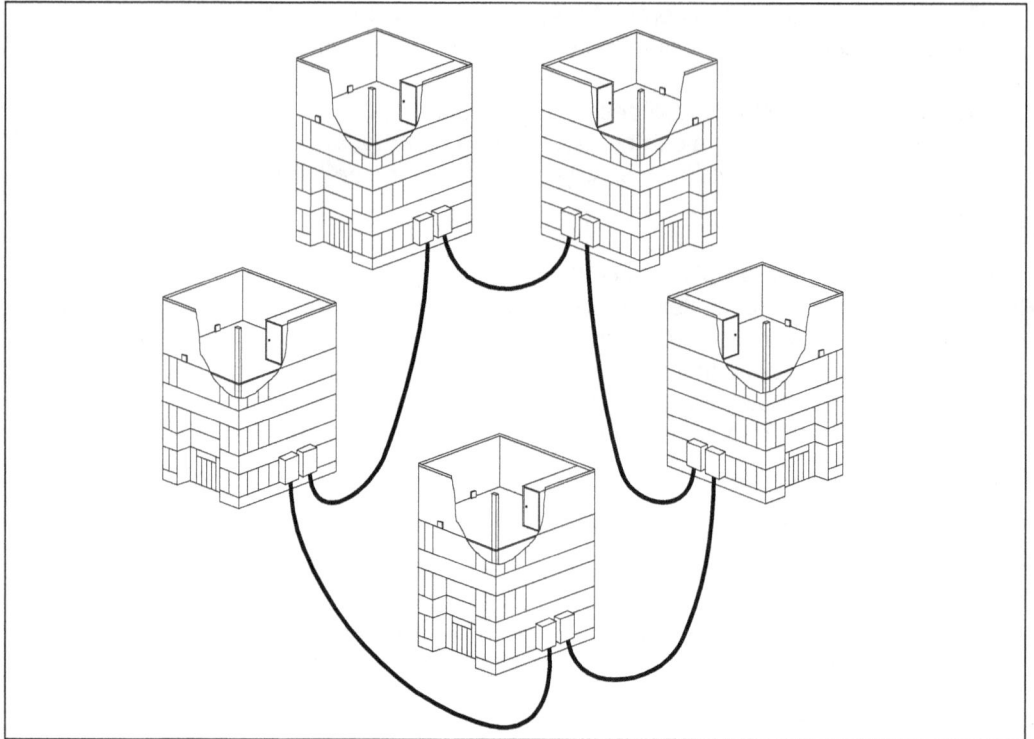

demands, such systems use a bit pattern, called a token. The token is circulated to each node allowing that node to capture the token and the right to transmit data.

Examples of ring topologies are:
- FDDI
- IBM Token Ring
- High-Speed Time Division Multiplexing

Bus Topology

Networkers based on a bus topology (Figure 14.6) also use a token passing scheme, or an access scheme known as carrier-sense multiple access with collision detection (CSMA/CD) or collision avoidance (CSMA/CA). Like a ring, messages on the bus are broadcast to all terminals. Since all the terminals tap into a single main trunk channel like branches on a tree, messages do not have to be repeated.

The most popular systems requiring bus topology are:
- Ethernet
- MAP (manufacturing automation protocol)

Fiber Optic Data Networks

As their data rates and transmission distances expand, networks are increasing their use of fiber optics. High-speed wide-area networks will use fibers to link smaller LANs or devices requiring high-speed connections. There are numerous network architectures being developed and tested with only a few being accepted. For these networks to connect varied devices uniformly, they must have standardized designs. The users receive the adaptability needed to connect different equipment and product behavior of hardware and software.

Figure 14.6

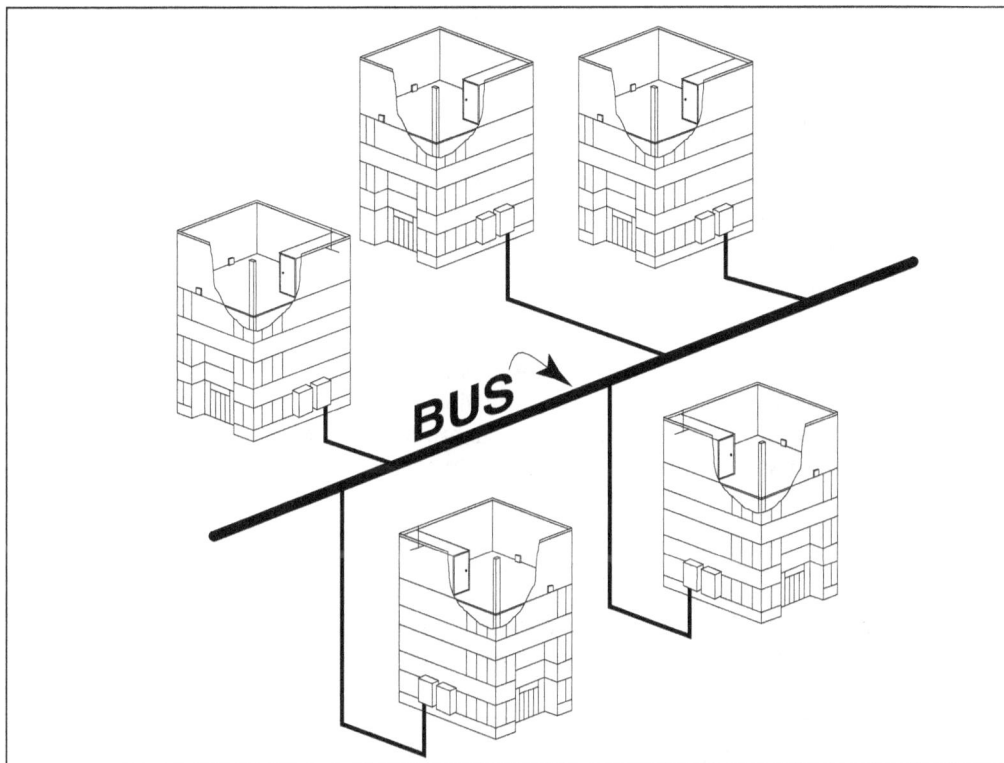

A decade ago, LANs used standard Ethernet and the IBM Token Ring network, transmitting about 10 Mbps. Many of these networks are still operating, but transmission rates have pushed continually upward for local area networks and are even higher for wide area networks interconnecting LANs. The 10-Mbps Ethernet was followed by the 100-Mbps Fast Ethernet and Gigabit Ethernets. Fibers are important elements of the networked versions of Fibre Channel and of the 100-Mbps fiber distributed data interface (FDDI) standard.

10-Mbps Ethernet

The original Ethernet standard codified as IEEE (Institute of Electrical and Electronics Engineers) standard 802.3 was the first LAN to gain considerable acceptance. It distributes variable length digital data packets at 10 Mbps to transceivers dispersed along a coaxial cable bus. Independent cables of up to 50 m long and containing four twisted wire pairs, run from the transceivers to individual devices (such as personal computers, file servers or printers). The network is able to serve up to 1,024 terminals.

An Ethernet network has no overall controller, therefore control functions are handled by individual transceivers. When a terminal is ready to send a signal, its transceiver checks for another signal going along the coaxial cable. If another signal is present, transmission is delayed. Otherwise, the terminal begins transmitting and continues until it finishes or detects a collision (the transmission of data at the same time by a second terminal). These collisions happen because it takes time (several nanoseconds a meter) for signals to travel along the coax. A delay of 6 ns/m would cause a collision if two terminals 300 m apart on the coax started sending within 1.8μs of each other. The terminal stops transmitting if it detects a collision and waits a random interval before trying again.

Address headers determine the destination for each data signal. All the transceivers on the network see every data signal, but the signals not directed to them are ignored. The only signals the transceiver relays to the terminal attached to it are those with the terminal's address.

The basic 10-Mbps Ethernet design has some significant variations. The original standard heavy coaxial cable allows transceivers to be up to 500 m apart, which is expensive. Substituting a lighter grade of coax limits transceiver spacing to 200 m, but this "thin" Ethernet is sufficient for a majority of purposes. As an alternative, using twisted wire pairs can carry signals up to about 100 m. Along with the data bus configuration, Ethernet is sometimes arranged in a star configuration. Cables radiate outward from a hub, which relays signals to other terminals.

Optical fibers can extend transmission distances beyond the limit dictated by the loss of coaxial cable to distances bound by other conditions, such as the time it takes signals to travel through the network. Frequently, a remote terminal or a point-to-point fiber link may connect two coaxial segments of an Ethernet with a central Ethernet. This allows a single Ethernet to connect terminals in different buildings, which is difficult with the 500 m limit of coax.

Maximum transmission distances are dependent upon whether the network is operating in half-duplex or full-duplex mode. In half-duplex, terminals either transmit or receive at any one time. In half-duplex mode, the maximum distance is 2 km for either multimode fiber or singlemode fiber. In full-duplex mode, they simultaneously send and receive data and multimode fiber allows cable runs to 2.5 km, and singlemode allows spacing to 15 km. (Light takes roughly $75\mu s$ to travel a 15-km fiber. This means the terminal on the end of a 15-km fiber lags $75\mu s$ behind the rest of the network.)

Fast Ethernet (100 Mbps)

Fast Ethernet standard was approved in 1995 and is a faster version of the original Ethernet. Fast Ethernet uses interface cards that operate at 100 Mbps, retaining the same transmission protocols and frame format as the 10-Mbps Ethernet. The same network configurations and cabling as 10-Mbps Ethernet are used. The primary change is the 10-Mbps Ethernet cards were replaced with Fast Ethernet cards. However, the faster speed limits coax runs to 100 m.

Specification for the Fast Ethernet limits half-duplex transmission to 412 m over single or multimode fiber with a travel time of $2\mu s$. Full-duplex transmission extends the maximum distance to 2 km for multimode fiber and 10 km for singlemode. These differences are caused by the differences in the nature of half- and full-duplex transmission.

Gigabit Ethernet (1 Gbps)

Gigabit Ethernet is the most recent move to higher speeds operating at 1 Gbps. It uses the same protocols and frame format as slower Ethernets but with only full-duplex transmission. Because of the high speed, the node spacing for Gigabit Ethernet is shorter. A special twinax cable (instead of a single metal wire, a coaxlike cable with a twisted pair at the center) is used for jumper cables running up to 25 m. Four twisted wire pairs in a Category 5 cable combines to span up to 100 m. This does not imply one pair in Category 5 cable – the 1-Gbps signal is split among four pairs.

The backbone for Gigabit Ethernet is fiber, with copper used only for short connections. There are different maximum distances for four types of fiber cable:

Short-wavelength (780-850 nm) in 62.5/125-μm fibers: 260 m
Short-wavelength (780-850 nm) in 50/125-μm fiber: 525 m
Long-wavelength (1300-1500 nm) in 62.5 - or 50-μm fiber: 550 m
Singlemode, long-wavelength: 3 km

Modal dispersion dominates the limit on multimode fibers; it is larger in the larger-core 62.5/125-μm fiber.

The Gigabit Ethernet standard is modeled after components developed for 1-Gbps transmission using the Fibre Channel standard, but it is not itself covered by Fibre Channel. Gigabit Ethernet could be the best choice for backbone networks and a logical choice for transmitting Internet protocol signals between networks.

Fibre Channel

Fibre Channel covers a wide range of signal transmission. In addition to covering the point-to-point transmission, Fibre Channel also covers switched networks and transmission around loops and other network topologies. Hubs connect nodes to form loops; switches are interconnected to make a fabric that functions somewhat like the phone system in routing signals between devices. Rather than allocating dedicated channels, Fibre Channel processes signals as frames.

Fibre Channel uses a 10-bit coding for each 8-bit byte. Fibre Channel transmits bits in the system in series rather than in parallel. Data rates can be specified either as megabits per second or megabytes (Mbytes or MB), which can be confusing. Overhead bits increase the bit rate by 6.25 percent. Table 14.1 lists the speeds in both formats.

Gigabit Ethernet and Fibre Channels allow transmission over both copper and fiber. Twisted pair, coax and twinax are the major alternatives for speeds of 1 Gbps and below. Fiber is specified at the higher rates. Transmission can be in the short or long-wavelength bands.

Fibre Channel protocol specifies limits on the numbers of terminals in a loop and the arrangement of hubs and switches as with Ethernet. Remember Fibre Channel as a high-speed digital transmission standard useful in networks in addition to point-to-point transmission.

Table 14.1

Assigned Number	Megabytes per Second	Megabytes per Second (with Overhead)
12.5	12.5	133
25	25	266
50	50	531
100	100	1062
200	200	2125
400	400	4250

Fiber Distributed Data Interface (FDDI)

Fiber distributed data interface (FDDI) network standard operates at 100 Mbps. The FDDI standard calls for the ring topology with two rings that can transmit signals in opposite directions to a series of nodes. It also specifies concentrator-type terminals allowing stars and/or branching trees to be added to the main FDDI backbone ring. Ordinarily, one ring carries signals while the other is kept in reserve in case of component or cable failure.

FDDI transmission is controlled by a scheme of "token passing" used in slower-speed token ring networks, covered by the IEEE 802.5 standard. Contrary to Ethernet, terminals do not compete for space to send signals, but pass around the loop an authorization code called a token. A node with a message to send receives the token, holds the token and sends the message, with a code identifying its destination. All other nodes ignore the message, which is canceled when it completes its path around the ring. The terminal that sent the message begins passing the token around the ring again. This is more methodical than Ethernet transmission, although it has limitations.

FDDI uses a 4 or 5 transmission code adding one extra bit for every four data bits, which means the actual data rate is 125 Mbaud for 100 Mbps of user data. This coding scheme balances transmission between on and off bits to enhance operating efficiency.

The standard was developed around fiber optic transmission. Copper wires can carry 100-Mbps signals short distances and wired versions of FDDI have been developed. (Sometimes called CDDI, with the C from copper substituted for the F from fiber.)

FDDI is a local area network, but oftentimes serves as a backbone network linking LANs operating at slower speeds. The nodes would then be gateways to other networks. Concentrators are devices attached to FDDI nodes to combine signals from many terminals, or to collect signals from a LAN for transmission to the FDDI network. With higher data rates and the growth of video and graphics-intensive applications, FDDI may be used more as a local area network.

Fibers

There are some specific details of the transmission equipment for FDDI standards:

- Multimode graded-index fiber with 62.5-(m core and 125-(m cladding is recommended, but 50/125 and 100/140 fiber can be used. Modal bandwidth should be at least 500 MHz-km at 1300 nm. Attenuation between nodes must not exceed 11 dB (or 7 dB for the low-cost FDDI standard).
- 1300-nm transmission, to take advantage of low fiber loss, dispersion and allow use of inexpensive LED sources.
- Pin photodiode detectors instead of 1300-nm APDs because of cost efficiency and reliability. Minimum power needed at the detector for a bit error rate of 1 in. 2.5 x 10 bits must be no higher than - 27 dBm.
- The FDDI standard also provides for the use of singlemode fiber with four combinations of laser transmitters and receivers, depending on transmission distance.

Other Fiber Networks

Other fiber optic networks are always being developed. A majority of these are in the research stage but look for some to emerge as standards. Numerous concepts are still evolving, such as basing data networks on the asynchronous transfer mode (ATM). As history shows, most of these ideas will not evolve, but a few may emerge as significant new systems.

TEST MODULE 14

Matching: Match the letter with the appropriate answer.

1. _____The original Ethernet standard codified as IEEE was first to gain acceptance.

2. _____A local area network that sometimes serves as a backbone network linking LANs operating at slower speeds.

3. _____A faster version of the standard Ethernet.

4. _____Covers a wide range of signal transmission and uses a 10-bit coding for each 8-bit byte.

5. _____Most recent Ethernet uses the same protocols and frame format as slower Ethernets but with only full-duplex transmission.

A. Star topology

B. FDDI - fiber distributed data interface

C. 100-Mbps Fast Ethernet

D. Fibre Channel

E. 10-Mbps Ethernet

F. Ring topology

G. Gigabit Ethernet

Section 15.0
Testing and Troubleshooting

Objective

Demonstrate how to find and correct system problems using accepted testing practices.

Outline

- Introduction to Testing
- System Configuration
- Fiber Attentuation Testing
- Connector and Splice Testing
- Connector Insertion Loss Testing
- Fundamentals of OTDR
- Visual Tracer
- Microscopes
- Fiber Optic Talk Set
- Problems in Fiber Optic Testing

Learning Activity

Assessment: Test Module 15

Lab Exercise: Hook up meters and test end to end with power units, meters and talk sets.

Introduction to Testing

A most important part of fiber optics is testing. In a fiber optic system, one needs to be able to determine if all the components are functioning properly. Testing is a necessary method to measure the transmission loss of an optical fiber path.

As discussed earlier, an optical signal is generated at one end of a fiber path by a laser or LED source. That signal undergoes a certain amount of loss as it moves from one end of the fiber path to another. This loss can come from the length or quality of the fiber itself, the number of connectors or from splices. After a certain point, however, fiber loss can indicate a defective link. Testing helps to identify the problem. This section describes the measure of optical transmission loss with a variety of equipment and tests. The following is the practical testing of fiber optic systems and components.

By fiber optic testing, this does not mean only testing optical fiber and cable, but also the connectors and splices used to join the optical fibers. We are talking about testing the sources and transmitters used to couple the signals into the fiber as pulses of light, and the detectors and receivers used to convert the light pulses back into electrical signals. We are talking about couplers and switches and other components that may be used for transferring signals among fibers. And finally, we are talking about the complete system and how it is tested and troubleshot in the field.
- Optical fiber and cable
- Connectors and splices
- Sources and transmitters
- Detectors and receivers
- Couplers and splitters
- Systems

System Configuration

The typical fiber optic system is shown in Figure 15.1. The transmitter consists of an electrical input that is converted into a light pulse by a source driver and an LED or laser, which actually converts the electrical signal into an optical pulse.

A connector is used to position the fiber in front of the source and fiber optic cables carry the signal from transmitter to receiver. A photodiode in the receiver converts the light pulse into an electrical pulse, which is then conditioned by an electronic circuit with a preamplifier and trigger into an electrical output that is compatible with the rest of the data transmission system.

Requirements

The characteristics of optical fiber and cable that need testing include:
- Attenuation or the actual loss of signal per unit length
- Bandwidth, or how fast a signal can be carried down the length of the fiber
- The numerical aperture that indicates how effective the fiber will be at coupling light in and out
- The diameter and concentricity, which are physical characteristics of the manufacture of the fiber, but critical for the alignment of the fiber when connectorization or splicing is necessary
- The cut-off wavelength on singlemode fiber, which is the wavelength below which the fiber is no longer carrying a singlemode
- The environmental and physical characteristics of the fiber, which indicates how it stands up under real-world operating conditions

Figure 15.1

TRANSMITTER

CABLE

SOURCE
DRIVER

CONNECTOR

INPUT

LED OR
LASER

RECEIVER

CABLE

PRE-AMP
TRIGGER

CONNECTOR

OUTPUT

PHOTODIODE

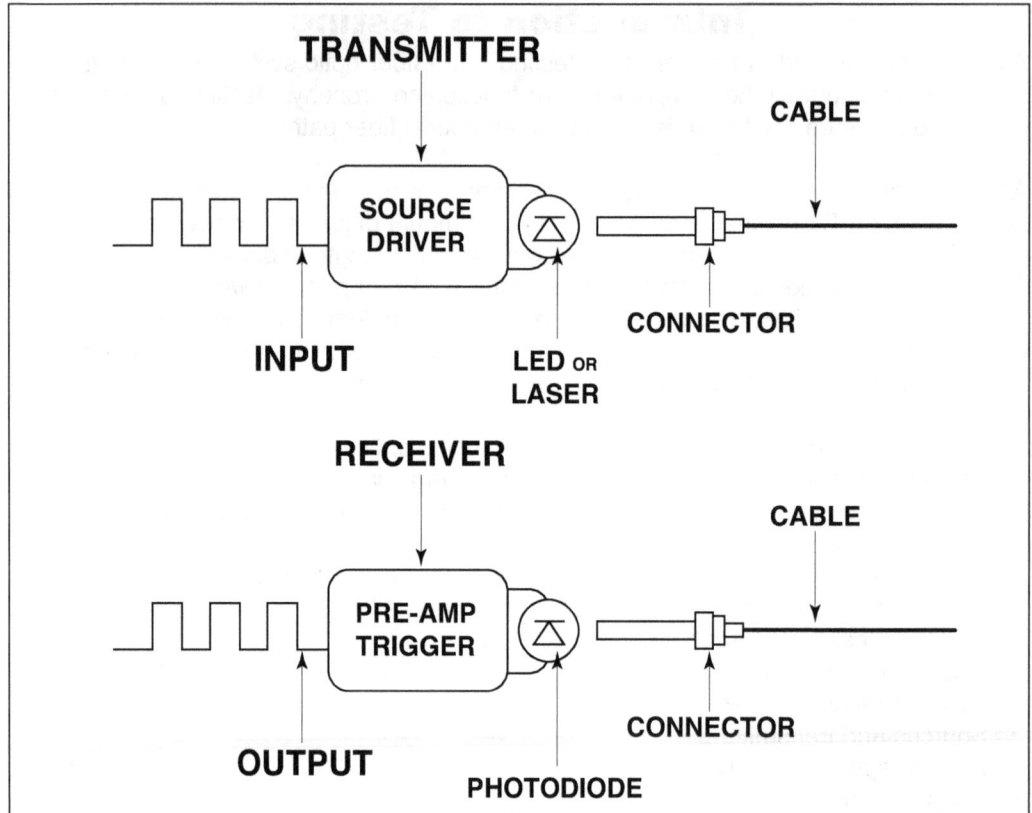

Fiber Loss Mechanisms

There are two major sources of loss in optical fibers. Scattering in the optical fiber occurs as light actually is scattered or bounces off molecules in the structure of glass itself. Scattering is less at long wavelengths, making fibers better transmission media at longer wavelengths.

The peaks are due to molecular absorption. Water and heavy metals, which are in the fiber as impurities or dopants, tend to absorb light at certain wavelengths. These absorption peaks tend to be quite strong, and one of the major factors in making optical fibers is to try to make those absorption peaks both small and in regions that are not overlapping where light is often transmitted.

The combination of these two curves is what you see in the standard loss curves for optical fiber. (See Figure 15.2.) And the points at which we transmit light tend to be areas that are in between the peaks and toward the longer wavelengths, since the scattering is a very strong function of wavelength. The wavelengths most commonly used with glass fibers are 820 to 850 nanometers (the so-called "short wavelength" region), and 1300 nanometers to 1500 nanometers (the "long wavelength" region), which will probably be used more in the future because of its extremely low loss.

Fiber Attenuation Testing

The standard method for testing the attenuation of optical fiber is the "cutback method" test. Here, (Figure 15.3) a source is used to put a signal into the optical fiber; modal conditioning is used in graded index or multimode fiber to establish a consistent launch condition to allow consistency of measurements. (Although modal conditioning is beyond the scope of this discussion, it is a very important topic for making measurements in multimode fiber because of the effects of modal conditioning on the values one will measure

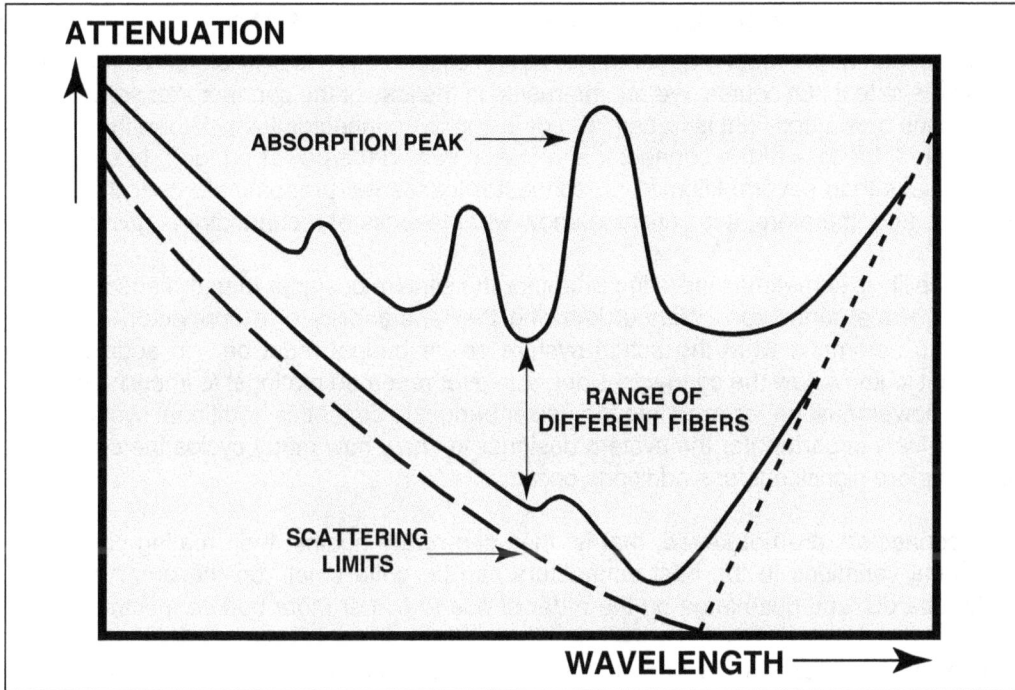

Figure 15.2

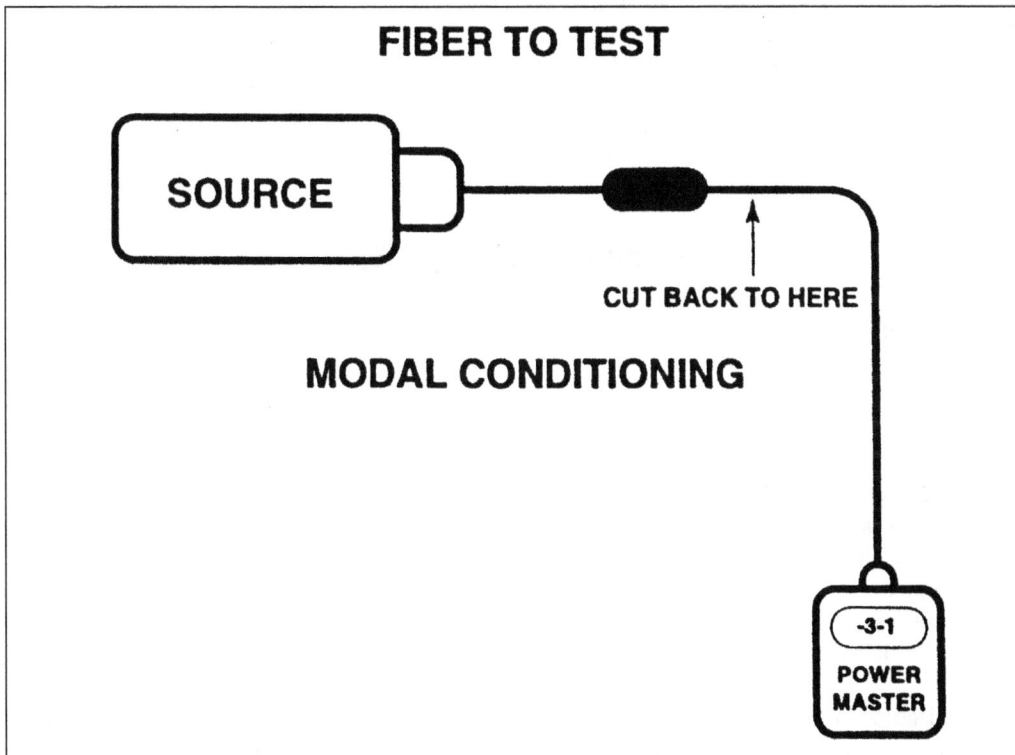

Figure 15.3

in this test.) Measure the amount of light that comes out there. The difference in the light at one end and the other - divided by the length of the fiber gives you the loss per unit length, or the attenuation of the fiber. This is the method used by all manufacturers for testing fiber.

Connector and Splice Testing

When we look at the testing of connectors and splicers, there is a whole series of potential factors to test. Of course, we are interested in the loss of the connector or splice. This determines how much light is transmitted or lost in the connector itself. When the system is designed, the loss of the connector is a critical item in the power budget. In short systems of less than several kilometers, connector losses will predominate over the actual fiber loss and, therefore, it is critical to know what the loss of a connector is going to be.

Repeatability is sometimes more important for the system designer than the absolute lowest loss of the connector. Without knowing the repeatability of a connector, it is very difficult to determine what the actual system power budget must be. In addition, it is important to know how the connectors degrade over repeated matings, to know how much excess power must be factored into the power budget to cover this additional system loss. It is also very important for the system designer to know how many cycles the connector will go before significant loss additions occur.

Most connectors are not keyed; that is, they can rotate against their mating connector. Rotational variations in the best connectors can be quite small, on the order of a few tenths of a dB, and quite large on the order of one to two or more bad connectors.

The rotational variations are very important for a given connector because if a service technician puts the connector back together in a different orientation, the power in the system can vary by a significant amount. Therefore, the rotational variation in a given connector is an important parameter for the user to know.

Obviously, the environmental and physical characteristics of the connector (how well it works over temperature, humidity, altitude and under physical stress) are very important because these will determine the type of connector that must be specified.

Connectors and Splices Summarized

- Loss
- Repeatability
- Degradation
- Rotational variations
- Environmental and physical characteristics

Connector Loss Factors

There are numerous factors that affect connector loss. For example, the end gap between the fibers will cause a loss. The light from the fiber exits in a cone, and as one moves the ends of the fiber apart, the amount of light picked up by the receiving fiber decreases as the cone of light overfills the core of the receiving fiber. End gaps, therefore, make good attenuators.

The end finish and dirt affects connector loss tremendously. A properly polished fiber, properly cleaned, will have significantly less loss than a poorly polished, dirty fiber. One of the things to remember is that an optical fiber is only the size of a human hair, and a piece of dandruff is bigger than the fiber. It is imperative to keep connectors clean.

The physical characteristics of the fiber, as well as the connectors, can produce a core offset either from the poor concentricity in the fiber, or the lack of coaxiality in the connector itself.

In either case, the mismatch of the two fibers will cause a significant loss of light at the termination. Neither of these are totally preventable, but both should obviously be minimized in all circumstances. See Figure 15.4.

Other connector loss factors include angular mismatch at the fiber ends, either due to the polished or cleaved end angle of the fibers being non-perpendicular, or the axial runout of connectors. The numerical aperture mismatch of two fibers can also cause loss. If one is transmitting from a higher numerical aperture into a lower numerical aperture, there will be significant extra loss because the high order modes exiting the fiber at larger angles will not be picked up by the receiving fiber.

A situation where the transmitting fiber has a lower numerical aperture than the receiving fiber will result in a lower loss because all of the light will be easily captured by the receiving fiber. Likewise, with core diameter mismatch, the situation works differently depending on whether the transmitting fiber is larger or smaller. A larger transmitting fiber will always have significantly higher loss when transmitting into a smaller fiber because the larger core will overfill the smaller receiving core. However, a situation where you have a smaller transmitting core and a larger receiving core will result in extremely small losses. See Figure 15.5.

Connector Concentricity Contributors
Here are some of the problems that cause concentricity errors in mated connectors. The tolerance pulled up in the fiber ferrules can result in a fiber that is significantly smaller than the hole in the ferrule. For example, in 125 micron O.D. fibers, the typical overall fiber size is between 121 and 126 microns, but because the ferrule has to be designed to fit worst-case fibers, (nothing is worse than having a fiber that won't fit in the ferrule), the typical hole size in the ferrule is on the order of 130 microns. Therefore, one can have as much as a 10 micron gap in the fit of the fiber to the ferrule.

While some of that will be compensated by the fact that one is putting epoxy in the hole with the fiber, and a well-coated fiber will be floating on the skin of epoxy, there is still sig-

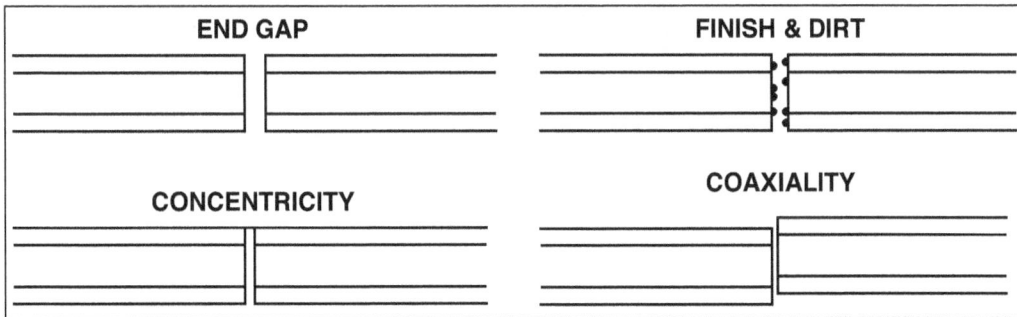
Figure 15.4

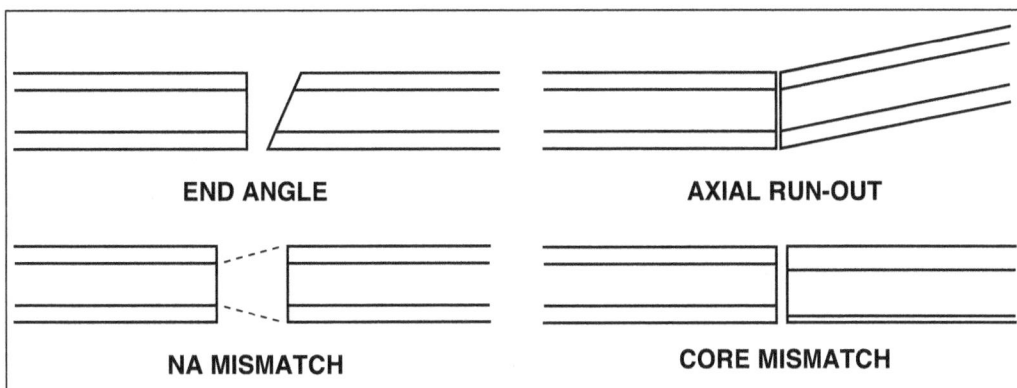
Figure 15.5

15.7

nificant room for misalignment. This is another reason to make sure when terminating connectors, that the fiber is thoroughly coated with epoxy and by pushing it in and out of the ferrule and fiber with epoxy. This will ensure the fiber is closer to the center of the ferrule.

When the ferrules of two connectors are put into splice bushings, the ferrules are always slightly smaller than the splice bushing; otherwise, obviously they would not fit. Here again, a few microns of tolerance build-up can lead one to a situation where the final cores of the fibers are significantly offset. See Figure 15.6.

Fiber Geometry Defects

Even if the two fibers are perfectly aligned, the physical geometry of the fiber itself can still cause misalignment. If the core and cladding are not exactly concentric or if they are both oval as opposed to perfectly round, this will result in core to core mismatches which will create loss. Needless to say, there is no such thing as a perfect fiber; no such thing as a perfect connector; no such thing as a perfect splice bushing. See Figure 15.7.

Connector Insertion Loss Test

How do manufacturers typically test connectors? The FOTP-34, or fiber optic test procedure 34, has become the EIA standard for testing fiber optic connectors. Basically, it looks like a fiber cutback test.

A source, some modal conditioning, a test fiber, and an optical power meter to make a measurement of the power coming out of the test fiber is needed. The test fiber here is

Figure 15.6

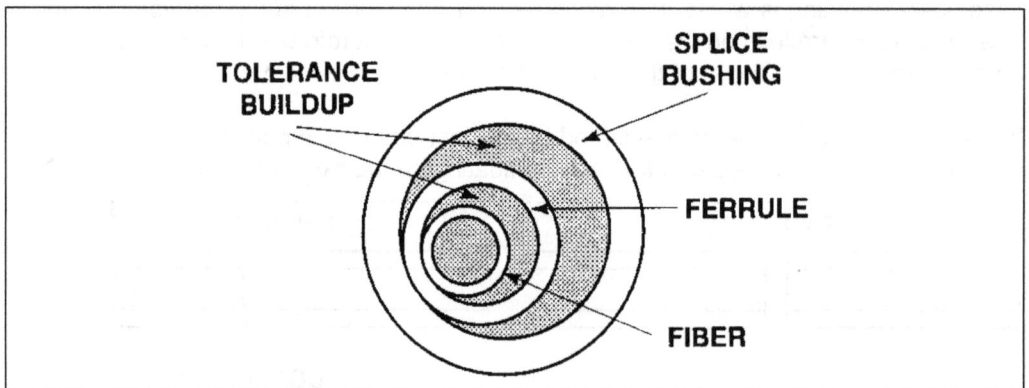

Figure 15.7

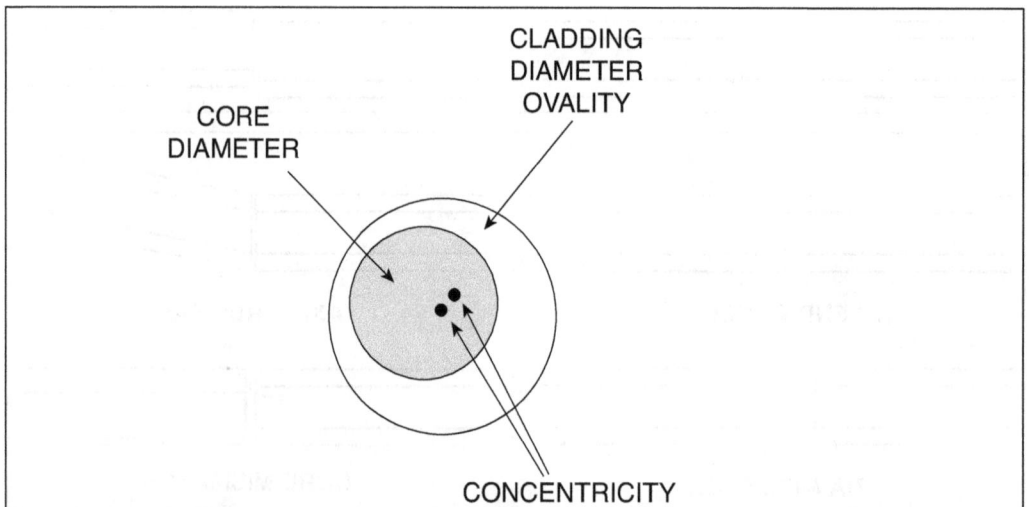

not long; it need only be a new meters. One measures the total amount of power out of the fiber, inserts a connector pair and measures the loss of the mated pair of connectors.

Over the years, companies have expressed the loss that is measured here in two ways. What one normally thinks of as connector loss is the total loss of a mated pair of connectors. There are still companies, however, that take the loss of a mated pair and divide it by two, attributing half the loss to each connector. (The standard may allow .50 dB loss in a mated pair and each connector would contribute roughly half that. One should always be careful when reading specs concerning the loss of a connector to ensure that one understands fully the way the test was done and the way the results were specified.) See Figure 15.8.

Cable Loss Test - Single and Double Ended
The fiber attenuation test by the cutback method and the basic connectors insertion loss test can be combined into the testing of connectorized cable because it is destructive.

A suggested method is to use a source with a short launch cable, approximately the exact size and type of fiber that is used in the cable to test and is terminated in an appropriate mating connector to the cable you're testing. The launch cable, therefore, gives you a standard launch condition against which cables are tested. The cable you wish to test is then mated to the launch cable to determine how well it mates with the launch cable connector and how much power is transmitted through the cable itself.

By measuring the amount of power out of the launch cable before mating the test cable and then the amount of power out of the test cable after mating to the launch cable, one determines the total amount of loss in the cable including the mated connector loss. As a matter of convenience, we often have people set the source power to give them zero dB referenced to a microwatt out of the launch cable. Then one can read on the optical/power meter in dB referenced to a microwatt the exact cable loss without having to worry about making calculations.

The following will demonstrate the cable loss test for both single and double-ended cables, equipment and suggested procedures.

The procedure used for testing connectorized cables requires launching a known amount of light from a standard test jumper cable into the cable under test and measuring the

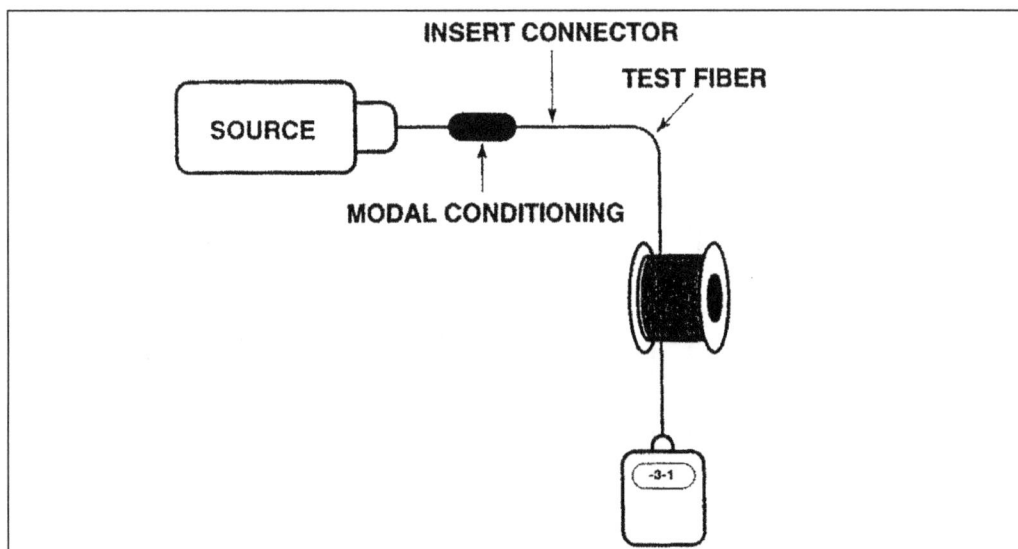

Figure 15.8

amount of light transmitted through the cable under test. The test cable is chosen to be of the type of fiber used in the cable under test and is connectorized with compatible connectors.

The launch cable provides a repeatable launch condition; consistent with the way the cable is used, i.e. attaching the other cables. It allows comparing all cables tested under similar test conditions, and allows one to make measurements with a minimum amount of operator-induced error.

This method tests the insertion loss of the launch connector and the cable. See Figure 15.9.

This is appropriate for systems where the cable goes directly from the source to the detector, as is common in many computer systems.

Basic Fiber Optic Cable Loss Test

The Test Kit
In these demonstrations, we will use a Fluke Networks (formerly Fotec) T302 Fiber Optic Test Kit similar to those used by many installers of fiber optic cables and systems. Any similar instruments are appropriate.

Contents of a test kit:
- Fiber optic power meter
- LED source
- Connector adapters
- Launch cables and splice bushings
- Microscope
- Connector cleaning wipes
- Attenuators
- Spares and supplies
- Carrying case

Figure 15.9

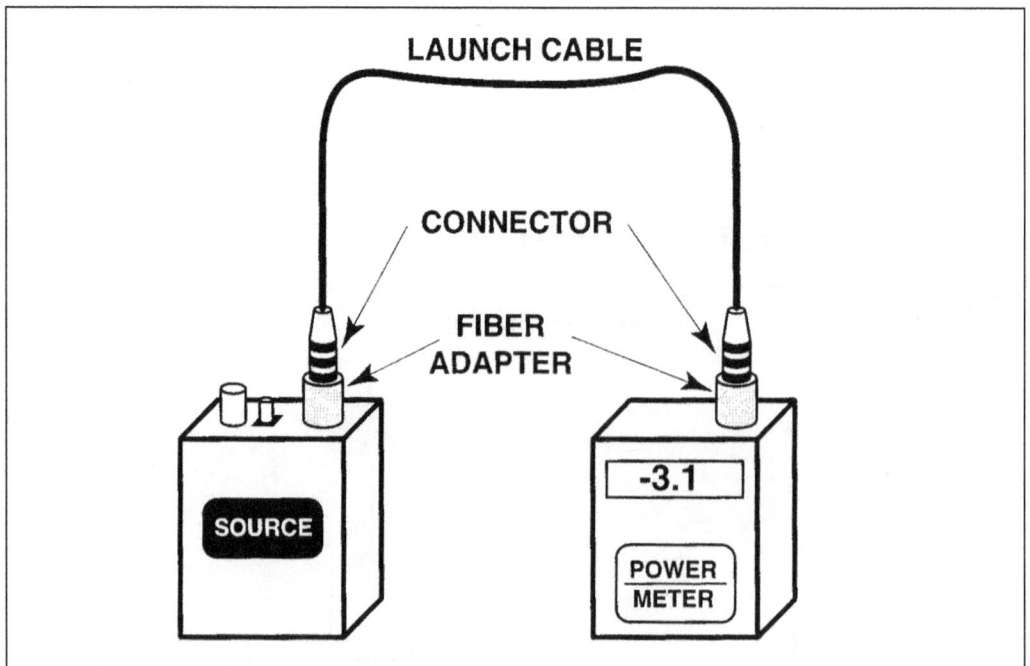

15.10

The Fluke test kit consists of a set of basic instruments for making fiber optic measurements. The fiber optic source, which uses an LED similar to those used in fiber optic systems, is used to inject a test signal into the fiber optic cable. A fiber optic power meter measures the optical power injected into the cable by the source and transmitted through the cable to determine the cable loss.

Connector adapters are required to adapt the test instruments to different connectors used on the fiber optic cables. Launch (and perhaps receive) test jumper cables are required to establish consistent test conditions in order to minimize measurement accuracy. A set of connector splice bushings, sometimes called "sleeves," will be required to connect the test cables to the cables you wish to test.

Many test kits include a simple portable microscope to allow inspection of the polished or cleaved end surfaces of fiber optic connectors. This allows diagnosing the problems with high loss connectors, which can lead to installing fewer bad connectors. The microscope should allow looking at the connector directly and at an angle to diagnose all common connector flaws.

And, of course, any test kit should include all AC adapters, extra batteries and other spares needed, especially for the benefit of the field technician who always needs it when and where it's unavailable.

Steps in setting up the instruments:
A. Before making any measurements, set up the instruments for testing. First, choose the correct connector adapter for the source and power meter.
B. Attach the adapter to the source.
C. Attach a connector adapter to the meter.
D. Attach the proper test jumper cable to the source. Make certain the connector is attached securely to prevent unwanted variations in the amount of power coupled into the test cable, since this will directly affect measurement accuracy.
E. If the source is battery-powered, make certain the battery is well charged. If not, or if working in the laboratory, connect the source to AC power through the proper adapter.
F. To test cables using a launch cable, calibrate the power launched from the cable. This is done by attaching the launch cable to a power meter and measuring its power output. If the source has an adjustable output level, it is preferable to adjust the output power to a convenient level that is easy to remember.
G. For LED systems, it is recommended setting the source to 1 microwatt. (-30 dBm or 0 dBu). Fotec meters are usually available with dBu range to facilitate making cable loss measurements.
H. Attach the launch cable to the optical power meter for calibration.
I. Turn on the optical power meter and set it on the dB scale.
J. Turn on the source.
K. Adjust the source to the desired value.

If there is any question about the cleanliness of the connectors on the launch or unknown cables, clean all connectors with optic pads.

To test an unknown cable, attach the transmit end of the cable to the launch cable using an appropriate splice bushing. Some connector manufacturers refer to these as sleeves.

Steps in setting up the instruments - continued:
A. Remember that connectors may have different losses measured in each direction due to connector defects, fiber NA and size variations. Therefore, test in the direc-

15.11

tion the cable will be used, or test both ways if the direction of transmission is unknown or unspecified.

B. For SMA connectors, attach the connectors by screwing them into the threaded splice bushing.

C. After attaching the second connector to the splice bushing, make certain both connectors are snugly but not too tightly, screwed into the bushing. If the "906" style SMA is used, be certain to use the plastic alignment bushing, as it has primary responsibility for mating alignment.

D. With the "906" SMA, tests can be performed without the metal barrel of the splice bushing, if care is taken to prevent bending the bushing or pulling the connector apart. While this can speed testing, it is a potential source of error and is recommended only for the experienced tester.

E. To measure the loss of the cable under test, attach the far (receive) end of the optical power meter and measure the total power transmitted through the cable.

F. Attach the cable firmly to the meter to prevent any variation in readings.

G. Here this 100 meter spool of cable shows a loss of 2.5 dB. Since we set the power out of the launch cable at 0dBu, we can read the loss directly. If we had set the launch power at another value, e.g., -20 dBm, we would have read -22.5 dBm on our fiber optic power meter and would have subtracted the launch power value to determine the loss of the cable.

H. Be careful loosening the launch cable connector and rotating the launch connector, we can measure only 2.2 dB. This variation of loss with rotation is normal, caused by eccentricities in the fibers and connectors of the two mated cables. Large rotational variations in loss are valid reasons for rejecting cables, as this leads to unreasonable variances in cable loss in actual application. See Figure 15.10.

Figure 15.10

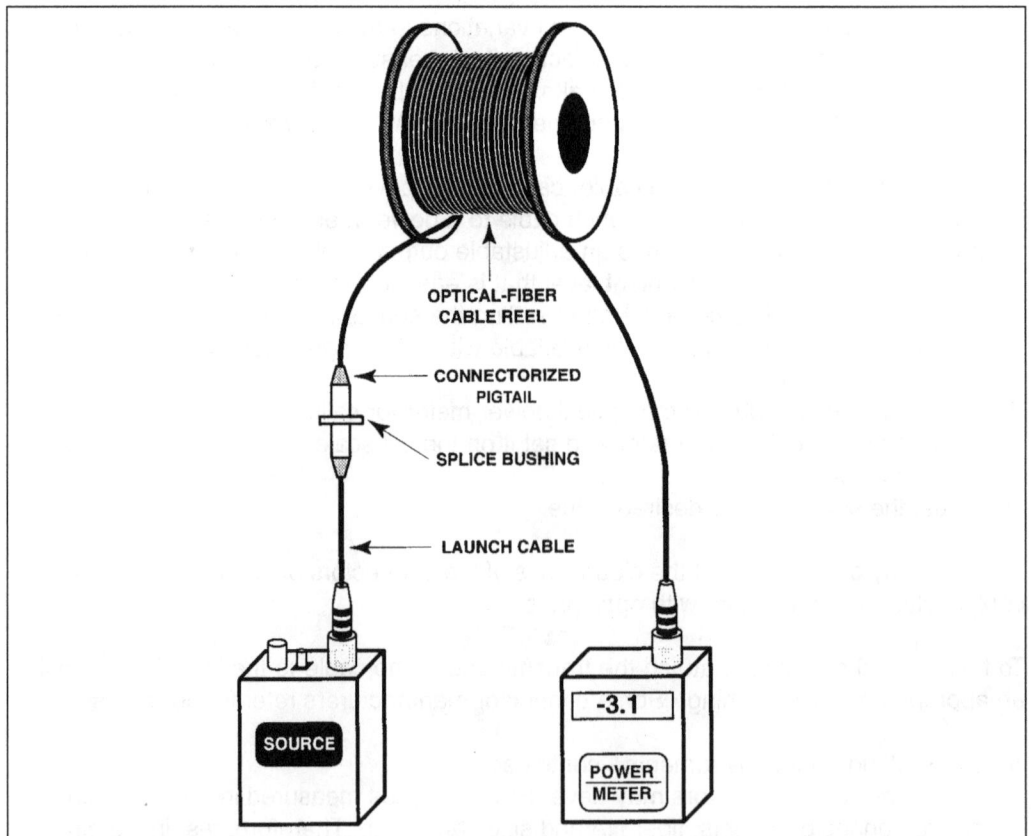

OPTICAL-FIBER
CABLE REEL

CONNECTORIZED
PIGTAIL

SPLICE BUSHING

LAUNCH CABLE

SOURCE

-3.1

POWER
METER

One can apply these basic tests to standard cables during installation in a way that will save considerable time in testing. By pairing cables and looping back to one end as shown here, one can test two fibers simultaneously from only one end of the cable.

Use a standard splice bushing to attach the two connectors, and remember to attach the source to the transmit fiber, if one is designated as such. The loss measured is the total of the insertion loss of both connectors and the attenuation of both fibers. Therefore, the loss will be about twice the expected value for one fiber of the cable.

The biggest advantage of this method is that it requires only one person at one end of the cable to perform the tests. It is extremely helpful in situations where many fibers are terminated in the single location, such as a computer room or patch panel.

Alternate Methods of Testing Cables

If the cable is used to connect with cables at both the launch and receive ends, a double-ended cable test may be more appropriate. Here, both a launch and receive test cable are used. Unlike the single-ended test that tests only the launch and connector, the double-ended test checks how well both connectors mate with the test cable connectors. The receive cable may be either a cable identical to the cable under test or it may be a "bucket" cable, with larger sized fiber, as is common on singlemode systems. See Figure 15.11.

Figure 15.11

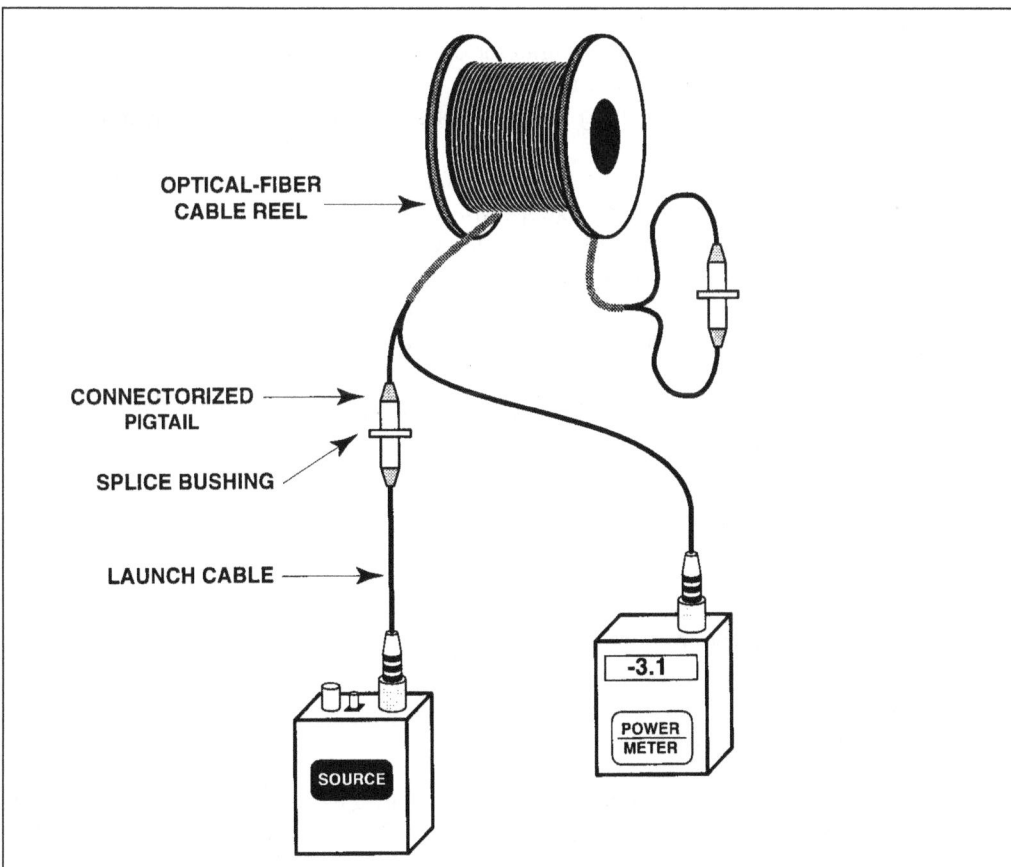

OPTICAL-FIBER CABLE REEL

CONNECTORIZED PIGTAIL

SPLICE BUSHING

LAUNCH CABLE

-3.1

POWER METER

SOURCE

15.13

Definition of Loss in dB

Fiber attenuation and cable loss are generally expressed in units of decibels, commonly abbreviated as dB. The formula for calculating dB is as follows:

$$dB = 10 \log (Pout/Pin)$$

Pin is the power launched from the test jumper into the cable under test, and Pout is the power transmitted through the cable and measured by the optical power meter.

If the power measurements are made in dB, just subtract the measured values of Pin and Pout to obtain the loss in the cable.

Definition of Cable Loss

$$Loss (dB) = 10 \log (Pout/Pin)$$

Note: All measurements are in watts (mW or uW). If dB measurements are used, merely subtract readings in dB.

A microscope can be used to diagnose connector faults. This makes it easier to know why connectors are bad and can help in perfecting the connectorizing process.

Most microscopes have a built-in illuminator and some means of holding the connector for viewing.
A. If the connector is dirty, or even suspected of being dirty, clean it with an optic pad before viewing. The microscope can show you how effective the optic pad really is, if you look at the dirty connector before and after cleaning.
B. To use the microscope, screw the connector onto the state or holding bracket.
C. Slip the state into the base of the microscope.
D. Turn on the illumination by opening the light housing, and view the connector end.

To maximize the accuracy of these measurements, remember to always treat connectors carefully. Keep covers on connectors, except when being tested or connected to the system. Don't drop them or touch the ends. Keep the cables from being kinked or even bent into a tight radius as they may introduce bending losses and affect calibration or measurements. Make certain all connections to the launch and receive cables are snug. Loose connections may have end gaps or angular misalignment losses.

Treat Cables and Connectors Carefully

· Keep covers on connectors.
· Don't drop or touch ends.
· Don't kink cables.
· Make connections snug.

Warning: Hazardous optical radiation may be present on optical fiber cables. Permanent eye damage may result from exposure!

· Don't look at fiber ends.
· Test for power with fiber optic power meter first.

And remember, hazardous light levels may be present in optical fiber systems, especially those with laser sources, or when being tested with OTDRs. Never look directly at the end of a fiber; always test the fiber first with a power meter before inspecting the fiber in a microscope to prevent potentially permanent eye damage.

Source and Transmitter Testing Requirements

Once you have a complete fiber optic system hooked up to the cables in the system, there are several things to know.

A. Whether the sources and receivers are working properly. For the sources and transmitters, the primary thing is the amount of output power the source has, and how much of that is coupled into the fiber.

B. The bandwidth and the wavelength of emission of the light. If it is a laser, the so-called threshold current and extinction ratio or efficiency are important.

C. The environmental and physical characteristics.

D. The stability and lifetime of the laser are very important although difficult to test.

Sources and Transmitters

- Output power (into fiber)
- Bandwidth
- Wavelength of emission
- Laser: threshold current, extinction ratio
- Environmental and physical characteristics
- Stability and lifetime

Semiconductor Failure Rates

Since lasers and LEDs are semiconductors, they have a high infant mortality and then a very low failure rate until they are worn out. The time scale on this graph (Figure 15.12) is basically unknown for most fiber optic sources and as a result, these devices tend to be extensively tested and burned in to try to determine how long they are likely to last in a given system application.

Detectors and Receivers Testing Requirements

For detectors and receivers, be primarily interested in the sensitivity. For example:

- how well they work with how low a level of received optical power
- the bandwidth or how fast they work
- how stable they are
- how much distortion would be introduced into an analog signal for analog receivers
- how well they work over a variety of environmental and physical conditions

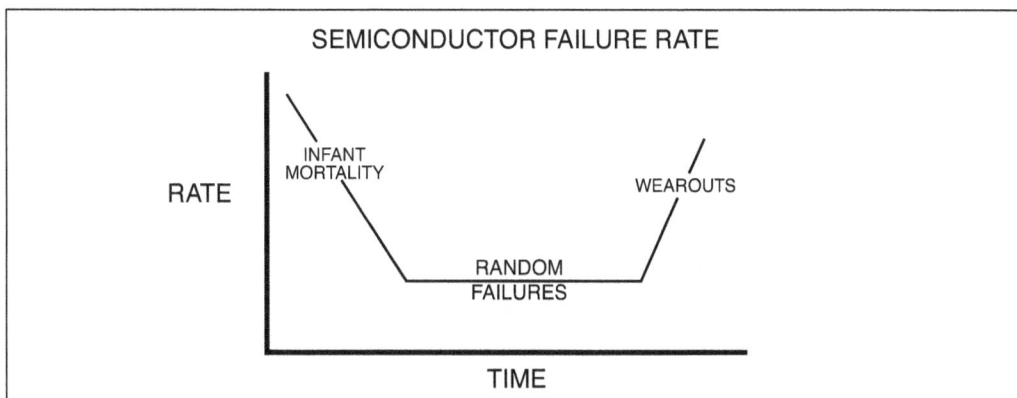

SEMICONDUCTOR FAILURE RATE

RATE

INFANT MORTALITY

RANDOM FAILURES

WEAROUTS

TIME

Figure 15.12

Detectors and Receivers
- Sensitivity
- Bandwidth
- Stability
- Distortion (analog)
- Environmental & physical characteristics

System Testing

For the complete system, be primarily concerned with how well it transmits data. So, therefore, be interested in such things as the bit error rate (BER) as a function of system margin, determined by the combination of receiver sensitivity, launched power, losses of the fibers, couplers, connectors and splices. The environmental and physical characteristics of the whole system are also important.

Fiber Optic System Testing

Data transmission (BER) as a function of system margin, determined by:
- Receiver sensitivity
- Launched power
- Losses in fiber, couplers, connectors and splices
- Environmental and physical characteristics

Bit Error Rate as a Function of Receiver Power

It is the BER of the fiber optic system that is of greatest interest. The BER, or the number of bits that are transmitted erroneously, turns out to be a function of the received optical power. As the power goes up, the bit error rate goes down because the signal to noise ratio improved, until the power at the receiver reaches a point at which the receiver saturates, and then the bit error rate goes up rapidly.

In a system that is designed to operate with a given bit error rate, there is a certain range of received optical power that is appropriate for system. Therefore, in the system testing application, the primary thing one wants to test is whether the power of the system at the receiver is adequate to provide proper system performance. See Figure 15.13.

Test Equipment Types and Applications

In order to perform the tests that have been described, one needs some specialized fiber optic test equipment. If you are mainly intending to make measurements of loss, you will need the optical loss test sets.

A real optical loss test set differs considerably from what was originally offered for fiber optic testing. These so-called "attenuation meters" did not provide for any control over the

Figure 15.13

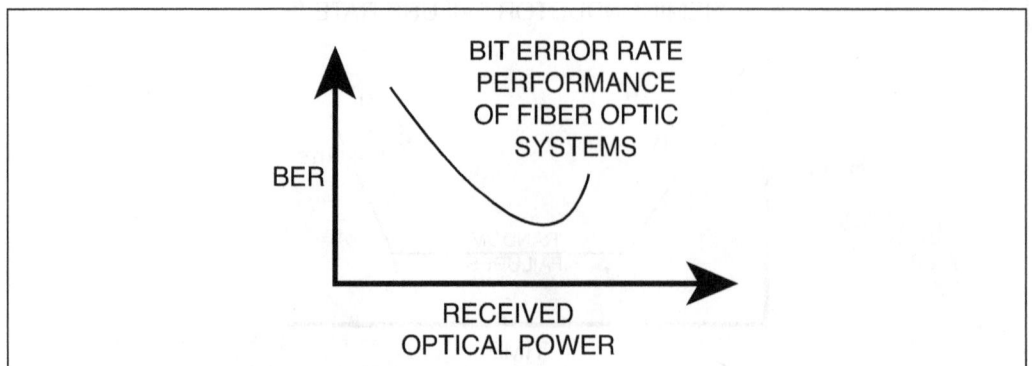

BIT ERROR RATE
PERFORMANCE
OF FIBER OPTIC
SYSTEMS

BER

RECEIVED
OPTICAL POWER

launch conditions of the sources used. As a result, they were not very accurate in making measurements. They were also generally designed by optical companies that did not have a good grasp of the needs of the field service or laboratory instrumentation users for high accuracy and ruggedness. Today, there are several optical loss test sets available, which are quite expensive but provide extremely good performance.

Fiber optic power meters are available from very inexpensive, handheld units through costly laboratory units which are interfaceable to computers. There are some good fiber optic fiber meters available, and there are some units that have performance deficiencies which make them extremely difficult to use.

These deficiencies include ergometric factors as simple as the incorrect placement of buttons and switches, as well as electrical performance features which provide for inaccurate measurements over the dynamic range that is often used for making measurements. Unfortunately, with fiber optic power meters, it is difficult to test these devices, and there are no independent calibration laboratories commonly providing calibration testing services.

The best method for choosing one of these instruments is to go to a reputable manufacturer or follow the recommendations of another satisfied user. For LED and laser sources, one is primarily interested in having a stable source, because if the light source varies, the measurements of loss would vary accordingly. These sources are available in a variety of different wavelengths of light, which are identical to those that are used in the actual transmission systems. If you are working on a system that is designed to transmit at 850 nanometers, you should use an 850 nanometer source to test your cables. Likewise for 1300 or 1550 nm systems used in telecommunications.

Attenuators are used to simulate the loss in long lengths of fiber so that one can test fiber optic transmission systems without having to have big spools of fiber sitting around to create the attenuation. There are several types of attenuators. For the person interested in testing systems, attenuators are quite valuable, even in the field service environment, because they allow you to insert additional loss into the system or use an attenuator in a loop-back mode between transmitter and receiver to simulate system operation for diagnostic work.

Optical time domain reflectometers are instruments that allow you to make loss measurements of optical fiber from only one end of the fiber.

Bandwidth testers allow you to determine the information carrying capacity of the optical fiber or how fast a signal can be transmitted reliably over the fiber.

Microscopes are widely used to determine whether or not connectors have been properly polished, or if fibers have cracks, chips or other faults. Microscopes can also be used to examine cleaved fibers, to find whether or not a cleave is appropriate for use in splicing. It is often recommended that microscopes be used whenever fiber optic cables are tested, because it is a quick and convenient way of determining whether or not the connectors are clean, whether or not they have been damaged by handling, and to determine whether or not they are appropriate for continued usage.

Fiber Optic Test Equipment
- Optical loss test sets
- FO power meters
- LED and laser sources

- Attenuators
- Optical time domain reflectometers (OTDRs)
- Bandwidth testers
- Microscopes

Application

Here are the types of tests that need to be performed, and the various types of fiber optic instrumentation required. The four types of instruments that are listed are: optical loss test sets, which can also be a power meter and a source; power meters alone; optical time domain reflectormeters; and attenuators.

If you want to measure the launch power in a fiber optic system, you only need a power meter and a test cable.

When performing a fiber loss test, you need an optical loss test set, a source and power meter or an OTDR.

For connector loss, an optical loss test set is the ideal way of handling this. But an OTDR can be used.

For receiving power, you only need a power meter. And, for receiver sensitivity, you can use a power meter and an attenuator to simulate cable loss and determine the bit error rate as a function of received optical power. Likewise, data transmission testing really only requires an attenuator to determine how well the system will transmit data with a fixed system loss. An OTDR is invaluable in determining faults.

In a situation where a fiber has been cut, for example underground or in the middle of a long cable run, an OTDR will often tell you precisely where the cable has been broken to facilitate a speedy repair.

Optical Time Domain Reflectometer (OTDR)

The OTDR is an invaluable instrument for fiber optic testing, but is not without its limitations. Basically, the OTDR works by looking at back-scattered light. Remember, in fiber loss mechanisms, the light is scattered by actually bouncing off molecules in all directions. A significant amount of that light is reflected back up the optical fiber toward the transmitter.

This phenomenon allows one to take a high powered laser, pulse it down the fiber and look at the light that comes back in a receiver to determine the amount of light that is being scattered and then attenuated on the round-trip through the fiber. You can, therefore, use this kind of instrument to infer, indirectly, the loss of the optical fiber. Remember, the fiber cutback test allows you to make direct measurements of the loss. With an OTDR, you are making an indirect measurement. See Figure 15.14.

Figure 15.14

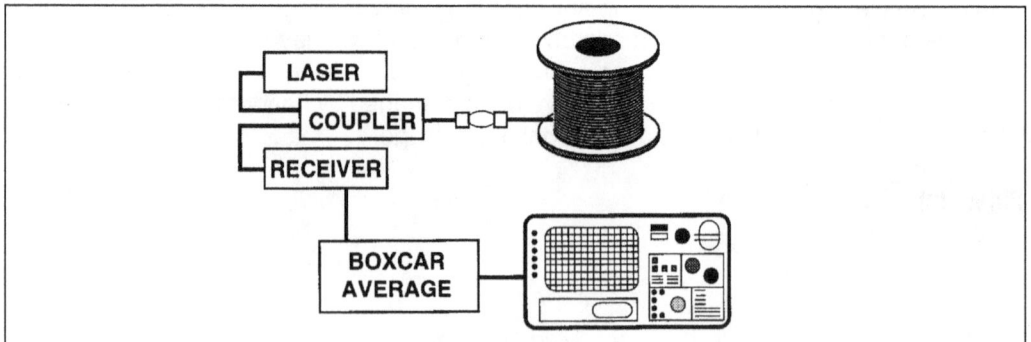

The OTDR display shows power in dB on the vertical axis and distance in the horizontal axis. The distance is actually converted from time, since the light travels in the fiber at a fixed speed. Starting from zero distance, the first.thing one sees is the laser pulse itself. This pulse blanks out the beginning of the display, so that typically, an OTDR cannot look at cables of less than 10 to 50 meters.

The scope of the curve you see between splices and connectors is the actual loss of the fiber, in dB per unit length. The steps show where splices and connectors are inserted and show the loss at the joint. At the end of the fiber, you hear only the noise where no more signal is being scattered back to the receiver. Splices show only a small dip in the curve, but connectors show a large reflection. This reflection is caused by light bouncing off the end of the fiber where there is an air gap. It is called fresnel loss, and it typically averages about 4 percent of the power that is going through the fiber. However, in a situation where the fiber has been cut, do not expect to see a large fresnel pulse reflection because the end of the fiber is unlikely a proper, clean and smooth cleave.

When you are installing fiber, it is always a good rule of thumb to cleave both ends of the fiber, because if you are going to look at it with an OTDR, it will give you the best possible signal to noise ratio. See Figure 15.15.

Advantages of the OTDR:
The initial popularity of the OTDR as a fiber optic tester was primarily due to its four simple advantages. An OTDR requires access to only one end of the fiber, a major advantage in the installation of long cable runs. If you are dealing with 1 to 2 km lengths of fiber, it is inconvenient to have a crew at both ends of the fiber in order to make loss measurements.

The OTDR graphically shows the fiber, splicer and connector loss. The skilled operator can very quickly interpret the readout of the OTDR to determine what is going on within the installed fiber optic cable. The OTDR also easily locates breaks or faults in the cable and shows how long the fiber is, or how long the distance is to the fault.

Advantages of the OTDR Summarized
- Requires access to only one end of the fiber
- Graphically shows fiber splice and connector loss
- Locates breaks or faults
- Shows fiber length or distance to fault

Disadvantages of the OTDR
These advantages of the OTDR are balanced by certain disadvantages. For example, the interpretation of the display requires an extremely skilled operator who understands the

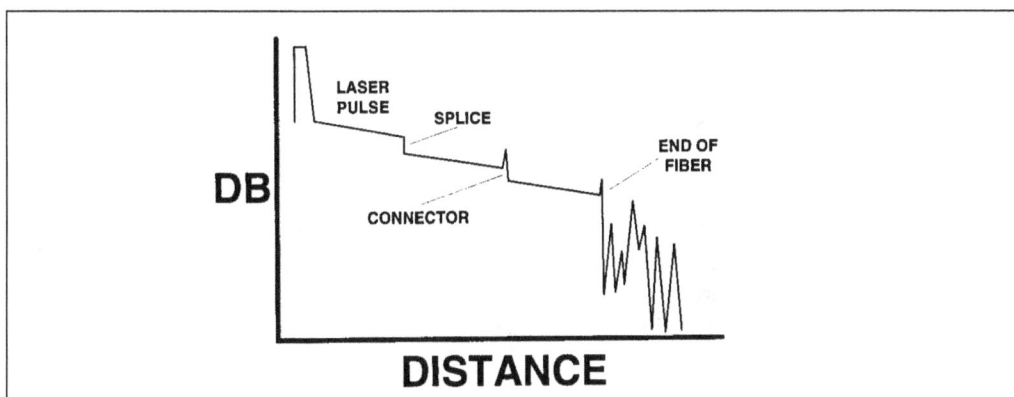

Figure 15.15

nuances of the way the OTDR works and why at times, the displays don't look the way they are supposed to look. Some OTDRs now have automatic event detection and do not require more experienced operators. Also, the OTDR is expensive, and those that give reliable results cost more than $10,000 - usually around $20,000.

The investment in a piece of equipment that is this complex and requires such a skillful operator, is only for those who have large amounts of installation work with long lengths of fiber where the OTDRs might prove to be worth its cost. But you also have to realize that OTDRs often give erroneous loss for connectors and splices. If one is merely taking a length of fiber and doing a connector insertion loss test, such as the FOTP-34 type test, as described earlier, you will be able to get good feedback on what the connector loss is. However, when installing actual connectors and splicing fibers, you will find that with two unmatched pieces of cable, the loss shown by the OTDR will not necessarily be the actual loss of the system in transmitting light.

This is because both numerical aperture and scattering mismatches in the fiber cause the OTDR to be fooled. An OTDR can even show gain instead of loss for connectors and splices. If you go and shoot the other direction with an OTDR, you will find that you can average it out and get a reasonable feeling for what the connector or splice loss is.

The OTDR also has limited distance capability and resolution. In a 1300 wavelength singlemode application, the OTDR is limited to about 15 or 20 km, which is quite short compared to the distance the systems are capable of. This is because there is not very much back scattering in the 1300 nm wavelength used in singlemode systems.

The OTDR has a resolution on the order of 0.1 to 0.2 dB, and since current state-of-the-art splicing provides splices at less than 0.1 dB loss, one very often cannot see splices on OTDRs. So, for singlemode systems, the OTDR is essentially relegated to being a fault-finding instrument. Splice loss is more commonly measured by loss test sets or by the local injection detection systems that are becoming available as accessories for fusion splicers.

Another problem which often occurs in using the OTDR with loose-buffered fiber optic cable, is the loose-buffered fiber is always slightly longer than the cable, and often by 1 or 2 percent. If you are using fiber optic cable that has the loose-buffered fiber construction, and you are trying to find the distance to a fault with an OTDR, be aware of the fact that there is a 1 to 2 percent error in the reading where the cable itself is always shorter than the fiber.

In spite of these disadvantages, the OTDR is an extremely powerful tool for testing fiber optic cables, and with appropriate training, the installer can use the OTDR to quickly troubleshoot problems and repair them.

Disadvantages of the OTDR Summarized
- Requires skilled operators to interpret display
- Expensive
- Often given erroneous loss (even gain) for connectors and splices
- Limited distance capability and resolution
- Insufficient accuracy for measuring state-of-art splice loss

Visual Tracer (Flashlight)
This device allows visual tracing and identification of fibers. The tracer uses a highly visible; red LED sources coupled to an ST or SMA connector to inject light into multimode fibers. There is adequate light to trace fibers through cable runs at about 20 dB. It is especially useful for identifying fibers in patch panels, splicing pigtails and finding broken wires.

Additionally, a single flashlight can check continuity of fibers. See Figure 15.16.

Microscope

Portable fiber optic inspection microscopes are low-cost visual inspection tools for use with fiber optic connectors and bare fibers. Application for these microscopes are:
  A. Visual inspection of fiber connectors
  B. Diagnosis of connector faults
  C. Inspection of cleaned fibers

Microscopes come in many different sizes, ranging from a portable handheld scope with a 100X magnification, to a table type up to 800X magnification that can be viewed on a monitor. (See Figure 15.17.)

Although transmission loss testing is the definitive test of the quality of an optical fiber connector, visual inspection will detect poor surface finish, scratches, unremoved particles, poor surface finish, scratches, unremoved particles, poor centering, cracks and every systematic polishing patterns. The microscope is often the best way of determining why a connector has high loss, and allows looking at the connector, both face on and at an angle, to allow best judgement of possible problems with the connector.

For bare fibers that have been cleaved for insertion in wet splices or non-polish connectors, the microscope allows inspection for breakover, lip, hackle, mist and other faults that affect coupling loss. Fibers may be rotated through a full 90 degrees for viewing end on and across the cleaved faced.

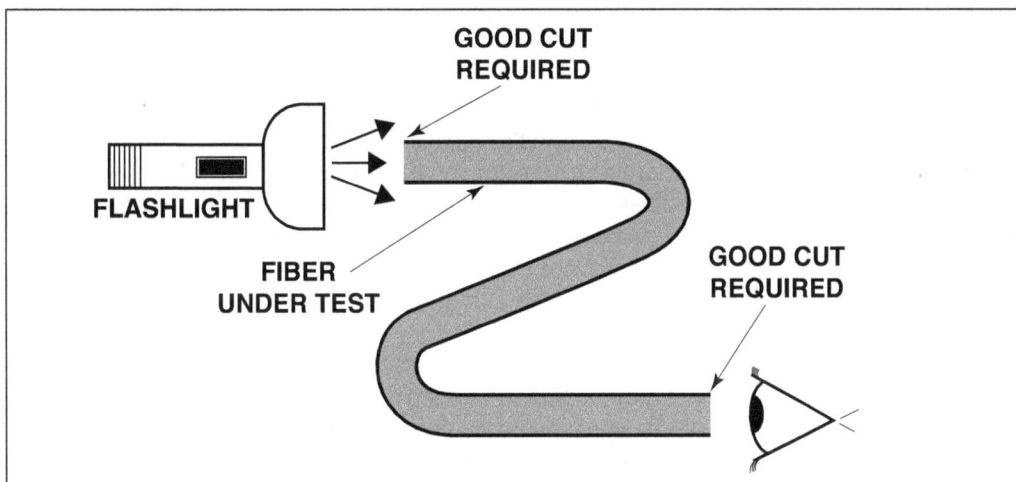

Figure 15.16

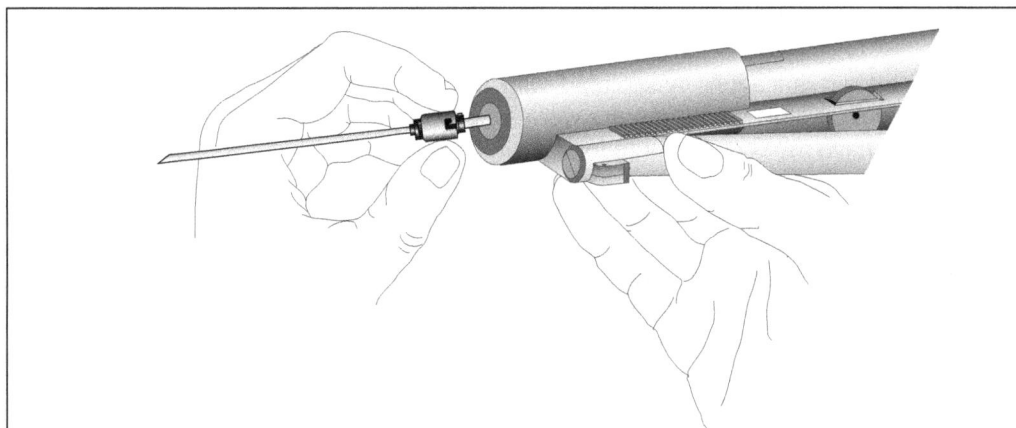

Figure 15.17

Fiber Optic Talk Set

Fiber optic talk sets are used in fiber cable operations very much like a butt-in set is used in copper cables. They provide temporary communications over a vacant fiber between work locations while the fiber cable is being installed, spliced, tested or restored.

A fiber talk set has a 50 dB of range in a full duplex set that is small, simple to use and costs around $500. Talk sets usually come in two versions: the local loop and LAN application type that talks 10 dB or about 20 km (13 miles); and a larger model at 30 dB that talks about 60 to 100 km, depending on your cable loss.

These units come in either full or half duplex. Full duplex is like a telephone where you can talk and listen at the same time. Half duplex is like a CB radio. You can talk and listen, but not at the same time. These units operate on batteries and are like a butt set, rugged and built for outdoor usage.

Additionally, there are units that can be both a test set and talk set. The handheld units function as a power meter and as a test set, as well as a talk set.

Problems in Fiber Optic Testing

Even though fiber optics has been around for many years, there are still some significant problems in fiber optic testing. For example, there are still not standardized test methods being used for the many different parameters that one needs to test. The EIA, the IEC, CCITT, even ANSI and SAE are all working on developing standards for fiber optics. The Department of Defense is participating in many of these groups. Although they have developed standards of their own, they are trying to adopt as many commercial standards as possible.

Most fiber optic measurements require making measurements of fiber optical power, and it has been found that there is very poor calibration accuracy for optical power. Part of this may be that the NBS standards for making measurements of optical power are not designed for the inherent difficulties of fiber optic measurements. In fiber, the light exits the fiber in an expanding cone, sometimes over-filling detectors. The light is at specific wavelengths in the infrared, which are not commonly used. The light signal is often not DC power, but is modulated, and sometimes at high bit rates.

Another problem with fiber optic testing is the unpredictable component specifications. Not all manufacturers test and specify components the same way. So, when testing components from one company, it may be difficult to compare the results with the specifications of that company and other companies. In this case, what is needed is people in the industry to standardize not only in the way that they test, but also in the way that they specify their components.

There are certain limitations associated with the test equipment that is used for fiber optics. Many fiber optic power meters were adapted from optical laboratory instruments, which do not have the requirements for dynamic range, accuracy or ruggedness needed in fiber optic instruments. Many are difficult to use, with too many confusing controls. Some are merely poor designs, with major performance flaws.

It is difficult to make consistent fiber optic power measurements; one is limited to a repeatability of + 1 percent. These inherent fiber optic repeatability limitations are:
- result of the unique characteristics of the light coming out of the fiber
- different types of fibers used
- different types of connectors

- varying wavelengths of light
- modulation of the light that is measured

Even the stress on test cables, caused by handling, creates a variance in transmitting power. All of these things provide a certain amount of difficulty in making repeatable measurements.

Problems in Fiber Optic Testing
- Nonstandard test methods
- Poor calibration accuracy
- Unpredictable component specifications
- Test equipment limitations
- Inherent fiber optic repeatability limitations
- Slow, expensive instruments

Designing Systems for Testability

The designer of the fiber optic system can design the system to make it more stable. This can be done by qualifying the components accurately, designing for worst-case margins, and, in some cases, designing electronic inputs and outputs in parallel with a fiber optic inputs and outputs. Then, you can determine if the electronics of the system are working properly before you have to determine whether the fiber optics system is working properly.

There are systems designed where there are actually parallel optical ports for testing as well as transmitting data. Although this is an expensive way of designing a system, for higher reliability systems, it may be an appropriate way of providing for test ports without having to compromise the transmission system itself by opening up the connectors for testing.

System Troubleshooting

Once the system is installed, troubleshooting the system is really not much different than troubleshooting a standard electronic cable system. You look for a high bit error in the transmission system which indicates a problem. You test the system in a loopback mode where it can self-test its own operation and indicate problem areas. If you have a start signal tracing, you go to the system receiver, disconnect the fiber optic connectors and measure the optical power to determine if it is adequate for the system.

If the optical power is low, that indicates that the problem is in the source, not in the cables. If the source power is high and the receiver power is low, then the problem is probably in the cable or connectors.

If there is some problem in the cable system which must be tracked down, it can be done with the source on, using the power meter in every accessible location in the cable system to determine if there are connectors or other components where the loss might have occurred. One can also use an OTDR and examine the cable from both ends to determine where there might be a high loss involved in the cable. If the cable is too short, use a long launch cable attached to the OTDR to allow examination of the system cable.

System Troubleshooting
- Look for high BER
- Test in loopback mode
- Signal trace:
    Receive power
    Source power
    Imply cable loss

TEST MODULE 15

1. What is the standard method for testing the attentuation of optical fiber ?
   a) Power meter
   b) Visual tracer
   c) Cutback method

2. What is used to diagnose connector faults?
   a) detectors
   b) OTDR
   c) microscope

3. What should you always use to test fiber before inspecting the fiber in a microscope to prevent permanent eye damage?
   a) Power meter
   b) Detector
   c) microscope
   d) OTDR

4. What is needed to measure the launch power in a fiber optic system?
   a) Cutback method
   b) Microscope
   c) Power meter
   d) Power meter and a test cable

5. Circle the advantages of the OTDR.
   a) Requires access to only one end of the fiber
   b) OTDR is expensive
   c) Shows fiber length or distance to fault
   d) Locates breaks or faults

LAB EXERCISE

Hook up meters and test end to end with power units, meters and talk sets.

Glossary & Acronyms

GLOSSARY OF TERMS

1BASE5
An implementation of the Institute of Electronic and Electrical Engineers (IEEE) StarLAN standard on a baseband medium (up to 1 Mbps). The maximum segment length is 1,640 ft. (500 m).

10BASE2
An implementation of the IEEE Ethernet standard on thin coaxial cable, a baseband medium of 10 Mbps. Maximum segment length 200 m.

10BASE5
An implementation of the IEEE Ethernet standard on twinaxial cable, a baseband medium of 10 Mbps. The maximum segment length is 1,640 ft (500 m).

10BASE-FL
An implementation of the IEEE Ethernet standard on 62.5/125-um fiber optic cable, a baseband medium of 10 Mbps.

10BASE-T
An implementation of the IEEE Ethernet standard on 24-AWG, unshielded, twisted pair wiring, a baseband medium of 10 Mbps.

802.3
Defined by the IEEE, these standards govern the use of the carrier sense multiple access/collision detection (CSMA/CD) network access method used by Ethernet networks.

802.5
Defined by the IEEE, these standards govern the use of the Token Ring network access method.

A-D
analog to digital conversion

Abrasion Resistance
Ability to resist surface wear.

Absorption
Loss of power in an optical fiber, resulting from conversion of optical power into heat and caused principally by impurities, such as transition metals and hydroxyl ions, and also by exposure to nuclear radiation.

AC
alternating current

Accelerated Aging
A test that simulates long-time environmental conditions in a relatively short time.

Acceptance Angle
The biggest possible angle between a ray and the center axis

3

Access Method
Set of rules by which networks arbitrate their use.

Active/Passive Device
A device, such as a Token Ring multistation access unit (MAU), that supplies current for the loop that is considered active. One that does not supply current is considered passive.

ACE
above ground cable enclosure

Adapter
The adapter is the hardware device that connects two dissimilar devices. Example: a synchronous data link control (SDLC) adapter connects a gateway PC to a modem for SDLC communications.

Address
A unique identifier assigned to networks and stations so each device can be separately designated to receive and reply to messages.

Adjunct Power
Power supplied to optional data or voice equipment in an equipment room, telecommunications closet, or work area, through separate power supplies.

Adjusted Main Ring Length (AMRL)
The equivalent electrical main ring length (EEMRL) minus the length of the shortest intercloset cabling path.

Administration Point
A central location at which communication circuits are administered; that is, rearranged or rerouted by means of cross-connections or interconnections to information outlets.

Administration Subsystem
The part of a cabling distribution system that includes the connecting hardware components where you can add or rearrange circuits. These components include cross-connects and interconnects, and their associated patch cords or jumper wire. See Administration Point.

Aerial Cable
Telecommunications cable installed on supporting structures such as poles, bridge hangers, building extension supports, etc. These cables are typically non-filled cables intended exclusively for aerial placement via a separate metallic "strand" cable, or with support strand embedded in the same outer sheath (figure 8 cable).

Aerial Distribution Method
The method of running cable between buildings in campus systems by going through the air; that is, building to pole, pole to pole, and/or building to building.

AF
audio frequency

AGX
above ground fiber cross-connect system

AIA
American Institute of Architects

Air Handling Plenum
A designated area, closed or open, used for environmental air circulation (return air).

Alarm Indicator
A device, or combination of devices, such as bells, lamps, horns, gongs or buzzers that respond to a signal from an alarm sensor (FED-std-1037A).

ALPETH
Aluminum-polyethylene, the primary sheath for aerial cable.

ALVYN
Aluminum-polyvinyl-chloride, the preferred sheath for riser back-bone cable where a flame-retardant sheath is required to meet National Electrical Code (NEC) standards.

AM
amplitude modulation

Ambient
Conditions existing at a test or operating location prior to energizing equipment (e.g.: ambient temperature).

American National Standards Institute (ANSI)
Organization responsible for the definition and maintenance of standards. ANSI is the principal group in the United States for defining relative standards. ANSI represents the United States in the International Standards Organization (ISO).

American Standard Code for Information Interchange (ASCII)
A 7-bit binary code standardized by American National Standards Institute (ANSI) for use by personal computers (PCs) and some mainframes to represent alphanumeric and graphical characters. An additional bit is included to form an 8-bit character (byte).

Ampere (A)
A standard unit of current. One ampere (1 A) of current is produced by one coulomb (1 C) of charge passing a reference point in one second (1 sec).

Amplifier
An electronic component used to increase the strength of a transmitted analog signal. Performance is measured in decibels (dB). Similar to a repeater in digital systems.

Amplitude
The relative value of a varying wave form.

Amplitude Modulation
One of three basic methods (see Frequency and Phase Modulation) of adding information to a sine wave signal in which the magnitude of the signal is varied to transmit information.

Analog
A format that uses continuous physical variables such as voltage amplitude or frequency variations to transmit and receive information.

Analog Signal
A nominally continuous electrical signal that varies in amplitude or frequency in response to changes in the physical quantity (such as sound) that it represents.

Annealing
A process of controlled heating followed by gradual cooling to relieve mechanical stresses. Annealing copper makes it more pliable.

APD
avalanche photo diode

API
Application program interface gives users the ability to write programs that communicate with a 3270 emulator, simulating operator keyboard actions. API programs often are written to automate host log-on procedures.

APPC
Advanced program-to-program communications consists of a set of IBM protocols that application programs running on different processors used to communicate with each other directly, without passing through the mainframe.

Appletalk
A proprietary local area network developed by Apple Computer to link Macintosh computers and peripherals, especially printers.

Application Layer
Layer 7 of the open system interconnect (OSI) model for data communications. It defines protocols for users or application programs.

APPN
Advanced peer-to-peer networking is a network architecture that allows mainframes, mini-computers and PCs to communicate as peers across LANs and WANs

Approved Ground
Only as specified in NEC (National Electrical Code Handbook). Refer to articles 250-24, 250-50, 250-71, 250-80, 250-81, 800-33 - 800-40 for compliant approved grounding methods and procedures. Refer to EIA/TIA 607 for standard telecommunications grounding.

Aramid Yarn
Strength elements that provide tensile strength and provide support and additional protection of the fiber bundles.

Architecture
The manner in which a system (infrastructure, hardware and software) is designed. Architecture usually describes how the system is constructed, how the components fit together, and the protocols and interfaces used to integrate these components. It also defines the functions and description of data formats and procedures used for communication between nodes and workstations.

ARCnet
A 2.5 Mbps baseband, token-passing network, designed by Datapoint Corporation, that supports up to 255 nodes.

Armor
Additional protective element beneath outer jacket to provide protection against severe outdoor environments. Usually made of plastic-coated steel, it may be corrugated for flexibility.

Array Connector
A connector that aligns and protects fibers from a ribbon fiber optic cable. A fanout array design can be used to connect ribbon fiber optic cables to nonribbon cables.

ASP
Aluminum-steel-polyethylene, the preferred sheath for filled cable.

Asynchronous Transmission
A data transmission technique controlled by start and stop bits at each end of a character and characterized by an undetermined time interval between characters.

Attachments
A general term to include straps, bolts, clamps or brackets used to support cable in an aerial distribution scheme.

Attachment Unit Interface (AUI)
Branch cable interface located between a media attachment unit and a data station.

Attenuation
The decrease in magnitude of power of a signal in transmission between points. For example, in fiber, expresses the total loss of an optical fiber consisting of the ratio of light output to light input. Attenuation is measured in decibels (fiber or copper) per kilometer (dB/km) at a specific wavelength or frequency. The lower the number, the better. Typical multimode wavelengths are 850 - 1300 nanometers (nm) and singlemode are at 1300 - 1550 nm. Copper is now characterized up to 350 MHz plus.

Attenuation Constant
A rating for a cable or other transmitting medium, which is the relative rate of amplitude decrease of voltage or current in the direction of travel.

Attenuation-Limited Operation
The condition in a fiber optic link when operation is limited by the power of the received signal (rather than by bandwidth or by distortion).

Audio
A term used to describe sounds within the range of human hearing. Also used to describe devices which are designed to operate within this range. (Ex: telephone 300 Hz-3400 Hz.)

AWG
American Wire Gauge

AWM
Appliance Wiring Manual

B or BUR
Buried cable

BDF
Building distribution frame. Primary location for administration of "backbone" cable for a particular building. May reside in same location as B.E.T. (BET) and/or MDF. While BDF implies that there is one distributing frame for a building, there may be IC/TC (IDF) locations that serve as cross-connect locations beyond the BDF. Usually located at, or near, telco entrance facilities (basement, mechanical cores, utility tunnels, etc.) may also be MPOE (minimum point of entry) (telco) location. Terminology varies - see ANSI/EIA/TIA 568A or latest revision, and EIA/TIA 569.

BET
Building entrance terminal, aka entrance facility. Usually the nearest location within a structure that permits termination and protection of telco entrance cable(s). May also serve as MDF and/or BDF. In some situations, the BET is co-located with, or serves as, PABX or key equipment room, MDF and/or BDF and may contain all associated power, battery and other communications equipment. See EIA/TIA 568A, 569. Terminology varies.

Backboard
A rigid support for mounting telecommunications terminating hardware, blocks, cross-connect components and wiring. May also be used to attach and support entrance and distribution cables, splice cases, etc. Typically 3/4 in. plywood anchored/fastened to existing wall. It is recommended that the plywood be fire-retardant, or be coated with a fire-retardant substance.

Backbone Cable
Typically considered to be horizontal or vertical distribution cable. Connects entrance facility to various floors or telecommunications closets. Some versions are shielded. Should be placed near central axis of building for protection and minimized risk of lightning strikes.

Back-End
Database server functions and procedures for manipulating data.

Background Process
When the PC is running more than one program at the same time, a background program or process is one that is sharing memory and CPU time with another active program, but is not interacting with the user. The background program may be waiting for input from the keyboard, but runs by itself in memory while another (foreground) program controls the keyboard and display.

Backscattering
The return of a portion of scattered light to the input end of a fiber; the scattering of light in the direction opposite to its original propagation.

Balanced Line
A cable having two identical conductors opposite in polarity and equal in magnitude and transmission characteristics, with respect to ground. See Balun.

Balloon Framing
A method of framing a structure with strength members, characterized by studs that run from basement to roof without interruption.

Balun
Balanced/unbalanced device used when interconnecting balanced circuits with unbalanced circuits, such as coaxial cables to balanced unshielded or shielded twisted pair (UTP-STP).

Band
A range of frequencies between two predetermined limits.

Bandwidth
The range of frequencies that can be used for transmitting information on a channel, equal to the difference in Hertz (Hz) between the highest and the lowest frequencies available on that channel. Bandwidth indicates the available frequency of a channel. Thus, the larger the bandwidth, the greater the amount of information that can pass through the circuit.

Bank
(a) A range of frequencies between upper and lower limits or (b) a group of tracks on a magnetic drum or magnetic disc or (c) telco carrier equipment (channel bank).

Barrier
A permanent partition installed in a cable raceway or housing that provides complete separation of the adjacent compartment.

Baseband Network
A network in which the entire bandwidth of the transmission medium is used for the signal. Unlike broadband, no characteristic modulation techniques are used.

Baseboard Raceway
A distribution method in which metal or wood channels, containing cables, run along the baseboards of a building. The front panel of the baseboard channel is removable, and information outlets may be placed at any point along the channel.

Basic Rate Interface (BRI)
ISDN standard interface to serve sources or destinations of relatively small capacity, such as terminals. Two "B" channels (64 kbps) and one "D" channel (16 kbps).

Batch
The batch method of processing occurs when data is collected over a period of time and then processed as a batch rather than as it becomes available.

Baud
Unit of data transmission speed meaning bits per second (500 baud = 500 bits per second). Less commonly used as technology permits much faster speeds.

Beamsplitter
An optical device, such as a partially reflecting mirror, that splits a beam into two or more beams and that can be used in fiber optics for directional couplers.

Bearing Wall
A wall that supports a load other than its own weight (upper floor, roof, etc.).

Bel
A unit that represents the logarithm of the ratio of two levels. The number of bels is equal to the logarithm 10 (P1/p2); 2 logarithm 10 (E1/E2); and 2 logarithm 10 (I1/I2). See dB.

Bend Loss
A form of increased attenuation caused by (a) having an optical fiber curved around a restrictive radius of curvature or (b) microbends caused by minute distortions in the fiber imposed by externally induced perturbations.

Bend Radius
Measure for copper cable or optical fiber bends. Refer to manufacturer's recommendations for specific minimum bend radius. Typically 4X, 6X or 10X the outside diameter dependent on specific performance characteristic limitations.

BGX
below ground fiber cross-connect system

BIC
building industry consultant

BICSI
Building Industry Consulting Services International

Binary Digit (BIT)
The smallest unit of information (data) and the basic unit in data communications. A bit can have a value of zero or one (a mark or space).

Binder (cable)
A tape, film or thread used for holding assembled cable conductors in place.

Bit
One binary digit.

Bit/s (BPS)
Bits per second. A measure of speed or data rate. Often combined with prefixes such as Kbps (kilo or thousands of bits per second) and Mbps (mega or millions of bits per second).

Bit Error Rate
The number of erroneous bits compared to the total number of bits transmitted over a fixed period of time.

Blue Field
The field used in telecommunications closets or equipment rooms to connect stations to the horizontal segment. Systems other than fiber distributed data interface (FDDI) can also be connected to the blue field. In twisted pair environments, the blue field terminates cable from the information outlet.

BNC Connector
Coaxial connector type used on many types of data communications equipment.

Bonding
A low-resistance path obtained by joining all current-carrying metallic elements to assure electrical continuity, and having the capacity to safely conduct any current introduced into the path.

Booster
A device or amplifier inserted into a line or cable to increase the voltage. Transformers may be employed to boost AC voltages. The term booster is also applied to antenna pre-amplifiers.

Braid
A group of textile or metallic filaments interwoven to form a tubular flexible structure which may be applied over one or more wires, or flattened to form a strap (as in bonding braid).

Breakout Cables
Multi-fiber compositions where each fiber is further protected by an additional jacket and optional strength elements.

Break Test Access
Method of disconnecting a circuit which has been electrically bridged to allow testing on either side of the circuit without disturbing cable terminations. Devices that provide break test access include: disconnect blocks, bridge clips, plug-on protection modules and patching devices.

Bridge
Linking together two or more networks by a device. A bridge is capable of providing logical routing of frames between rings based on routing information contained in the frames.

Bridging Connection
A parallel connection through which some of the signal energy in a circuit may be withdrawn, usually with imperceptible effect on the normal operation of the circuit.

Broadband
Denotes transmission facilities capable of handling frequencies required for high-grade communications. Broadband infers the use of carrier signals as opposed to direct modulation. Characteristically used for simultaneous multi-channel transmission.

Broadcast
The address for all nodes in a network or the message sent to all nodes.

Brouter
A device that can route specific protocols and bridge others, thus combining the capabilities of a bridge and a router.

Brown Field
The field used in telecommunications spaces to terminate campus backbone cables.

Buffer or Buffering
(a) Protective material extruded directly on the fiber coating to protect it from the environment (tight buffered); or (b) extruding a tube around the coated fiber to allow isolation of the fiber from stresses in the cable (buffer tubes).

Buffer Tubes
Extruded cylindrical tubes covering optical fiber(s) used for protection and isolation.

Building Core
That portion of any building devoted to stairwells, elevators, rest rooms, utility, mechanical, electrical, HVAC and telecommunications cabling/equipment.

Building Entrance Area
The area inside a building where cables enter and may be connected to riser/backbone cables and where electrical protection is provided. The network interface, as well as protectors and other distribution components for campus backbone subsystems, may also be located here. See BDF, BET.

11

Building Footing
The concrete base under the foundation of a building, in which copper wire may be laid to form an electrical ground.

Bundle
Many individual fibers contained within a single jacket or buffer tube. Also, a group of buffered fibers distinguished in some fashion from another group in the same cable core.

Buried Cable
A gel-filled, mechanically protected cable that is direct buried in a trench in such a fashion that it cannot be removed without excavation (FED-std-1037A). As a general reference to types, characteristics and makeup, refer to AT&T Outside Plant Systems, Issue 3 or later. Not to be confused with underground cable (in ducts).

Buried Distribution Method
The method of running cable underground between buildings in campus systems by burying the cable in a trench.

Bus
(a) data path shared by many devices or (b) a linear network topology in which all workstations are connected to a single cable. On a bus network, all workstations receive all transmissions; only the workstation that the information is addressed to will use the information. Contrast with ring and star.

Busbar Wire
Minimum #6 AWG copper wire in telecommunications.

Bus Interface Unit (BIU)
The data circuit equipment that provides physical access to the bus.

Bus Topology
A local area network (LAN) topology in which endpoints connect to a single cable or fiber, or set of wires or fibers, at any point.

Butyl Rubber
A synthetic rubber with good electrical insulating properties.

Byte
A group of eight bits makes a byte. Typically a 16 bit "word" is itself divided up into two bytes for handling. A byte is usually the smallest addressable unit of information in a data store or memory.

C
Symbol designation for capacitance and Celsius..

Cabinet
An enclosure that may house connection devices, terminated cables, splices, apparatus, wiring and equipment. Typically affords security and/or protection from prevailing conditions such as weather, vandalism or accidental damage.

Cable Assembly
(a) Optical fiber cable that has connectors installed on one or both ends (also applies to copper). General use of these cable assemblies include the interconnection of optical fiber cable systems and opto-electronic equipment. If connectors are attached to only one

end of a cable, it is known as a pigtail. If connectors are attached to both ends, it is known as a jumper or patch cord. (b) The method by which a group of insulated conductors is mechanically assembled or twisted together.

Cable Attenuation
The measure of the loss in electrical strength encountered by signals sent through cable.

Cable Bend Radius
Cable bend radius during installation infers that the cable is experiencing a tensile load. Free bend infers a small allowable bend radius since it is at a condition of no load.

Cable Budget
The equivalent electrical cable, as an example, length between the Token Ring adapter card (DB9 Connector) and the media interface connector (MIC) on the multistation access unit (MAU) plus adjusted main ring length (AMRL).

Cable Grip
A device that slips over the end of a cable and connects to a winch or hand line to assist in pulling cable during installation.

Cable Listings
National Electrical Code (NEC)

Article 800:
MPP	Multipurpose Plenum
MPR	Multipurpose Riser
CMP	Plenum Rated Communications Cable
CMR	Riser Rated Communications Cable
CM	General Purpose Not Used In Plenums Or Risers
CMX	Residential And Restricted Commercial Use

Article 770:
OFC	Optical Fiber, Conductive
OFCP	Optical Fiber, Conductive, Plenum
OFCR	Optical Fiber, Conductive, Riser
OFN	Optical Fiber, Non-Conductive
OFNP	Optical Fiber, Non-Conductive, Plenum
OFNR	Optical Fiber, Non-Conductive, Riser

Refer to appropriate articles of the NEC handbook for details on ratings and specific approved applications.

Cable Plant
The cable plant consists of all the optical elements, for example, fiber, connnectors, splices, etc. between a transmitter and a receiver.

Cache
Memory location set aside to store frequently accessed data for improved system performance.

CAD/CAM
(a) computer-aided design/computer-aided manufacturing or (b) computer-aided drafting/computer-aided management.

Campus
The buildings and contiguous property of a complex, such as a university, college, industrial park, military establishment, municipality or health care facility, to name a few.

Campus Backbone Cable
The communications cable that is part of the system and runs between buildings. Typical methods of installing campus backbone cable: in-conduit (underground conduit), direct-buried (in trenches), aerial (on poles), and in-tunnel (in steam tunnels).

Campus Cable Entrance
The point at which campus backbone system cabling (aerial, direct-buried, or underground) enters a building.

Capacitance
The property in a system of conductors and dielectrics that permits the storage of electrical charges whenever a difference in potential exists between the conductors. Capacitance is undesirable in copper cable because it interferes with signals by opposing the desired flow of current.

Capacitive Reactance
The opposition to alternating current due to the capacitance of a cable or circuit. It is measured in ohms and is equal to $1/6.28fC$ where f is the frequency in Hz and C is the capacitance in farads.

Capacitor
Two conducting surfaces separated by a dielectric material. The capacitance is determined by the area of the surface, type of dielectric, and spacing between the conducting surfaces.

Carbon Block
A surge-limiting device that is grounded by arcing across the air gap when the voltage of a conductor exceeds a predetermined level. If the current flow across the gap is large or persists for a length of time the protector mechanism will operate and the protector will become permanently grounded.

Carrier
A company which provides network transmission services or (b) a continuous electrical signal capable of being modified to carry information. The carrier carries no information until some component of the signal (amplitude, frequency or phase) is changed. These changes convey the information.

Carrier Sense Multiple Access with Collision Avoidance (CSMA/CA)
Network access method using contention similar to carrier sense multiple access/collision detection (CSMA/CD) used by LocalTalk Networks. Unlike CSMA/CD, in this method, the sending node responds with a clear-to-send signal before transmission begins.

Carrier Sense Multiple Access/Collision Detection (CSMA/CD)
Network access method in which nodes contend for the right to send data. If two or more nodes attempt to transmit at the same time, they abort their transmission until a random time period of microseconds has transpired and then attempt to resend.

Cascaded Stars
Topology in which a centralized multiport repeater serves as the focal point from many other multiport repeaters, also known as hierarchical star.

Category 1-5 Cabling
(Structured Wiring)

        Cat 1   POTS Voice And Low-Speed Data.

        Cat 2   ISDN, Low-Speed Data, 4 Mbps Token Ring

        Cat 3   Cables/connecting hardware with transmission characteristics
                     up to 16 MHz

        Cat 4   Cables/connecting hardware with transmission characteristics
                     up to 20 MHz

        Cat 5   Cables/connecting hardware with transmission characteristics
                     up to 100 MHz and beyond

CATV (Community Antenna Television/Cable Television)
A method of delivering high-quality television reception by transmitting signals from a central antenna throughout the community, via fiber and coaxial cable. CATV is a broadband transmission facility which generally uses a 75 ohm coaxial distribution drop cable to carry numerous frequency-divided TV channels simultaneously.

CB
Citizens band

CCTV
Closed-circuit television

CDDI
Copper distributed data interface is the term used for a copper cable on which a high-speed (100 Mbps) data is run. Called FDDI over fiber (fiber distributed data interface).

Ceiling Distribution Systems
Distribution systems that use the space between a suspended or false ceiling and the structural floor of the story above for placing the cable. Methods include zone, poke-through, conduit, raceway and cable trays.

Cellular Floor Method
A floor distribution method in which cables pass through floor cells, constructed of steel in concrete, that provide a ready-made raceway for distributing power and communications cables (separately).

Cellular Polyethylene
Expanded or foam polyethylene, consists of individual closed cells of inert gas suspended in a polyethylene medium, resulting in a desirable reduction of the dielectric constant.

Center Wavelength (Laser)
The nominal value central operation wavelength. It is the wavelength defined by a peak mode measurement where the effective optical power resides.

Center Wavelenth (LED)
The aveage of the two wavelengths measured at the half amplitude points of the power spectrum.

Centralized Cabling
A cabling topology used with centralized electronics connecting the horizontal cabling with intrabuilding backbone cabling in the telecommunications closet.

Central Member
The center component of a cable. It serves as an anti-buckling element to resist temperature-induced stresses. Sometimes serves as a strength element. The central member material is either steel, Fiberglas, or glass-reinforced plastic.

Central Office
Facility where common carriers originate subscribers circuits and where the switching equipment that interconnects those circuits is located.

Central Office Local Area Network (CO LAN)
A LAN data switching system typically located at the local telephone company's central office and used for providing advanced features or service to customers, Ex: centrex features.

Central Processing Unit (CPU)
(a) personal computer's (PC's) primary microprocessor chip or (b) host computer, centralized PABX or other central electronic intelligence.

Characteristic Impedance
A frequency-dependent resistance that quantifies the complex opposition to current flow offered by a transmission line.

Chromatic Dispersion
Spreading of a light pulse caused by the difference in refractive indices at different wavelengths.

CICS
The Communications Information Control System is a mainframe database application designed for remote access via the systems network architecture (SNA).

Circuit
A two-way communication path between electronic devices.

Circuit Mil
A term used to define cross sectional areas using an arithmetic shortcut in which the area of the round wire is taken as diameter in mils (.001") squared.

Cladding
The low refractive index material that surrounds the core of an optical fiber, usually silica.

Cleave
The process of separating an optical fiber by a controlled fracture of the glass, for the purpose of obtaining a fiber end, which is flat, smooth, and perpendicular to the fiber axis.

Client
A node that requests network services from a server.

Client-Server Computing
A technique by which processing can be distributed between nodes requesting information (clients) and those maintaining data (servers).

Closed Architecture (Proprietary System)
An architecture that is compatible only with hardware and software from a single vendor. Contrast with Open Architecture.

Closet
Typically a location for hardware, conduit, power panels and electronics, such as multiplexers and concentrators. See Telecommunications Closet (TC).

Cluster
A collection of terminals or other devices in a single location.

COAM
customer owned and maintained

Coating
A protective layer of material over the cladding of an optical fiber.

Coax Connection
Coax connection consists of the link established between a PC and a cluster controller via coaxial cable.

Coaxial Cable
A cable with one transmission conductor (inner conductor) and an outer conductor/braid/shield insulated from one another by a dielectric foam.

CODEC (Coder/Decoder)
Equipment used to transform analog voice signals to digital signals (coder) and digital signals to analog signals (decoder). May be in digital PABX or in the device/instrument itself.

Coherent Light
Parallel, narrow band of light of the same wavelength and phase.

Coil Effect
The inductive effect exhibited by a spiral-wrapped shield, especially above audio frequencies.

Collapsed Backbone
A local area network configuration wherein bridging and routing functions are located at the main cross-connect and accessed via concentrators at the horizontal cross-connects.
Color-Coded Cable
Cable having color-coded insulation on the conductor to aid identification.

Common Carrier
A private communications utility company or a government regulated organization that furnishes services to the general public. It is typically licensed or regulated by a state or federal government agency.

Communication Network
Hierarchy of stations capable of intercommunications, but not necessarily on the same channel or circuits.

Communication Power Pole

A raceway placed between the ceiling and floor used in conjunction with a ceiling distribution system for the purpose of distributing communication and power service to a work area. Also called utility column, ceiling drop pole or power pole.

Communications System

A collection of individual communications networks, transmission systems, relay stations, tributary stations, and terminal equipment capable of interconnection and inter-operation to form an integral whole. These individual components must serve a common purpose, be technically compatible, employ common procedures, respond to some form of control and, in general, operate in unison.

Composite Cable

A cable construction technique that combines multiple cables or media in a single over-jacket.

Computer Peripherals

The auxiliary devices under control of a central computer, such as card punches and readers, high-speed printers, magnetic tape units and optical character readers.

Concentrator

A type of station defined in the fiber distributed data interface (FDDI) station management (SMT) standard. Concentrators can be connected to other concentrators to form a single-ring/tree offshoot from the main dual ring. Concentrators are used to divide a data channel into two or more channels of average lower speed, dynamically allocating space according to demand, to maximize data throughput at all times. Also called an intelligent time division multiplexer (TDM), asynchronous time division multiplexer (ATDM), or statistical multiplexer.

Concentric Stranding

A group of uninsulated wires twisted together and containing a center core with subsequent layers spirally wrapped around the core to form a single conductor.

Concurrency

This term refers to the ability of some emulation programs to allow DOS programs to run in the background while 3270 emulation runs in the foreground.

Conductivity

The ability of a material to allow electrons to flow, measured by the current per unit of voltage applied. It is the reciprocal of resistivity.

Conductor

A medium such as copper wire that can carry electrical current.

Conduit

A pipe, usually metal, that runs underground, from floor to floor, or along a floor or ceiling to protect cables. In the riser backbone subsystem when riser telecommunications closets are not aligned, conduit is used to protect cable and to provide the means for pulling cable from floor to floor. In the horizontal subsystem, conduit may be used between a telecommunications closet and an information outlet in an office or other room. Conduit is also used for campus distribution, where it is run between buildings and intermediate manholes and is made of PVC occasionally encased in concrete. Multiduct conduit may also be used.

Conduit Sizing
All INC (intra and interbuilding network cabling) conduits should be sized at a minimum of 4 inches. When placing new conduits, size installation to include one spare 4" conduit for future. All INC cores and sleeves should meet same criteria (min. 4 in.).

Connecting Block
A flame-retardant plastic block containing metal wiring terminals (clips) that establish an electrical connection between the cable and the cross-connect wire.

Connecting Hardware
A device used to terminate cable with connectors and adapters that provides an administration point for cross-connecting between cabling segments or interconnecting to electronic equipment.

Connector
A device to connect and disconnect copper wires or fibers in cable to equipment or to other wires or fibers. Copper wire and fiber optic connectors most often join transmission media to equipment or cross-connects.

Connector Panel
A panel designed for use with patch panels; it contains either 6, 8, or 12 adapters pre-installed for use when field-connectorizing fibers.

Connector Panel Module
A module designed for use with patch panels; it contains either 6, 8, or 12 connectorized fibers that are spliced to backbone cable fibers.

Contention
Network access method in which nodes compete for transmission by sending signals at will.

Controller
(a) A device used to control the input/output operations between the host computer and a group of terminals.

Control Unit Terminal (CUT)
CUT refers to the connection of a single-session terminal device to a cluster controller.

Controller
(b) In an SNA environment, the controller refers to IBM Cluster Controllers, devices that interface display terminals and printers to the host computer.

Convector Area
An area allocated for heat circulation and distribution. Convector areas, typically built into a wall, can be used as a satellite location only if a more suitable area is unavailable.

Convergent Light
Rays coinciding in one point.

Cord
A flexible insulated cable (stranded vs. solid conductors).

Core
(a) The central transmission area of a fiber. The core always has a refractive index higher than that of the cladding or (b) section of building dedicated to utilities, HVAC, mechanical, electrical, etc. See building core.

Corona
The ionization of gasses about a conductor that results when the potential gradient reaches a certain value.

Coulomb
A quantity of electrically transferred by a current of one ampere (1 A) in one second.

Coupler
A multiport device used to distribute optical power.

Coupling
The transfer of energy between two or more cables or components of a circuit.

CPE
Customer remises (or provided) equipment. Equipment residing on customer sites such as, PABX systems, key systems, data devices, etc. This term is frequently interchanged with "station equipment" in protection practices.

CPUC
California Public Utilities Commission

Critical Angle
That angle of incidence, at which total reflection is obtained.

Cross-Connect
System component where communication circuits are administered (that is, added or rearranged using jumper wire or patch cords). In 110 connector systems, jumper wire or patch cords are used to make circuit connections. In fiber optic connector systems, fiber optic patch cords are used. The cross-connect is located in an equipment room or telecommunications closet.

Cross-Connect Field
Copper wire or fiber terminations grouped to provide cross-connect capability. The groups are identified by color-coded sections of backboards mounted on the wall in equipment room or telecommunications closets, or by designation strips or labels placed on the wiring block or unit. The color-coding identifies the type of circuit that terminates at the field. See EIA/TIA 606.

Cross Pinning
A wiring configuration that permits two DTE devices to two DCE devices to communicate, when straight-through pinning (568A/568B) cannot be used.

Crossover
A conductor which connects to a different pin number at each end. See Cross Pinning, above.

Crosstalk
Undesired signals in one circuit as a result of inductive coupling from another circuit. See Near End Crosstalk (NEXT).

CRT
cathode ray tube

CSA
Canadian Standards Association. Compare with EIA/TIA.

Current, Alternating (AC)
An electric current that periodically reverses direction of electron flow. The rate at which a full cycle occurs in a given unit of time (generally a second) is called the frequency of the current as in 60 cycle (hertz) AC.

Current, Direct (DC)
Electrical current whose electrons flow in one direction only. It may be constant or pulsating as long as its movement is in the same direction.

Current Loop
A two wire transmit/receive interface.

Cut-Down
A method of securing a conductor to a wiring terminal. The insulated conductor is placed in the terminal groove and pushed down with a special tool. As the conductor is seated, the terminal cuts through the insulation to make an electrical connection, and the spring-loaded blade of the tool trims the conductor flush with the terminal. Also called punch-down.

Cutoff Frequency
The lowest possible frequency having a propagation mode.

Cut-through Resistance
The ability of a material to withstand mechanical pressure without damage.

CXC
Coaxial cable, CSA (Canadian Standards Association) cable designation.

Cyclic Redundancy Check (CRC)
A coded sequence of information allowing error checking and correction.

DAC
dual-attached concentrator

Daisy Chain
A cabling practice no longer recommended, where devices were connected from one to another in a chain configuration.

Data Circuit-Terminating Equipment (DCTE)
General terminology for data communications equipment such as a modem. A device that terminates a data communications session and provides encoding or conversion if necessary (for example, modems or printers).

Data Communication
The movement of encoded information by means of electric transmission systems via one or more data links according to protocol.

Data Communications Equipment (DCE)
A device that establishes, maintains, and terminates a data communications session and provides encoding or conversion if necessary.

Data Concentrator
A device that permits a common transmission medium to serve more data sources than there are channels currently available within the medium.

Data Connector
IBM refers to the multistation access unit's (MAU's) port connector as a data connection. Certain applications use the term media interface connector (MIC) instead of data connector.

Data Link Control (DLC)
DLC is the protocol used by IBM or other vendors (Token Ring).

Data Link Layer
Layer 2 of the open system interconnect (OSI) model; it defines protocols governing data packetizing and transmission into and out of each node.

Data Packet Switch
System-common equipment that electronically distributes information among data terminal equipment connected to a data transmission network. The switch distributes information by means of information packets addressed to specific terminal devices.

Data Terminating Equipment (DTE)
General terminology for data equipment such as terminals and host computers. DTE can also stand for data terminal equipment. See DCTE.

Data Transmission
The sending of data from one place to another by means of a signal over a channel using various media.

DB9
A standardized connector with nine pins typically for Token Ring and serial connections.

DB15
A standardized connector with 15 pins typically for Ethernet transceivers.

DB25
A standardized connector with 25 pins typically for parallel-to-serial connections.

DC
direct current

Decibel (dB)
The standard unit for expressing transmission gain or loss and relative power ratios. The decibel is one tenth the size of a Bel, which is too large a unit for convenient use. Both units are expressed in terms of logarithm to the base 10 of a power ratio used primarily for attenuation and crosstalk measurements in telecommunications.

Decibel/kilometer (dB/km)
A unit of measurement for fiber optic attenuation.

Delay Line
A transmission line or equivalent device designed to delay a wave or signal for a specific length of time.

DEMARC
Demarcation point is the point of interface that readily identifies division of loop or circuit responsibility. Other terms frequently applied (properly or improperly): SNI (subscriber network interface), MPOE (minimum point of entry), EF (entrance facility) and others.

Demodulation
The process of extracting the information signal from an analog carrier signal. The reverse of modulation.

Demountable Walls
Metal walls that can be disassembled and moved to other locations. They contain vertical and horizontal slots through which cable can be run, aka: modular panels.

Deviation
Change in the direction of propagation

Dielectric
A nonconducting or insulating material that prevents passage of electric current and resists inductive coupling.

Dielectric Breakdown
Any change in the properties of a dielectric that causes it to become conductive. Normally a catastrophic failure of an insulation because of excessive voltage.

Dielectric Cable
A nonconducting cable, such as a fiber cable, without metallic members.

Dielectric Constant
Also called permitivity. That property of a dielectric which determines the amount of electrostatic energy that can be stored by the material when a given voltage is applied to it. Example: the ratio of capacitance of a capacitor using the dielectric to the capacitance of an identical capacitor using a vacuum as a dielectric.

Dielectric Heating
The heating of an insulating material when placed in a radio-frequency field, caused during the rapid polarization reversal of molecules in the material.

Dielectric Loss
The power dissipated in a dielectric as the result of the friction produced by molecular motion when a alternating electric field is applied.

Dielectric Strength
A measure of the maximum voltage that the insulation of a particular cable can withstand without breakdown.

Diffuse
The phenomenon when the molecules of two materials in contact mix.

Digital

A data format that typically uses two physical levels to transmit information. It is a discrete or discontinuous signal, on-off, zero-one, etc. MLT uses three levels instead of two.

Digital PBX (DPBX); also DPABX

A PBX (see Private Branch Exchange) designed to switch digital signals. Telephones used with a DPBX must digitize the voice signals, but computers and terminals may communicate directly through the DPBX, which functions as a point-to-point local area network through a central processor.

Digital Signal

A signal that represents information by a series of fixed, encoded, rectangular pulses, usually consisting of two possible voltage levels. Each voltage level indicates one of two possible values or logic states, such as on or off, open or closed, or true or false. See Digital, above.

Direct-Conduit Method

A ceiling distribution method in which cables are run in conduit from a centralized location directly to the desired information outlets.

Direct Distance Dialing (DDD)

The Bell System used for the telephone service in North America that enables a user to dial long-distance calls directly without operator assistance.

Disk/Disc

An electromagnetic storage medium for digital data.

Disk Server/Disc Server

A disk storage device which provides multiple users with access to their own assigned section of the disc. Compare with file server.

Dispersion

The cause of bandwidth limitations in a fiber. Dispersion causes a broadening of input pulses along the length of the fiber. Three major types are: (1) modal dispersion caused by differential optical path lengths in a multimode fiber; (2) chromatic dispersion caused by a differential delay of various wavelengths of light in a waveguide material; and (3) waveguide dispersion caused by the light traveling in both the core and cladding materials in singlemode fibers.

Display (Example:)

The display represents the PC monitor when it is running 3270 emulation. One PC monitor can emulate more than one 3270 terminal display.

Distortion

Any undesired change in a wave form or signal.

Distributed Architecture

A network that uses a shared communications medium (such as star, bus or ring LAN) and uses shared access methods.

Distribution Block/Frame

Centralized connection equipment where telephone or data terminal cabling is terminated and cross-connections are made.

24

Distribution Field
The cross-connect or interconnect field used to further distribute the cabling from one point in the network to another. Distribution fields are color coded by function (EIA/TIA 569).

Distribution Network
Part of the local exchange network, comprising small cables between subscribers' distribution points (DPs) and cabinets, remote line units (RLUs) or other flexibility points.

Distribution Panel
Wall or rack mounted panel that permits accessible physical "patching", or cross-connecting of voice/data circuits and/or devices.

DLC
digital loop carrier also, SLC or subscriber line carrier

Drain Wire
An uninsulated wire in contact with a shield throughout its length. Used for terminating the shield, typically at the equipment end only.

Drop Cable
In a CATV system, the transmission cable from the distribution device to the structure being served.

DSX
digital signal cross-connect

Dual-Attached Station (DAS)
A station defined in the fiber distributed data interface (FDDI) standard that has two dual-fiber-optic connections (MIC A and MIC B) that allow it to be connected to the dual FDDI ring.

Dual-Fiber Cable
A type of fiber cable that has two single-fiber cables enclosed in a jacket of extruded polyvinyl chloride (PVC), with a rip cord for pulling back the jacket to access the fibers.

Dual Homing
The optional connection of dual-attached stations (DASs) to concentrators to increase reliability of the DAS network attachment.

Dual-Media Access Control (DMAC)
A station with two media access control (MAC) entities, which allow for logical connectivity to each of the dual FDDI rings. A DMAC station can independently send and receive data from both rings and thus has an available bandwidth of 200 Mbps from the network.

Ducts
Various pathways or conduits ranging from PVC to metallic to clay/tile. Example: the main feeder channels in which communication cable is routed between buildings in a campus environment.

Dumb Terminal
A dumb terminal is a device that can send and receive data but cannot process data.

Duplex
(a) In data communications, a circuit used to transmit/receive signals simultaneously in both directions, or (b) in general, two receptacles or jacks in a common housing which accepts two plugs.

E
Voltage also, EMF (electromotive force).

Earthing
British terminology for zero-reference ground.

Elastomer
Any material that will return to its original dimensions after being stretched or distorted.

Electromagnetic
Referring to the combined electric and magnetic fields caused by electron motion through conductors.

Electromagnetic Coupling
The transfer of energy by means of a varying magnetic field, also inductive coupling.

Electromagnetic Interference (EMI)
The interference in signal transmission or reception caused by the radiation of electrical and magnetic fields. See RFI, Radio Frequency Interference.

Electron Volt
A measure of the energy gained by a electron passing through an electric field produced by one volt.

Electronic Industries Alliance (EIA)
Governing agency for established standards and published test procedures. See CSA.

Electrostatic
Pertaining to static electricity, or electricity at rest.

Electrostatic Coupling
The transfer of energy by means of varying electrostatic fields. Capacitive coupling.

EMC
Electromagnetic compatibility or conformity.

Emergency Power
An alternate electrical supply source, separate and distinct from the primary electrical utility (generators, batteries, rectifiers, etc.). Typically referred to as "back up", "hot standby", standby, emergency generators, etc.

EMF
Electromotive force (voltage).

Emulator
An emulator is a computer system or component that functions as another type of computer system or component. (See Printer Emulator.)

Energy Dissipation
Loss of energy from a system due to the conversion of work energy into an undesirable form, usually heat. Dissipation of electrical energy occurs when current flows through a resistance.

Entrance Facility
Typically denotes any or all of the following: telecommunications space, equipment, support hardware, cables, connectors, blocks, protectors, sleeves, splices, or other items specific to the DEMARC or MPOE where telco responsibility ends and the customer's begins. aka: telecommunications service entrance.

EPDM
Ethylene-propylene-diene monomer rubber. A material with proven electrical insulating properties.

EPR
Ethylene-propylene copolymer rubber. Another material with proven electrical insulating properties.

Equilay
More than one layer of helically laid wires with the length of the lay the same for each layer.

Equipment Closet
A termination point for customer premise cabling designed to offer access to common equipment. Closets generally serve a specific area. See Telecommunications Closet (TC).

Equipment Room (ER)
Area dedicated to housing equipment associated with telecommunications systems such as PABX, data, power, key and or peripheral components. (Also PABX Room, MDF.)

Equipment Subsystem
The part of a premises distribution system that includes the cable and distribution components in an equipment room and that interconnects system-common equipment, other associated equipment, and cross-connects.

Equivalent Electrical Cable Length (EECL)
The length of cable that provides the same electrical loss (dB/MHz) as the transmission path from A to B.

Equivalent Electrical Lobe Length (EELL)
The sum of the equivalent electrical cable length (EECL) of all the components in the path from the connector to the media interface connector (MIC).

Equivalent Electrical Main Ring Length (EEMRL)
The equivalent length of cable obtained by adding up the equivalent electrical cable length (EECL) of all the components in the intercloset cabling path. This should not include the short 2-ft. multistation access unit (MAU) patch cords that connect MAU ring segments in a closet.

Ethernet
A baseband local area network used for connecting computers and terminals, etc., within the same. Ethernet was marketed (and trade marked) by Xerox. It is the basis for the IEEE Standard 802.3.

EV
electron volt

Extrinsic Loss
In a fiber interconnection, that portion of loss that is not intrinsic to the fiber but is related to imperfect joining, which may be caused by the connector of splice.

F
frequency

Fall Time
The time needed for a pulse to fall from 90 percent to 10 percent of its maximum value.

Fan Out Cable
Multifiber cable constructed in the tight buffered design. Designed for ease on connectorization and rugged applications for intra- or inter-building requirements.

Farad
A unit of capacity that will store one coulomb of electrical charge when one volt of electrical pressure is applied.

FAS
Fire alarm and signal cable, CSA (Cabling Standards Association) cable designation.

Fasteners
A general term referring to the screws and anchors used to secure attachments to wood, concrete, brick or stone in a cable distribution scheme.

Fault Management
One of five categories of network management defined by the International Standard Organization (ISO). Detects, isolates and corrects network faults.

Fault Tolerance
The ability of a system to perform fault management and continue operating in the event of system failure.

FDI
Feeder distribution interface. Fiber distribution interface. Not to be confused with FDDI.

FDT
fiber distribution terminal

Federal Communications Commission (FCC)
A board of commissioners that regulates all electronic communications systems originating in the United States, including telephone systems.

Feedback
Energy that is extracted from a high-level point in a circuit and applied to a lower level. Positive feedback reduces the stability of a device and is used to increase the sensitivity or produce oscillation in a system. Negative feedback, also called inverse feedback, increases the stability of a system as the feedback improves stability and fidelity.

Feeder Cable
(a) In telecommunications, main distribution cable or trunk cable, or (b) in CATV, the transmission cable from the head end (signal pickup) to the trunk amplifier.

FEP
Fluorinated ethylene-propylene. A thermo-plastic material with superior electrical insulating properties and chemical and heat resistance, aka: TEFLON (DuPont).

Ferrous
Composed of and/or containing iron. A ferrous metal exhibits magnetic characteristics.

Ferrule
A mechanical fixture, typically a rigid tube, used to protect and align a fiber in a connector. Generally associated with fiber optic connectors.

FET
field effect transistor

Fiber
Thin filament of glass. An optical waveguide consisting of a core and a cladding that is capable of carrying information in the form of light.

Fiber Distributed Data Interface (FDDI)
An American National Standards Institute (ANSI) standard for a fiber-based physical and data line protocol that operates at a 100-Mbps data transfer rate.

Fiber Optic Cable
A transmission medium consisting of a core of glass or plastic surrounded by a protective cladding, strengthening material, and outer jacket. Signals are transmitted as light pulses, introduced into the fiber by a light transmitter (either a laser or light-emitting diode {LED}). Some of the advantages offered by fiber optic cable are low data loss, high-speed transmission, greater bandwidth, small physical size, light weight, and freedom from electromagnetic interference or electrical ground problems. Common types are single, dual, multifiber and ribbon.

Fiber Optic Connectors
Connectors designed to connect and disconnect either single or multiple optical fibers repeatedly. Fiber optic connectors are used to connect fiber cable to equipment and interconnect cables.

Fiber Optic Cross-Connection
Fiber optic apparatus for terminating cable in couplings. Designed for high-density cross-connection fields, the apparatus can terminate up to 72 fibers on each shelf, with up to nine shelves in a bay frame. Single shelves can also be wall mounted. Cross connections are handled with fiber optic patch cords.

Fiber Optic Interconnect
An interconnection unit used for circuit administration and built from modular cabinets. It provides interconnection for individual optical fibers but, unlike the fiber optic cross-connect panel, it does not use patch cords. The fiber optic interconnect provides some capability for routing and rerouting circuits, but is usually used where circuit rearrangements are infrequent.

Fiber Optics
The technique of conveying light or images through glass or plastic fibers. Coherent fiber optics should actually be called aligned fiber optics because the fibers are all the same length and are held in a constant spatial relationship.

Field
An area through which electric and/or magnetic lines of force pass.

File Server
A mass storage device that allows files to be accessed by several computers.

Fillers
Nonconducting components assembled with the insulated conductors or optical fibers to impart roundness, flexibility, tensile strength, or a combination of all three, to the cable.

Fire-Rated Doors
An assembly of various materials and types of construction used in wall openings to retard the spread of flames, gases and smoke.

Fire-Rated Poke-Through
A cable distribution device which is fitted through a predrilled core hole and allows cables to be fed from the floor below. A compartment at or above the floor surface is used to provide voice, data, power and video connections while maintaining the fire integrity of the floor.

Fire Resistance Rating
(Expressed in hours, or fractions of hours) Rating of designs or assemblies that show an acceptable resistance to fire (full scale); and, descriptions of materials or assemblies that withstand the passage of flame and heat transmission when exposed to fire under specified test and performance criteria. Reference to NFPA/NEC articles for specific rating.

Fire Shield
A material, device or assembly of parts in, or between, cabling systems, to prevent propagation of flames from one cable system to an adjacent cabling system.

Fire Wall
A wall that helps prevent fire spreading from one contained area to another and that runs from structural floor to structural ceiling.

Firebreak
A material, device or assembly of parts installed in a cable system to prevent the spread of fire along a cable system (not to be confused with fire barrier penetration). See Fire Shield.

Fireproof
A property in material such as masonry, block, brick, concrete or gypsum board that does not support combustion even under accelerated conditions.

Firestop Zoning
A unique group of architectural structure or assembles that prevents the passage of fire or toxicity from one contained area to another, thus reducing the possible spread of combustion through the fire barrier.

Firestop, Firestopping
(a) A material, device, or assembly of parts installed after penetration of a fire-rated wall, ceiling or floor area to prevent passage of flame, smoke or gases through the rated barrier. Refer to NFPA specifications for the intended application(s).
(b) The use of special devices and materials to prevent the outbreak of fire within telecommunications utility spaces and to block the spread of fire, smoke, toxic gases and fluids through openings, cable apertures and along cable pathways. The techniques used are often mandated by local building codes.
Note: Classifications are available under the rating criteria of ASTM E814.

Rating/Achievement:

F
Withstands the fire test for the rating period without: Permitting flames to pass through the firestop flame occurring on any element of the unexposed side of the firestop (auto-ignition) developing any opening in the firestop that permits a projection of water beyond the unexposed side during the hose strength test.

T
Meets the criteria of an "F" rating and prevents the transmission of heat during the rating period so that the temperature rise is not more than 325 degrees Fahrenheit on any exposed surface, thermocouple or penetrating item.

FITL
fiber in the loop

FLC
fiber loop converter

Flex Life
The ability of a cable to bend many times before breaking (stranded).
Flexibility
The ability of a cable to bend in a short radius.

Floor Box
A cast iron, stamped steel or nonmetallic box placed in the concrete floor (prior to pouring the concrete slab) of a building, which is fed via conduit and used to house voice, data, power or video connections.

FM
frequency modulation

Foam Polyethylene
Cellular polyethylene.

Foreground Process
When the PC is running more than one program at the same time, a foreground process or program is the one that currently controls the display screen and keyboard, and the one the user is interfacing with.

Frame
A metallic structure for supporting connectors, protectors, patch panels, blocks, etc.

Frequency
The number of cycles completed by a signal in one second (1 sec); expressed in Hertz (Hz). Example: 100 MHz.

Frequency Division Multiplexing (FDM)
A technique for combining many signals on a single circuit by dividing the available transmission bandwidth by frequency into narrower bands, each used for a separate communication channel.

Frequency Modulation (FM)
One of three basic methods (see also Amplitude and Phase Modulation) of adding information to a sine wave signal in which its frequency is varied to impose information on it.

Frequency, Power
Normally, the 60 Hertz power available in the United States

Frequency Response
The characteristic of a device denoting the range of frequencies over which it may be used effectively.

Fresnel
French physicist (1788-1827), one of the founders of the theory that light is an electromagnetic wave motion.

Fresnel Reflection Losses
Reflection losses that are incurred at the input and output of optical fibers due to the difference in refraction index between the core glass and immersion medium.

Front-end Processor (FEP)
An FEP is a computing device that handles communications tasks such as data transfer for the mainframe.

FTTC
fiber to the curb

FTTH
fiber to the home

Furniture System
Furniture walls combined with furniture units such as desks, work surfaces and file cabinets (modular furniture).

Furniture Wall
Hollow metal partitions with vertical and horizontal slots through which cable can be run. These walls can be covered with fabric, wood, veneer or other material and are usually from 48 to 72 inches high. See above.

Fuse
A UL-Approved overcurrent device with a circuit opening fusible part that is heated and severed by the passage of overcurrent through it.

Fusible Links
Short lengths of fine-gauge wires inside metallic sheath cable that melt to interrupt an electrical circuit and to prevent overheating or damage to wiring and equipment.

Fusion
A process for splicing optical fibers. See also mechanical splicing. Refer to technical specifications to determine which is most effective for intended applications.

FWHM
Full width half maximum, the width of a pulse at half its maximum.

Gain
The increase of voltage, current or power over a standard or previous reading. Usually expressed in decibels.

Gas Tube
A surge-limiting device similar in operation to a carbon block except that it has specially configured electronics with a more precise narrow gap (also available with a wide gap) and a sealed gas composition. The gas tube results in a more accurate and precise operating voltage range and extended service life under conditions of repeated operation.

Gas Tube Protector
An overvoltage protector featuring metallic electrodes which discharge in a gas atmosphere within a ceramic, glass or synthesized envelope.

Gateway
A device that converts protocols between dissimilar communications systems.

Gauge
A measure of a conducting wire's physical size, usually referred to as AWG. See also American Wire Gauge (AWG).

Gigahertz (GHz)
A unit of frequency equal to 1 billion Hertz.

GND or GRD
Abbreviation for ground.

Graded Index Fiber
A fiber design in which the refractive index of the core is lower toward the outside of the fiber core and increases toward the center of the core. The refractive index bends the rays inward and allows them to travel faster in the lower index of refraction region. This type of fiber provides high-bandwidth capabilities.

Gray Field
The cross-connect field used in telecommunications closets to terminate horizontal cables that ties two closets together. As in IC to TC (EIA/TIA 568A/569).

Ground

An electrical connection to the earth, generally through a bonding conductor or ground grid. Also a common return to a point of zero potential, such as the main grounding busbar. See EIA/TIA 607.

Grounding

A conducting connection, whether intentional or accidental, between an electrical circuit or equipment and the earth, or to some conducting body that serves in place of the earth.

Grounding Conductor

The conductor used to connect electrical equipment to a grounding electrode. Also, bonding conductor.

Grounding Electrode

A conductor or group of conductors (usually a rod, or grid) in direct contact with the earth, providing a connection to the earth.

Ground Loop

A completed circuit between conductors created by random contact. An undesirable circuit condition in which interference is created by ground currents when grounds are connected at more than one point.

Ground Potential

The potential of the earth. A circuit, terminal, or chassis is said to be at ground potential when it is used as a reference point for other potentials in the system.

Half Duplex Transmission

Data transmission over a circuit capable of transmitting in either direction, but not simultaneously.

Handshaking

A preliminary procedure, usually part of a communications protocol, to establish a connection between devices.

HDSL

high bit rate digital subscriber line

Headerduct

The main or feeder duct for bringing cable from telecommunications closets to distribution ducts in cellular and underfloor duct systems.

Heat Coil

The device which grounds a conductor when the conductor's current time limits are exceeded. Heat coils are suitable for sneak protection if they are located at the building entrance terminal (BET) aka entrance facility (EF).

Henry (H)

The standard unit of inductance. The inductance of a current is one Henry when a current variation of one ampere (1 A) per second is present at one volt (1 V).

Hertz

The unit of frequency, one cycle.

HF
high frequency

HiCAP
high capacity circuit (DS-1 and above)

Hierarchical Computer Network
A computer network in which processing and control functions are performed at several levels by computers specially tailored for the functions performed.

High Frequency
The band from 3 to 30 MHz in the radio spectrum, as designated by the Federal Communications Commission.

Home-Run Method
A distribution method in which individual cables are run directly from the telecommunications closet to each information outlet in a star configuration.

Horizontal Length (HL)
The cable distance from the work area outlet (WAO) to the specific field of the cross-connect (maximum 295 ft.) in the telecommunications closet (TC).

Horizontal Cabling
The portion of the cabling system extending from the work area outlet to the TC. The cross-connect facilities in the telecommunications closet are considered part of the horizontal cabling.

Host Computer
The central computer in a data communications system which provides the primary data processing functions such a computation, data base access, special programs or programming languages.

Hub
A concentrator or repeater in a star topology at which node connections originate. Hubs can be either active or passive.

Hum
A term typically used to describe the 600 cycle per second noise present in some communications equipment. Usually, hum is the result of undesired coupling to a 60 cycle source or to the defective filtering of ripple output of a rectifier.

Hybrid Cable
A cable containing two or more different types of cable, such as copper and fiber optic.

Hybrid Ring Control (HRC)
A Fiber Distributed Data Interface (FDDI) standard that allows the multiplexing of data packets and circuit-switched traffic on the FDDI local area network (LAN).

I
Symbol used to designate current (ampere).

IC
Intermediate cross-connect (EIA/TIA 568A/569). Formerly IDF.

ICEA
Insulated Cable Engineers Association

IF
intermediate-frequency

IFC or IFOC
Interfacility or intrafacility fiber optic cable

ILD
injection laser diode

IM
intensity modulation

Impedance
The total opposition that a circuit offers to the flow of current at a particular frequency. It is a combination of resistance (R) and reactance (X) and is measured in ohms.

Impedance, Characteristic
In a transmission cable of infinite length, the ratio of the applied voltage to the resultant current at the point the voltage is applied. Or the impedance which makes a transmission cable seem infinitely long, when connected across the cable's output terminals.

Impedance, High
Generally, the area of 25,000 ohms or higher.

Impedance, Low
Generally, the area of 1 through 600 ohms.

Impedance Match
A condition whereby the impedance of a particular circuit cable or component is the same as the impedance of the circuit, cable or device to which it is connected.

Impedance Matching Sub
A section of transmission line or pair of conductors cut to match the impedance of a load. Also called matching sub.

Impedance Matching Transformer
A transformer designed to match the impedance of one circuit to that of another.

INC
Intrabuilding network cable. Also, interbuilding network cable. Cabling that comprises infrastructure beyond the telco MPOE. See Backbone Cable. MPOE.

Index-Matching Fluid
A fluid with an index of refraction close to that of glass that reduces reflections caused by refractive-index differences.

Index of Refraction
The ratio of light emissions in a vacuum relative to a given transmission medium.

Inductance
The electrical property of a circuit that induces change in existing current.

Induction Heating
Heating a conducting material by placing it in a rapidly changing magnetic field. The changing field induces electric currents in the material and dissipation accounts for the resulting heat.

Information Outlet (IO) aka: Work Area Outlet (WAO)
A connecting device designed for a fixed location (usually a wall or floor in an office) on which horizontal cable pairs terminate and which receives an inserted plug. It is the point where the horizontal system meets the work area. Although such devices are also referred to as phone jacks, the term information outlet or WAO encompasses the integration of voice, data and other communication services that can be supported via a premises distribution system.

Infrared Light
Light in the wavelength range 750-1000 nm heat radiation.

Infrastructure
See INC, backbone cable, riser. Encompasses all wire, cable, equipment, power, hardware, devices and associated components or labor to maintain operating systems beyond the telco MPOE (minimum point of entry).

Inhomogenity
Irregularity in e.g. the composition or geometry

Injection Laser Diode
Sometimes called a semiconductor diode. A laser in which the stimulation occurs at the junction of n-type and p-type semiconductor materials.

Innerduct
Flexible conduit originally produced for protection of optical fiber cables. See manufacturer's standards for sizing and placement.

Input
A signal (or power) which is applied to a piece of electric apparatus, or the terminals on the apparatus to which a signal or power is applied to generate intelligent signal processing.

Insertion Loss
A measure of the attenuation of a device determining the output of a system before and after the signal is inserted into the system.

Institute of Electrical and Electronics Engineers Inc. (IEEE)
The standards group that develops standards for Token Ring, fiber distributed data interface (FDDI), 10BASE-T/FL, etc.

Insulation
A material having high resistance to the flow of electric current. Thin conducting wires are covered with color-coded insulation for protection and ease of identification.

Insulation Displacement
The type of wire terminals that require no insulation removal; when the conductor is correctly attached. The insulation is displaced (pierced) to form a connection.

Insulation Resistance
The measure of the ability of an insulation material to resist the flow of current through it; usually measured in Megohm-feet.

Insulation Stress
The molecular separation (pressure) caused by a potential difference across an insulator. The practical stress on insulation is expressed in volts per mil.

Integrated Services Digital Network (ISDN)
Integrated voice and data network based on digital communications technology and standards interfaces.

Integrated System
A telecommunications system that moves analog and digital traffic over the same network, including voice, data and video.

Intelligent Hub
A hub that performs bridging and routing functions in a collapsed backbone environment.

Intelligent Terminal
An input/output device, remote from its main computer, which contains an integral microprocessor capable of performing some amount of information storage or processing.

Interactive
Interactive, also referred to as on-line or real-time processing, occurs when the user is directly connected to the host and receives immediate response to requests or inquiries.

Intercloset Cables (more accurately, Backbone Cables)
Cables that connect telecommunications closets.

Interconnect
A circuit administration point, other than a cross-connect or information outlet, that provides capability for routing and rerouting circuits. It may not use patch cords. Typically it is a jack-and-plug device used in smaller distribution arrangements or to connect circuits in large cables to those in smaller cables.

Interface
The location where two systems or a major and a minor system meet and interact with each other.

Interface EIA Standard RS232 B/C/D
Standardized method adopted by the EIA to insure uniformity of interface between data communications equipment and data processing terminal equipment.

Interference
Disturbance of an electrical, electromagnetic or RFI nature that introduces undesirable responses into other electronic equipment.

Intermediate Cross-connect (IC)
In telecommunications, it is the space between MC and TC, if required.

Intermediate Frequency
A frequency to which a signal is converted for ease of handling. Receives its name from the fact that it is an intermediate step between the initial and final conversion or detection stages.

International Standards Organization (ISO)
The organization responsible for the open systems interconnect (OSI) standards among others.

International Telegraph and Telephone Consultative Committee (CCITT) aka: ITU-International Telecommunication Union(s)
A standards organization that, among numerous other activities, specializes in the electrical and functional characteristics of switching equipment. The CCITT sets standards to ensure compatibility between data communications equipment (DCE) and data terminating equipment (DTE).

Interoperability
The ability to operate and exchange information in a multivendor/multiproduct network.

Ionization Voltage
The potential at which a material ionizes. The potential at which an atom gives up an electron.

IPCEA
Insulated Power Cable Engineers Association
IR Drop
The value of a voltage drop in terms of current (ampere=I) and resistance (R).

IRS
ignition radiation suppression

Isolated Ground
A separate ground which is insulated from the equipment or building ground. Not recommended. See EIA/TIA 607.

Isolation
The ability of a circuit or component to reject interference.

IXC, IEC
Interexchange carrier. Ex: AT&T, MCI, US Sprint, Interdata.

Jack
A receptacle used with a plug to make electrical contact between communications circuits. Jacks and their associated plugs are used in a variety of connecting hardware applications including adapters, information outlets and equipment connections.

Jacket
The flexible covering of a cable, used to bind and protect the color-coded conductors inside.

Joule
A unit of energy, work or quantity of heat equal to 0.4342 foot-pounds. One Joule is the energy expended when a force of one Newton is applied over a displacement of one meter in the direction of the force.

Jumper
Optical fiber cable that has connectors installed on both ends.

Jumper Wire
Typically, a short length of copper used to route a circuit by linking two cross-connect termination points.

Junction Box
A connection point in a duct system that allows access to cables running in the ducts.

Keying
A mechanical mounting on connectors that can be used to prevent improper connection of station posts, as in "keyed" jacks/plugs.

Kilo
Metric prefix = one thousand.

KPSI
Tensile strength in thousands of pounds per square inch.

KV
kilovolt (1000 Volts)

KVA
kilovolt amperes

KW
kilowatt = 1000 watts

L
Symbol for inductance.

Label AA
Administration label for use on fiber optic cross-connect equipment which defines the fiber distributed data interface (FDDI) dual-ring assignments to the first four color-coded fibers in a building cable as follows:
fiber 1 (blue) - secondary input
fiber 2 (orange) -   primary output
fiber 3 (green) -   primary input
fiber 4 (brown) -   secondary output

Label BB
fiber 1 (blue) - secondary output
fiber 2 (orange) -   primary input
fiber 3 (green) -   primary output
fiber 4 (brown) -   secondary input

Label CC
fiber 1 (blue) - input
fiber 2 (orange) -   output
fiber 3 (green) -   input
fiber 4 (brown) -   output

LAN NetView Management Utilities for OS/2
This software management package allows an administrator to view statistics and config-

ure DOS, Windows, OS/2 or Macintosh workstations or servers from an OS/2 management console. It operates with LAN servers and NetWare networks.

Laser
(light amplification by stimulated emission of radiation) In telecommunications, a device which produces light at a narrow range of frequencies, to generate signals used in fiber optic communications systems.

LATA
Local access transport area. Regional TELCO Service, Interlata: See IXC.

Lay
Pertaining to wire and cable, the axial distance required for one cable conductor or conductor strand to complete one revolution about the axis around which it is cabled.

Lay Direction
The direction of the progressing spiral twist in a cable while looking along the axis of the cable away from the observer. The lay direction can be either left or right.

Lays
The twists in twisted pair cable. Two single conductors are twisted together to form a pair; by varying the length of the twists, or lays, the potential for signal interference between pairs is reduced due to canceling effect of emanated fields.

Lead-in
The cable that provides the path for RF energy between an antenna and a receiver or transmitter.

Leakage
The undesirable loss of signal over the surface of, or through, an insulator.

Leased Line
A private telephone line rented for the exclusive use of a leasing customer, without interexchange switching arrangements.

LED
See Light Emitting Diode.

Level
Typically, a measure of the difference between a quantity or value and an established reference.

LF
Low frequency

Light Emitting Diode (LED Source)
A semiconductor device that emits light formed by the P-N junction. Light intensity is roughly proportional to electrical current flow.

Lightguide Building Cable (LGBC)
Alternative term used for a fiber cable in which individual optical fibers are stranded around central members. For interior use.

Lightguide Cross-Connect (LGX) Distribution System
A component of fiber optic connecting hardware. This component accommodates 24-216 fiber terminations. Also referred to as an LGX shelf or frame.

Lightguide Interconnection Unit (LIU)
A component of fiber optic connecting hardware. This component accommodates 12, 24 or 48 fiber terminations. Also referred to as an LIU. Not to be confused with Line Interface Unit (PABX).

Lightpack Cable
A cable core design that allows bundles of optical fiber in a cable core without central strength members.

Line Driver
A signal converter that conditions the digital signal transmitted by an RS232 interface to ensure reliable transmission beyond the standard RS232 limit and often up to several miles; it is a baseband transmission device. Also called a baseband modem, limited distance modem or short haul modem.

Line rop
A voltage loss occurring between any two points in a power or transmission line. Such loss, or drop, is due to the resistance, reactance or leakage of the line. See Voltage Drop, IR Drop.

Line Equalizer
A reactance (inductance and/or capacitance) connected in series with a transmission line to alter the frequency-response characteristics of the line.

Line Level
Refers to the output voltage level of a piece of electronic equipment. Usually expressed in decibels.

Line Voltage
The value of the potential existing on a supply or power line.

Link
The communications circuit or transmission path connecting two points.

Link Budget
Optical loss budget that determines the maximum distance allowable between stations. Loss and dispersion factors are included.

Load
A device that consumes power from a source and uses that power to perform a function.

Loaded Line
A transmission line that has lumped elements (inductance or capacitance) added at uniformly spaced intervals. Loading is used to provide a given set of characteristics extending a transmission line.

Loading
See Loaded Line.

Lobe

In Token Ring networks, a 2-pair circuit connecting the adapter's DB9 connector to the media interface connector (MIC) on the multistation access unit (MAU).

Lobe Length (LL)

The distance from the connector on a Token Ring adapter card to the media interface connector (MIC) on the multistation access unit (MAU). If this path has no intermediate connection points and is made up of one cable type, its effect on the Token Ring's cable budget will contribute to the cable loss.

Local Area Network (LAN)

A data communications network consisting of host computers or other equipment interconnected to terminal devices, such as personal computers, often via twisted pair or fiber cables. LANs allow users to share information and computer resources. Typically, a LAN is limited to a single organization, department or geographic site.

Logical Unit (LU)

A device attached to the PC that is configured as a part of the network. These devices include displays and printers.

Logical Unit Address (LUA)

The host system address that identifies printers and displays attached to a PC.

Long-wire Antenna

Any conductor length in excess of one-half of a wavelength. In a television installation, a horizontal run of unshielded lead-in will act as a long-wire antenna and introduce additional signal on top of the regular antennae, signal causing "ghosts."

Loop

The cable pair which connects the customer to the switching center (ex: central office, main PABX). This path is called a loop because it is generally two wires out to the customer which are electrically tied together through the terminal set (device, instrument) when the device goes off-hook, creating a continuous path, or "loop."

Loopback

A type of diagnostic test in which a transmitted signal is returned to the sending device after passing through a data communications link or network. This test allows the comparison of a returned signal with the transmitted signal.

Loose Tube Cable

Type of cable design whereby coated fibers are encased in buffer tubes offering excellent fiber protection and segregation.

Loss

The portion of energy applied to a system that is dissipated and performs no useful work, or contributes to system impairments.

Low Frequency

A band of frequencies extending from 30 to 300 KHz in the radio spectrum, designated by the Federal Communications Commission.

M

Mutual inductance. See Inductance.

MA
Milliampere; one-thousandth of an ampere.

Macrobending
Macroscopic axial deviations of a fiber from a straight line, in cotrast to microbending.

Mainframe
Typically, a large-scale computer normally supplied complete with peripherals and software by a single, large vendor, often with a closed architecture. (Not recommended.) See OSI.

Main Ring Length (MRL)
The sum of the length of the intercloset cables, not including the short 2-ft. multistation access unit (MAU) patch cords connecting multiple MAUs together, that form a Token Ring. In multiple-closet MAU rings, the signal must travel through multiple MAUs, the cables connecting closets, the short 2-ft patch cords connecting the MAU segments in a closet and the lobe length (LL) without exceeding the maximum attenuation limit. See Horizontal Length (HL).

Main Terminal
See MPOE, equipment room, BDF.

Mbps
Megabits per second. One million units (bits) of information per second. As in binary language 0 or 1.
MC
Main cross-connect. Formerly MDF. See EIA/TIA 569.

MDF
Main distribution (distributing) frame, no longer used. Replaced by MC (main cross-connect). Location within the building that serves as the main cross-connect point between telco entrance cables, backbone or distribution cables (intra- and inter- building cables). May also serve as, or be co-located with main telecommunications equipment room, housing PABX, key, data, power or other associated equipment. May also serve as B.D.F. within the building. See MPOE, BDF, BET, entrance facility, network interface, etc.

MDPE
Abbreviation used to denote medium density polyethylene. A type of plastic material used to make cable jacketing.

Mechanical Splicing
Joining two fibers together by permanent or temporary mechanical means (vs. fusion splicing or connectors) to enable a continuous signal.

Media Access Control (MAC)
Refers to both the media access portion of the interface standard and the hardware and firmware which implements this portion of the standard.

Media Interface Connector (MIC)
A port connector also known as a data connector on a multistation access unit (MAU) in a Token Ring environment; also a dual-fiber connector for fiber distributed interface (FDDI).

Mega
Prefix = million.

Megabaud (Mbaud)
One million baud.

Megabit (Mb)
One million binary bits.

Megabyte (MB)
One million binary bytes.

Megahertz (MHz) One Million Hertz (Cycles).
A bandwidth-length product rating: Bandwidth is found by multiplying length by band-width-length equation.

Metal Oxide Varistor (MOV)
An electronic component that provides protection from voltage surges by absorbing them and then dissipating the energy as heat.

Metropolitan Area Network (MAN)
An extended LAN operating within a metropolitan area and providing integrated services for real-time data, voice and image transmission.

MFD
Microfarad; one-millionth of a farad. See Capacitance.

MGN
Multiground neutral system. A utility power system where the neutral conductor is continuously present along with the phase conductors. The neutral conductor is connected to earth periodically along its route, typically four times per mile. (This is not permitted in premises wiring systems, since the neutral conductor is only allowed to be grounded at the service entrance, or at the source of a separately derived system.) Not typically a telecommunications concern. See EIA/TIA 607.

MHO
Unit of conductance equal to the reciprocal of the unit of resistance (ohm).

MHz
Megahertz. Analog Frequency Spectrum Unit, one million cycles per second.

Micro
Prefix meaning one-millionth.

Microbending
Curvatures of the fiber which involve axial displacements of a few micrometers and spatial wavelengths of a few millimeters. Microbends cause loss of light and consequently increase the attenuation of the fiber.

Microfarad
One-millionth of a farad. This is the common unit for designating capacitance in electronics and communications.

Microinch
One millionth of an inch.

Micron
A micrometer; One-millionth of a meter.

Microphonics
Noise caused by mechanical excitation of a system component. In a single-conductor microphone cable, for example, microphonics can be caused by the shield rubbing against the dielectric as the cable is flexed.

Mil
One-thousandth of an inch.

Milli
Prefix meaning one-thousandth.

Minicomputer
A small or medium scale central computer designed to be accessed by dumb terminals. Compare with Microcomputer and Mainframe.

MIPS
Millions of instructions per second. A measure of processing power.

Modal Bandwidth
Bandwidth limited by modal dispersion inherent in multimode fiber optic cable.

Modal Dispersion
Dispersion resulting from the different transit lengths of different propagating modes in a multimode optical fiber.

Mode
A variable wave traveling in an optical fiber.

Mode Field Diameter
The diameter of the one mode of light propagating in a singlemode fiber. The mode field diameter replaces core diameter as the practical parameter in singlemode fiber.

Modem
A modulator/demodulator unit used for data transmission. It converts digital data into analog tones when transmitting over standard voice-grade telephone lines and reverses this process when receiving.

Modem Eliminator
A device used in place of the pair of modems normally needed to connect a local terminal and computer. It allows DTE to DCE data and control signal connections not easily achieved by standard cables and connectors.

Modular Jack
A female telecommunications interface connector. Modular jacks are typically mounted in a fixed locations and may have 4, 6 or 8 contact positions. Not all positions need be equipped with contacts. See also Telecommunications Outlets.

Modular Plug
A male telecommunications interface connector. Modular plugs may have 4, 6 or 8 contact positions. Not all positions may be equipped with contacts.

Modulation
Altering the characteristics of a carrier wave to convey information. Modulation techniques include amplitude, frequency, phase, plus many other forms of on-off digital coding.

Molding Raceway Method
A cable distribution method in which hollow moldings support cables. Small sleeves of pipe can be placed in the wall behind the molding to allow cable to pass through the wall.

Mono Filament
A single strand filament as opposed to multiple braided or twisted filaments.

Monochromatic
Consisting of a single wavelength. In practice, radiation is never perfectly monochromatic but, at best, displays a narrow band of wavelengths.

MPOE
Minimum point of entry. Frequently, this location is coincident with the BET, BDF and/or in close proximity to, or co-located with, the main telecommunications equipment room or MDF (aka network interface, BDF, BET, MDF, equipment room). See MC, MDF.

Multifiber Cable
An optical cable containing two or more fibers, each providing a separate information channel.

Multimode Fiber
An optical waveguide in which light travels in multiple modes. Typical core/ cladding size (measured in micrometers) is 62.5/125.

Multiplexer (MUX)
A MUX alternates the access of several data communication devices to a single communication line, such that the line is shared among the devices, but each functions as if it had sole access to the line.

Multimode Fibers
Optical fibers that have a large core (25 to 300 um) and that permit nonaxial rays or modes to propagate through the core. 62.5/125 mc is the common standard core for premises distribution systems.

Multiplexing
The process of combining multiple signals, usually by time-division multiplexing (TDM) on a high frequency carrier, to optimize the use of available transmission media. Also, see FDM.

Multistation Access Unit (MSAU)
A concentrator or transceiver for connecting nodes to a transmission medium.

Multiuser Outlet
A telecommunications outlet used to serve more than one work area, typically in open-system furniture applications.

Mutual Capacitance
The capacitance between two conductors when all other conductors, including the shield, are bonded to ground.

MV
Millivolt; one-thousandth of a volt.

MVS
Multiple virtual storage - An IBM mainframe operating system.

MW
Milliwatt; one-thousandth of a watt.
Nanometer
A unit of measurement equal to one billionth of a meter.

Nanosecond
One billionth of a second.

NAP
Network access point. See MPOE, Network Interface (NI), MDF, etc.

National Electrical Code (NEC)
A nationally recognized safety standard for the design, construction and maintenance of electrical circuits. The NEC, sponsored by the National Fire Protection Association (NFPA), generally covers electrical wiring within buildings.

NEMA 6P
National Electrical Manufacturers Association - Waterproof rating.

Neoprene
A synthetic rubber with resistance to oil, chemical and flame. Also called polychloroprene.

NetBIOS
The Network Basic Input/Output System in an IBM-developed LAN interface standard. Application programs on personal computers use NetBIOs for peer computer communications. Computers use this standard to communicate with remote resources attached to the LAN.

Network
An interconnection of computer systems, terminals or data/voice communications facilities.

Network Architecture
A formalized definition of the structure and protocols of a computer network.

Network Communication Cable (NCC)
Network communication cable, often called NCC, is generally used in the riser backbone subsystems. The cable consists of 24-AWG, annealed-copper conductors insulated with color-coded polyvinyl chloride (PVC) in twisted pairs, encased in an outer PVC jacket whose frictional properties permit it to be pulled in conduit. This type of cabling used to be referred to as direct inside wire (DIW). Not rated Category 5.

Network Connectivity
The topological description of a network which specifies the interconnection of the transmission nodes in terms of circuit termination locations and quantities.

Network Interface (NI) or Subscriber Network Interface (SNI)
The location of all connections at which any network channel, service or tariffed offering

is properly terminated in terms of: design, installation, maintenance parameters and a physical interface is provided for connection to the network.
Note: network interface can also mean the point of interconnection between one network and another, or a portion thereof (FED-STD-1037A).

NEXT
near end cross talk

Nibble
One half byte (4 bits).

Node
(a) in PABX architecture, a distributed location or remote cabinet/processor. (b) In Token Ring networks, a work station/device.

Noise
In a cable or circuit, any extraneous signal which tends to interfere with the signal normally present in or passing through the system. See NEXT.

NRZ
no return to zero

Null Modem
A device which allows the connection of two DTE devices by emulating the physical connections of a DCE device.

Numerical Aperture
The number that expresses the light-gathering point of an optical fiber.

NVP
Nominal velocity of propagation. Speed electrons travel relative to a percentage of the speed of light in a vacuum. Example: .70c expresses 70 percent of the speed of light.

Nylon
An abrasion-resistant thermoplastic with good chemical resistance.

Ohm
The unit of measurement of the volume resistivity of a cubic meter of material, as determined by measuring the DC resistance between any two opposite faces of the cube. For soil measurements, the resulting reading in ohms is the earth's resistivity for that soil. When earth resistivity is expressed in ohm/centimeters, convert to ohms by dividing by 100.

Ohm's Law
Stated $E = IR$, $I = E/R$ or $R = E/I$, the current, I in a circuit, is directly proportional to the voltage E, and inversely proportional to the resistance, R.

ONU
optical network unit

Open Architecture
An architecture that is compatible with hardware and software multiple vendors. See OSI.

Open System Interconnect (OSI)
A collection of international protocol standards for data networking. Multivendor/multi-product applications.

Optical Fiber
A thin filament of glass. Optical waveguide consisting of cladding and a core capable of carrying information in the form of light.

Optical Time Domain Reflectometer (OTDR)
An instrument that characterizes cable loss by measuring the backscatter and reflection of injected light as a function of time. It is useful for estimating attenuation and for locating splices, connections, anomalies and breaks.

Optical Waveguide Fiber
A transparent filament of high refractive index core and low refractive index cladding that transmits light.

OSP
outside plant

Outer Protection
An outer layer of material, composed of armored wire or metallic tape, covering the sheath of the cable. Specified when additional mechanical protection is required due to external factors such as gophers, squirrels, rocks or other site specific requirements. Also known as armored cable.

Output
The useful power or signal delivered by a circuit or device.

Overfloor Duct Method
A distribution method that uses metal or rubber ducts to protect and conceal exposed wiring across floor surfaces.

Ozone
Extremely reactive form of oxygen, normally occurring around electrical discharges and present in the atmosphere in small but active quantities. In sufficient concentrations, it can break down certain rubber insulations under tension (such as a bent cable).

Packet(s)
Groups of bits, including address, data and control elements, that are switched and transmitted together.

Packet Switching
A data transmission method whereby data is transmitted in packets through a network to a remote location. The packet switch sends packets from different data conversations along the best route available in any order. At the other end, the packets are reassembled to form the original message which is then sent to the receiving computer. Because packets need not be sent in a particular order, and go any route as long as they reach their destination, packet switching networks can choose the most efficient route and send the most efficient number of packets down that route before switching to another route.

Pair
Two wires, grouped (usually twisted) together and marked with reciprocal color coding.

Parallel Circuit
A circuit in which the identical voltage is presented to all components, with current dividing among the components according to the resistance or the impedance of the components. Side by side versus end to end.

Parallel Transmission
A method of transmission in which all bits of a character are sent simultaneously over separate lines to a high-speed printer or other locally attached peripheral. Contrast with Serial Transmission.

PASP
Polyethylene-aluminum-steel-polyethylene, the preferred sheath for protection of cable against damage by lightning, mechanical means or rodents.

Passive Device
A component of the broadband system which is not supplied with activating power.

Patch Cord
A short length of stranded copper wire or fiber optic cable with connectors on each end used to join communication circuits at a cross-connect.

Patching
Connecting circuits by means of cords with plugs inserted into appropriate jacks.

Patch Panel
A device, usually located in a telecommunications closet, in which temporary or semi-permanent connections can be made between incoming and outgoing lines. Used for modifying or reconfiguring a communications system or for connecting devices such as test instruments to specific lines.

PCC
Premises communication cable, CSA (Canadian Standards Association) cable designation.

PCE
pole mount cable enclosure

Peak
The maximum instantaneous value of a varying current or voltage.

Pedestal
An enclosure, usually mounted on the floor, which is used to house voice/data jacks or power outlets at the point of use. Also referred to as a monument, tombstone, above floor fitting or doghouse.

Peer-To-Peer Communications
The ability of programs and devices to communicate directly with one another, without passing communications through the mainframe.

Periodicity
Uniformly spaced variations in the insulation diameter of a transmission cable that results in reflections of a signal.

Peripheral Equipment
Equipment which itself has no on-line role but works closely with on-line equipment, e.g., printers, modems, sorters, etc.

Personal Computer (PC)
A computer for personal, single-user use, as opposed to mainframes or mini-computers, which are shared by many users.

Phase
An angular or sinusoidal relationship between two alternating quantities of energy.

Phase Modulation
One of three basic methods (see also Amplitude and Frequency Modulation) of adding information to a sine wave signal in which its phase is varied to impose information on it.

Phase Shift
A change in the phase relationship between two alternating quantities. See Phase.

Photodetector (Receiver)
Converts light energy to electrical energy. The silicon photo diode is most commonly used for relatively fast speeds and good sensitivity in the 0.75 um to 0.95 um wavelength region. Avalanche photodiodes (APD) combine the detection of optical signals with internal amplification of photo-current. Internal gain is realized through avalanche multiplication of carriers in the junction region. The advantage in using an APD is its higher signal-to-noise ratio, especially at high bit rates (transmission speeds).

Photodiode
A semiconductor diode that produces current in response to incident optical power and is used as a detector in fiber optics.

Photon
One quantum electromagnetical energy

PHY
Physical layer of the Fiber Distributed Data Interface (FDDI) standard. Also used to refer to the actual hardware used to implement the physical layer (PHY entity).

Physical Unit (PU)
Devices on an SNA network, such as controllers, that are physically rather than logically addressed by the host.

Pickup
Any device which is capable of transforming a measurable quantity of intelligence (such as sound) into relative electrical signals (e.g., a microphone) inductive or direct coupled.

Pico
Prefix meaning one-millionth of one-millionth.

Picofarad (pF)
A unit of capacitance used to designate capacitance unbalance of the two wires of a pair relative to ground. One picofarad equals one trillionth of a farad.

Pigtail
Optical fiber cable that has a connector installed on one end.

Pin
A conductor on a plug or connecting device/apparatus.

Pin Diode
A device used to convert optical signals to electrical signals in a receiver.

Plastic
High polymer substance, including both natural and synthetic products, yet excluding rubber that is capable of density flowing under heat and pressure.

Plasticizer
A chemical added to plastic to make them softer and more pliable.

Plenum
A return air space inside buildings through which environmental air is handled.

Plenum Cable
Cable specifically designed for use in a plenum. Plenum Cable has insulated conductors often jacketed to give them low flame-spread and low smoke-producing properties.

Plug
A device used for connecting conductors to a jack. It is typically used on one or both ends of equipment cords or on wiring for interconnects or cross-connects.

PMD
Physical medium dependent. Determines the specifications for transmitters and receivers, cables, connectors and bypass switches.

Point-to-Point Transmission
An uninterrupted connection between two pieces of equipment (link/channel).

Poke-Through Method
A ceiling distribution system method that involves drilling a hole through the floor from the ceiling space below and poking cables through to terminals. Not Recommended.

Polybutadene
A type of synthetic rubber often blended with other synthetic rubbers to improve their properties.

Polyethylene
A thermoplastic material having excellent electrical properties.

Polymer
A substance made of specific repeating chemical units or molecules. The term polymer is often used correctly or incorrectly, in place of plastic, rubber or elastomer.

Polypropylene
A thermoplastic similar to polyethylene but stiffer and having a higher softening temperature.

Polyurethane
Broad class of polymers noted for excellent abrasion and solvent resistance. Can be in solid or cellular form (formed/expanded).

Polyvinyl Chloride (PVC)
A multipurpose thermoplastic used for wire and cable insulation and jackets.

Polyvinylidene Difluoride (PVDF)
A fluoropolymer material that is resistant to heat and is widely used in the jackets of plenum cable.

Ports
Terminations in equipment systems at which various types of communication devices, switching equipment, and other devices are connected to the transmission network.

POTS
plain old telephone service

Potting
Sealing by filling with a substance to exclude moisture intrusion.

Power
The amount of EMF/energy per unit of time. Usually expressed in watts.

Power Arrestor
A protection device used on power lines to limit ground surge voltage due to lightning, while simultaneously interrupting power "follow-on" (the discharge of normal power).

Power/Communication Pole
A raceway placed between the ceiling and floor used in conjunction with a ceiling distribution system for the purpose of distributing communication and power service to a work area. Also called Utility Column or Ceiling Drop Pole.

Power Loss
The difference between the total power delivered to a circuit, cable or device and the power delivered by that device to a load.

Power Ratio
The ratio of power appearing at the load to the input power, usually expressed in dB.

Preform
A glass structure from which an optical fiber waveguide may be drawn.

Prefusing
Fusing with a low current to clean the fiber end. Precedes fusion splicing.

Premises Distribution System (PDS)
The transmission network inside a building or group of buildings that connect various types of voice and data communication devices, switching equipment, and information management systems together, as well as to outside communications networks. It includes the cabling and connecting hardware components and facilities between the point where building wiring connects to the outside network lines, back to the voice and data terminals in the office or other work locations. The system consists of all the transmission media and electronics, administration points, connectors, adapters, jacks, plugs, and support hardware between the building's side of the network interface and the terminal equipment required to make the system operational.

Premise Wiring/Structured Wiring (Premise Cabling/Structured Cabling)
The entire wiring system on the user's premises used for transmission of voice, data and video.

Prewiring
Wiring installed before walls and ceilings are enclosed or finished, and in anticipation of future use or need. It is more cost effective to prewire all potential locations at the time of a major installation, rather than return on a repeated basis to keep adding locations one (or more) at a time.

Primary Coating
The plastic coating applied directly to the cladding surface of the fiber during manufacture to preserve the integrity of the surface.

Primary Power
Voltages >300V RMS to ground.

Primary Rate Interface (PRI)
ISDN standard interface comprising 23 "B"+1 "D" channel for North America, and 30 "B"+1 "D" Channel for Europe. See Basic Rate Interface (BRI) and Integrated Services Digital Network (ISDN).

Primary Ring
One of two rings of the dual counter-rotating ring architecture. This ring is designated as the primary ring because it is the default ring that single-media access control (SMAC) stations connect to unless instructed to do otherwise.

Printed Circuit
A copper foil circuit formed on one or both faces of an insulating board to which circuit components are soldered. The copper foil pattern serves to connect components and is produced either by etching or plating.

Printer Emulation
A feature of 3270 emulation that makes the printer attached to a PC appear to be an IBM 3270 printer.

Private Branch Exchange (PBX) or Private Automatic Branch Exchange (PABX)
A private telephone switching system, usually located on a customer's premises connecting a common group of lines from one or more central offices to provide service to a number of individual phones. Now used interchangeably with PABX (private automatic branch exchange).

Propagation Delay
Time required for a signal to pass from the input to the output of a device. See NVP.

Protector
Device used to limit damaging foreign voltages and currents on metallic telecommunications conductors and equipment.

Protector (Cable)
Cable protectors limit the voltage between the conductors and shield of a cable. Standard cable protectors are equipped with 6-mil carbon electrodes. Standard cable protectors limit voltages to 800V DC which may not address today's low-voltage requirements.

Protector (Unit)
A device to protect against either overvoltage, overcurrent or both. The unit may contain carbon electrodes, gas tubes, solid state components, heat coils, fuses or a combination thereof. Units may be integrated or have plug-in/screw-in elements depending on the

application and design. Used with, or in, protector blocks, protected terminals, connecting blocks and central office connectors as well as PABX and other devices/systems.

Protector Module
A device that limits voltage between telecommunications conductors and ground. The protector is equipped with 3 mil carbon electrodes or equivalent gas tubes. Typical line protectors limit voltage to 350V DC. See Protector (Cable).

Pseudo Random NRZ
A wave form of binary signals that may be used in a computer system. It is called NRZ, non-return to zero, because the voltage does not return to zero. Also, non return to zero, inverted (NRZI).

Public Data Network
A network established and operated for the specific purpose of providing data transmission services to the public through IXCs/IECs.

Public Switched Network
Any common carrier network that provides circuit switching between public users, such as the public interactive telephone network.

Pulling Tension
The amount of pull, measured in pounds or foot-pounds, placed on a cable during installation.

Pulse
A current or voltage which changes abruptly from one value to another and back to the original value in a finite length of time. Used to describe one particular variation in a series of wave motions.

Pulse Code Modulation
The most common method of representing an analog to digital signal, such as speech, by sampling at a regular rate and converting each sample to an equivalent digital code.

Pulse Spreading
The dispersion of an optical signal with time as it propagates through an optical fiber.

Punch Down
See Cut Down.

PVC
Polyvinyl chloride, widely used in cable sheaths/jackets and conduits.

QA/QC
Quality assurance/quality control. Should be clearly defined in contract terms and conditions.

QFLC
quad fiber loop converter (four T-1s)

Quad Fiber Cable
A type of fiber optic cable that has four single cables enclosed in an extruded jacket of polyvinyl chloride (PVC), with a rip cord for pulling back the jacket to access the fiber.

Quick Clip
An electrical contact used to provide an insulation displacement connection to telecommunications cables.

R
Symbol for resistance (ohms).

Raceway
Examples of raceways include, but are not limited to:
1. Conduit (rigid or flexible, metallic or nonmetallic) EMT, a thinwall electrical metallic tubing
2. Sleeves, slots, cores or auxiliary channels (gutters)
3. Baseboard (concealed) systems
4. Underfloor systems. Cellular floor systems
5. Cable trays, troughs, ladder racking
6. Busways, surface raceways, lighting fixture raceways and latch duct, nonmetallic or metallic. There are other raceways/ pathways that may need to be considered.

*Note: Flexible metallic (flex) conduit should never be a consideration, recommendation or option for communications.

Rack
A vertical or horizontal open support, usually made of aluminum or steel, that is attached to a floor, ceiling or wall. Cables are laid in and fastened to the rack and connected to the equipment(s).

Radio Frequency
The frequencies in the electromagnetic spectrum that are used for radio communications.

Raised Floor Method
A floor distribution method in which square, steel or wood-laminated plates resting on aluminum locking pedestals are attached to the building floor. Also called access floor, since each plate can be removed for easy access to cables below.

Random Access Memory (RAM)
A semiconductor storage device in which data can be entered, read and erased. RAM is the fastest form of data storage and retrieval, however, the data is erased when the power is turned off.

Rayleigh
English physicist (1842-1919), nobel prize winner

Rayleigh Scattering
The scattering of light that results from small inhomogeneities in material density of composition.

RBOC (Regional Bell Operating Company)
One of the seven Bell operating companies that were formed during the divestiture of AT&T.

RCDD
Registered communications distribution designer. Certification program for telecommunications industry personnel planning to consult or design. Requires documented experience, knowledge and testing to obtain status.

Reactance
A measure of the combined effects of capacitance and inductance on an alternating current. The amount of such opposition varies with the frequency of the current. The reactance of a capacitor decreases with an increase in frequency; the opposite occurs with inductance.

Real Time
A form of information processing where output is generated nearly simultaneously with the corresponding input. Used mostly where the results of the computation are used to influence a process while it is occurring.

Receiver
An electronic fiber package that converts light energy to electrical energy in a fiber optic system.

Reflectance
Reflectance is the ratio of power reflected to the incident power at a connector junction or other component junction, usually measured in decibels or dB. Reflectance is stated as a negative value, e.g., -30 dB. A connector that has a better reflectance performance would be a -40 dB connector or a value less than -30 dB. The term return loss, back reflection and reflectivity are also used synonymously in the industry to describe device reflections, but stated as positive values.

Reflection
The change in direction (or return) of waves striking a surface. For example, electromagnetic energy reflections can occur at an impedance mismatch in a transmission line, causing standing waves.

Refractive Index
The ratio of light velocity in a vacuum to its velocity in the transmitting medium.

Refraction
The bending of a beam of light at an interface between two dissimilar media or a medium whose refractive index is a continuous function of position (graded index medium).

Regenerator
A device which restores a signal to its original undistorted quality, including amplitude, waveshape and timing.

REH
remote electronic hubs

Remote Terminal
A terminal that is physically removed from the host but connected to it by a communications link such as a phone line. Ex: distributed node, PABX.

Repeater
A device inserted at intervals along a circuit to boost and amplify a signal being transmitted. Repeater may also regenerate a digital signal - squaring it and cleaning it up - but not changing it. Regenerating the signal removes noise and thus reduces the likelihood of error.

Requirements Survey
The systematic study of a building or campus of buildings to determine the needs for voice and data telecommunications equipment and distribution media. This is normally done prior to designing the system for the site.

Resistance
The property of a conductor that determines the current produced by a given potential difference. It impedes the flow of current and results in the dissipation of power as heat. Resistance is measured in ohms.

Resonance
An ac circuit condition in which inductive and capacitive reactance interact to cause a minimum or maximum circuit impedance.

Response Time
The time it takes a system to react to a given input. The response includes the transmission time, the processing time, the time for searching records and the transmission time back to the originator.

Retractile Cord
A cord having specially treated insulation or jacket so it will retract like a spring. Retractibility may be added to all or part of a cord's length.

Return Loss
Noise or interference caused by impedance discontinuities along the transmission line at various frequencies. Return loss is expressed in decibels.

RF
radio frequency

RFI
(a.) request for information. (b) radio frequency interference. A disturbance in the reception of signal transmission due to conflicting undesired signals, either through induction, radiation or, less frequently, unbalanced line conditions or poor circuit design. See EMI.

RFP
request for proposal

RFQ
request for quotation

RG/U
RG is the military designation for coaxial cable, and U stands for universal.

Ribbon Fiber Cable
A cable that contains one to 12 ribbons, with each ribbon having 12 fibers for a cable size range of 12 to 144 fibers. Ribbon fiber cables are designed for use in large distribution systems where small cable size and high pulling strength are important.

Ribbon Riser Cable
An optical fiber, nonconductive, riser (OFNR)-rated premises cable containing optical fibers possibly in ribbons.

Ring Circuit
Ring-shaped cable network with half of the circuits going in each direction so a cut at any point will still leave all locations with limited access to the central office.

Ring Conductor
A telephony term used to describe one of the two conductors in a cable pair used to provide telephone service. This term was originally coined from its position as the second (ring) conductor of a tip-ring-sleeve switchboard plug.

Ring In (RI)
Inward port of connecting multistation access units (MAUs).

Ring Out (RO)
Outward port of connecting multistation access units (MAUs).

Riser
The conduit or path between floors of a building into which telephone and other utility cables are placed to bring service from one floor to another.

Riser Backbone System
The part of a premises distribution system that includes a main cable route and structure for supporting the cable from an equipment room (often in the building basement) to the upper floors, or along the same floor, where it is terminated on a cross-connect in a riser telecommunications closet, at the network interface, or at distribution components of the campus backbone subsystem.

Riser Telecommunications Closet
The closet where riser backbone cable is terminated and cross-connected to either horizontal cable or to other riser backbone cable. The riser telecommunications closet houses cross-connect facilities, and may contain auxiliary power supplies for terminal equipment located at the user work area.

RMS
root mean square

Rope Strand
A conductor composed of groups of twisted strands.

Router
A router can be used to connect networks with similar protocols (802.5 Token Ring LANs) or dissimilar open system interconnect (OSI) model protocols (802.5 Token Ring LANs and X.25 packet switching networks). Routers are more sophisticated than bridges and can be used to prevent some of the speed mismatch, security and reliability problems that occur in large networks.

RS232-C
A set of standards specifying various electrical and physical characteristics for interfaces between computers, terminals and modems. The RS-232-C standard was developed by the Electronics Industries Association (EIA), and defines the mechanical and electrical characteristics for connecting DTE and DCE data communications devices. It defines what the interface does, circuit functions and their corresponding connector pin assignments. The standard applies to both synchronous and asynchronous binary data transmission. The traditional RS-232-C plug is functionally equivalent to CCITT V024/V.28.

RS-442 and RS-443
Both are EIA recommended standards for cable lengths that extend the RS-232-C 50 foot limit and describe the electrical characteristics of balanced-voltage and unbalanced-voltage digital interface circuits.

RS-449
An EIA recommended standard for the mechanical characteristics of two connectors (a 37-pin connector and a 9-pin connector). Designed for higher speeds not widely used yet.

Rubber (Wire Insulation)
A general term used to describe wire insulations made of thermosetting elastomers, such as natural or synthetic rubbers, neoprene, butyl rubber and others.

SAI
service area interface

Satellite Cabinet
Surface-mounted or flush-type wall cabinets for housing circuit administration hardware. Satellite cabinets, like satellite telecommunications closets, supplement riser telecommunications closets by providing additional facilities for connecting horizontal cables from information outlets in user work areas. Sometimes referred to as a satellite location. No longer recommended terminology. See TC.

Satellite Telecommunications Closet
A walk-in or shallow wall closet that supplements a riser telecommunications closet by providing additional facilities for connecting riser backbone to horizontal cables from information outlets. Also referred to as a satellite location. No longer recommended terminology. See TC.

SBR
A copolymer of styrene and butadene. Also GR-S or Buna-S. Most commonly used type of synthetic rubber.

Scattering
A property of glass that causes light to deflect from the fiber and contributes to optical attenuation.

Schedule 40
PVC (polyvinyl chloride) conduit typically used for underground entrance facilities. Refer to local governing codes and conditions.

SDIP
standardized dedicated inside plant
Secondary Power
Power operating at <300V RMS to ground. See Primary Power.

Secondary Protection
Supplemental. Auxiliary. A secondary protector installed in series with the indoor communications cable between the primary protector and the equipment. The secondary protector must provide overcurrent protection which will safely fuse at currents less than the current carrying capacity of listed:

Indoor communications wire and cable; telephone set line cords, patch cords, etc.; communications terminal equipment having ports for external wire line circuits.

Secondary Ring
One of the two rings of the dual counter-rotating ring architecture. Designated the secondary ring because single-media access control (SMAC) stations connect in the default mode to the primary ring. The secondary ring is used for data transport by dual-media access control (DMAC) stations, SMAC stations commanded to move to the secondary ring, or in the case of ring wraps, by all stations.

Semiconductor
In cable industry terminology, a material possessing electrical conductivity that falls somewhere between that of conductors and insulators. Usually made by adding carbon particles to an insulator. Not the same as semiconductor materials such as silicon, germanium, etc., used for making transistors and diodes.

Sensitivity
For a fiber optic receiver, the minimum optical power required to acheive a specified level of performance, such as BER.

Separator
Pertaining to wire and cable, a layer of insulating material such as textile, paper, etc., which is placed between a conductor and its dielectric, between a cable jacket and the components it covers, or between various components of a multiple-conductor cable. It can be utilized to improve stripping qualities, flexibility, or can offer additional mechanical or electrical protection to the components it separates.

Serial Transmission
A method of transmission in which data is sent one bit at a time, in contrast with a parallel transmission, in which multiple bits (usually eight) are sent simultaneously.

Series Circuit
A circuit in which the components are arranged end to end to form a single path for current.

Service Clearance
The space encompassing the equipment, or unit, which is required to permit proper working room for operating, inspecting and servicing equipment. This space should adequately allow:
> Doors to be fully opened.
> Component drawers to be pulled out, or racks opened.
> Allows safe work operations. Min. 36".

Session
A logical connection with a host system. The session begins when you establish the communications link and ends when you terminate emulation and return to DOS.

Sheath
A common term for the jacketing of twisted pairs in multipair cable.

Sheave
The grooved wheel or pulley used to assist in pulling cable through a bend in the routing; especially used in underground installations between manholes.

Shield
The metallic layer that surrounds insulated conductors in shielded cable. The shield may be the metallic sheath of the cable or the metallic layer inside a sheath.

Shield Coverage
The physical area of a circuit or cable actually covered by shielding material often expressed as a percentage.

Shield Effectiveness
The relative ability of a shield to screen out undesirable interference. Frequently confused with the term shield coverage.

Shielding
A metallic layer used to reduce EMI, RFI, noise, emissions or absorption. Also, the reduction of undesirable effects on circuits caused by electrostatic fields (FED-STD-1037A).

Signal
Any visible or audible indication which can convey information. Also, the information conveyed through a communications system.

Signal to Noise Ratio
The ratio of the received optical power, with fill signal averaging, divided by the noise floor for the detector.

Silicone
General Electric trademark for a material made from silicone and oxygen. Can be in thermosetting elastomer or liquid form. The thermosetting elastomer form is noted for high heat resistance.

Simplex Transmission
Data transmission over a circuit capable of transmitting in one preassigned direction only.

Single-Attached Stations (SAS)
A station type defined in the fiber distributed data interface (FDDI) stations management (SMT) standard, an SAS has a single physical medium dependent/physical layer (PMD/PHY) interface. SASs are less expensive than dual-attached stations (DASs) because of other reduced electronics and optics, but they do not directly gain the reliability advantages of dual rings.

Single-ended
Unbalanced, such as grounding one side of a circuit or transmission line.

Single-Fiber Cable
A plastic-coated fiber surrounded by an extruded layer of polyvinyl chloride (PVC), encased in a synthetic strengthening material, and enclosed in a PVC sheath.

Single Media Access Control (SMAC)
Fiber distributed data interface (FDDI) stations with a single media access control (MAC) entity. SMAC stations can logically send and receive data over a single ring of the dual-ring architecture at any given time. SMAC stations have access to 100 Mbps bandwidth from the network.

Singlemode Fiber
A fiber wave guide in which only one mode will propagate. The fiber has a very small core diameter of approximately 8 um. It permits signal transmission at extremely high bandwidths and is generally used with laser diodes.

Sintering
Melting of powder into solid state

Sinusoidal
Varying in proportion to the sine of an angle or time function. Ordinary alternating current is sinusoidal.

Skew Rays
A ray that does not intersect the fiber axis. Generally, a light ray that enters the fiber core at a very high angle.

Skin Effect
The tendency of alternating current to travel on the surface of a conductor as its frequency increases.

Slab on Grade
A concrete floor place directly on the soil without a basement or crawl space.

SLC
subscriber loop carrier or subscriber line concentrator

Sleeve
A metallic section of conduit (typ.) that extends above the floor line (after coring) at least one (1) inch, and extends into the space below as required by local fire codes. Review local codes to determine exact dimensions and firestopping practices.

Slot
An opening, usually rectangular (typ. 6" x 9"), through floors, ceilings or walls that accommodate placement of cable and wiring. See Firestopping. Refer to NFPA Articles, and local ordinances.

SMA 905/906 (Subminiature Type A)
A threaded type fiber optic connector. The 905 version is a straight ferrule design, whereas the 906 is stepped ferrule design. See EIA/TIA standards for connector 568SC.

SNA
System network architecture is a synchronous communications protocol designed by IBM that provides an interface between components in a computer network.

Sneak Current
A foreign current flowing to ground through terminal wiring and equipment that is driven by a voltage that is too low to cause a protector to operate.

Sneak Current Protection
The use of devices to protect against sneak currents either by interrupting the current (fuses) or grounding the conductor (heat coils).

SONET
Synchronous optical network; provides broadband connectivity for optical networks on a global scale.

Source
The device (usually LED or laser) used to convert an electrical information-carrying signal into a corresponding optical signal for transmission by an optical fiber.

Source Routing
A bridge uses source routing when the route to be followed is carried within each frame by the source stations. The source station acquires and maintains information by a search process, allowing parallel bridges to exist and to share traffic between the same two rings.

Spectral Bandwidth
Frequencies that exist in a continuous range and have a common characteristic. A spectrum may be inclusive of many spectrums (e.g. electromagnetic radiation spectrum includes the light spectrum, radio spectrum, infrared spectrum, etc.).

Speed of Light (c)
186,000+ mi. per second. See NVP.

Splice
The physical joining of two or more copper conductors or optical fibers to form a continuous circuit/conductor.

Splined Ceilings (Variation of False Ceilings)
A ceiling construction method where ceiling tiles are usually 12 in. x 12 in. and are lipped and locked together using runners or T-bars, and thin strips of metal called splines. Cable is run in the area above the tiles. Aka: "Locktile" ceilings.

Splice Closure
A container used to organize and protect splice trays. Typically used in outside plant environments.

Splice Tray
A container used to secure, organize and protect spliced fibers.

Stabilized Light Source
An LED or laser diode that emits light with a controlled and constant spectral width, central wavelength, and peak power with respect to time and temperature.

Stalpeth
Steel-aluminum-polyethylene, the primary sheath for underground cable.

Standard Protection
Also, primary protection. the minimum basic protection required on all exposed facilities to comply with NEC requirements.

Standing Wave Ratio (SWR)
A ratio of the maximum amplitude to the minimum amplitude of a standing wave stated in current or voltage amplitudes.

Star Topology
A network interconnection scheme in which one central location has links to all other nodes, which have no direct connections to each other. See examples in workbook.

Static Charge
An electrical charge that is bound to an object. An unmoving electrical charge.

Station Field
The field used in telecommunications closets or equipment rooms to connect horizontal cabling to stations via work area outlets. Station fields are blue.

Station Management (SMT)
The portion of the fiber distributed data interface (FDDI) standard that specifies ring monitoring, ring recovery, logical topologies and administration of station entities.

Stay Cord
A component of a cable, usually of high tensile strength, used to anchor the cable ends at their points of termination and keep any pull on the cable from being transferred to the electrical conductors.

Step-Index Fiber
An optical fiber in which the core is of a uniform refractive index with a sharp decrease in the index of refraction at the core/cladding interface.

STP
Shielded twisted pair typically (today) refers to 150 ohm, 22 AWG, 2 pair STP-A. See EIA/TIA 568A.

Strain Gauge
A device for determining the amount of strain (change in dimensions) when a stress is applied.

Straight-Tip (ST) Connector
(Orig. AT&T). A fiber optic connector used to join single fibers together at interconnects or to connect them to fiber optic cross-connects. See EIA/TIA 568SC recommended standard connector.

Stranded Cable
A strong woven-steel cable used to support cable in aerial distribution systems. The cable is lashed to the stranded cable during installation.

Strength Member
That part of a fiber optic cable composed of aramid yarn, steel strands or Fiberglas filaments that increase the tensile strength of the cable.

Stub Cable
A short cable (usually 25 ft. or less) that extends from a cable terminal, protector or block and is used to splice incoming cable connections to such devices.

Subminiature D Connector
A family of multipin data connectors used in RS-232-C communications. The connectors are available in 9, 15, 25 and 37 pin configurations. Sometimes referred to as DB9, DB15, DB25 and DB37 connectors respectively.

Support Hardware
The racks, clamps, cabinets, brackets, trays and other equipment that provides the physical means to attach the transmission media and connecting hardware to walls and ceilings, or in outside plant, ducts, manholes, vaults, poles, pullboxes, etc.

Surge
A temporary and relatively large increase in the voltage or current in an electric circuit or cable. Also called transient.

Surge Suppression
The process by which transient voltage (surges) are prevented from reaching sensitive electronic equipment.

Surface Raceway
A cable distribution method in which channels containing cables are run along or within the baseboards of a building.

Suspended Ceilings
A ceiling construction method where ceiling tiles are suspended by wires and T-bar. Cable is run in the area above the tiles. Aka: false ceilings.

Sweep-test
Pertaining to cable, the frequency response is verified by generating an RF signal whose frequency is swept repeatedly through a given frequency range at a rapid constant rate. The cable response is observed on an oscilloscope. The structural return loss sweep-test measures the magnitude of internal cable reflections.

Synchronous Data Link Control (SDLC)
The protocol IBM defined for synchronous data communications within SNA networks.

Synchronous Transmission
Transmission in which the data characters and bits are transmitted at a fixed rate with the transmitter and receiver being synchronized. Compare with Asynchronous Transmission.

System-Common Equipment
The equipment on premises that provide functions common to terminal devices such as telephone, data terminals, integrated workstation terminal and personal computers. Typically, the system-common equipment is the private automatic branch exchange (PABX) switch, data packet switch or central host computer.

T1
A digital transmission link with 1.544 Mbps bandwidth. T1 operates on two twisted pairs and can handle 24 voice channels, each digitized at 64 Kbps. More voice channels are available with advanced digital encoding techniques.

T1 Carrier
The digital transmission system which transmits data at 1.544 Mbps. See T1.

Tap
In communications, tap is an electrical connection permitting signals to be transmitted onto or off a bus. The link between the bus and the drop cable that connects the workstation. In power, a tap is an intermediate point in an electric circuit where a connection may be made.

Telecommunications
The transmission and reception of electrical or optical signals by copper wire, optical fiber or electromagnetic means. Encompasses all forms of transmitted intelligence, wire or wireless.

Telecommunications Closet (TC)

A space (formerly known as floor closet, IDF, satellite closet or other terms), in a building that is set aside to provide a safe, secure and environmentally suitable area for the installation of cables, wires, telecommunications equipment and/or termination and administration systems.

Telecommunications Service Entrance

The point where regulated telecommunications cables enter the building or property. See Entrance Facility.

Terminal

A device that provides a user interface to the host.

Terminal Block

An assembly, premanufactured to accept 22-26 AWG conductors, allowing termination and cross-connect administration, test points and/or fusing and protection where required.

TFE

Tetrafluoroethylene. A thermoplastic material proven to have excellent electrical insulating properties with chemical and heat resistance.

Thermal Rating

The temperature range in which a material will perform its function without undue degradation.

Thermoplastic

A plastic material that softens and flows when heated and becomes firm when cooled. This process can be repeated.

Thermoset/Thermosetting

A plastic material that is crosslinked by a heating process known as curing. Once cured, thermosets cannot be reshaped.

Thin Ethernet

An Ethernet LAN or IEEE 802.3 LAN which uses smaller diameter coaxial cable than standard Ethernet. Thinnet, cheapnet, etc.

TIA

Telecommunications Industry Association.

TIC

Token Ring Interface Coupler - A device that provides the connection between a host computer or front-end processor and a Token Ring LAN.

Tight-Buffered Cable

Type of cable construction whereby each glass fiber is tightly buffered by a protective thermoplastic coating to a diameter of 900 micrometers. Increased buffering provides ease of handling and connectorization.

Time Division Multiplexing (TDM)

A technique for combining multiple signals on a single circuit by interleaving bits or bytes of data from successive channels.

Timesharing
A method of computer operation that allows many users to use one computer. Due to the power and speed of the computer, it appears as though the users are served simultaneously, when in fact they are being served in sequence.

Tinsel
A type of electrical conductor comprised of a number of tiny threads, each thread having a fine, flat ribbon of copper or closely spiraled about it. Used for small size cables requiring flexibility and extra-long life.

Tip Conductor
A telephony term used to describe the positive conductor of a pair. This term was originally coined from its position as the first (tip) conductor of a tip-ring-sleeve switchboard plug.

TNC
A threaded connector used to terminate coaxial cables. TNC is an acronym for threaded Neill-Concelman.

Token Passing
A special data sequence that is continuously sent around the ring. The term token represents permission to transmit from one station to its downstream neighbor.

Token Ring
A baseband LAN access method with a ring topology supporting a four- to 16-Mbps data rate. Network access is controlled by passing a token to attached devices. Only the device in possession of the token can communicate over the network at one time.

Topology
The physical or logical configuration of a local area network (star, ring, bus). Used in telecommunications to describe the ways in which different facilities and services are cabled, and how the addition of new devices can most readily be accomplished (prewired).

Transceiver
A single device capable of both transmitting and receiving information.

Transducer
A device for converting mechanical energy to electrical energy.

Transfer Impedance
For a specified cable length, transfer impedance relates a current on one surface of a shield to the voltage drop generated by this current on the opposite surface of the shield. Transfer impedance is used to determine shield effectiveness against both ingress and egress of interfering signals. Cable shields are normally designed to reduce the transfer of interference-hence, shields with lower transfer impedance are more effective than shields with higher transfer impedance.

Transient
An abrupt change in voltage, of short duration, which may cause signal impairments, loss of memory or physical damage to equipment. See Surge.

Transition Point (TP)
An enclosure used to house the transition connection between flat undercarpet cabling and the traditional round cable feeding from the equipment closet.

Transmission Distance
The actual length of the path from the transmitter of one node to the receiver of the next downstream node. The maximum transmission distance is determined by the maximum signal loss (attenuation limit) that can be withstood between any transmitter and receiver.

Transmission Electronics
Any of the various devices used with different transmission media to convert from one transmission method to another. Transmission electronics may typically include multiplexing equipment and asynchronous data units (ADUs).

Transmission Line
An arrangement of two or more conductors or an optical fiber used to transfer signal energy from one location to another.

Transmission Loss
The reduction in power between any two points in a telecommunications system.

Transmission Media
The various types of copper wire and fiber optic cable used for transmitting voice, data or video signals.

Transmitter (optical fiber)
The electronic package that converts electrical energy to light energy in a fiber optic system, laser or LED.

Transport Control Protocol/Internet Protocol (TCP/IP)
A common network layer and transport layer data networking protocol.

Triboelectric Noise
Noise generated in a shielded cable due to variations in capacitance between the shield and conductor as the cable is flexed.

Trunk
Typically, a communication link between two switching systems. The term switching includes equipment in a central office (of the telephone company) and PABXs. A tie trunk connects PABXs. Central office trunks connect a PABX to the switching system at the central office and/or link other central offices together.

Turn-Key
A contractual arrangement in which one party designs and installs a system and turns over, in its entirety, to another party who will operate the system, after acceptance.

Twinaxial Cable
Two insulated conductors inside a common insulator, covered by a metallic shield and enclosed in a cable sheath. See Coaxial.

Twin-lead
A transmission line having two parallel conductors separated by insulating material. Line impedance is determined by the diameter and spacing of the conductors and the insulating material is usually 300 ohms for television receiving antennas. Aka: flat lead.

Twisted Pair
Two or more insulated wires twisted together. The twists, or lays, are varied in length to reduce the potential for signal interference between pairs. In cables greater than 25 pair,

the twisted pairs are grouped and bound together in a common binder group. Twisted pair is the most common type of transmission media. Formerly referred to as direct inside wire (DIW). See Unshielded Twisted Pair (UTP).

Two-Wire Circuit
A circuit in which information signals in both directions are carried by the same two-wire path or loop.

UHF
Ultrahigh frequency. The spectrum extending from 300 to 3000 MHz as designated by the Federal Communications Commission.

UL Approved
Tested and approved by the Underwriters Laboratories Inc.

Unbalanced Circuit
A two-wire circuit with legs which differ from one another in resistance, capacity to earth or to other conductors.

Unbalanced Line
A transmission line in which voltages on the two conductors are unequal with respect to ground (e.g., a coaxial cable).

Undercarpet Wiring
A cable distribution method which uses flat cables placed beneath carpeting to provide voice, data video and power services to open office workstations. Difficult to install, administer or troubleshoot.

Underfloor Duct Method
A floor distribution method using a series of metal distribution channels, often embedded in concrete, for placing cables. This method uses one or two levels, depending on the complexity of the system. Sometimes referred to as underfloor raceways.

Underground Cable
Not "direct buried" but in support structures beneath the surface of the earth as in conduit, duct, ductbanks or other approved structures that isolate the cable from direct contact with earth and/or foreign power/EMI/RFI influences. Considered exposed to elements such as lightning, commercial power "hits", etc. (FED-STD 1 037A).

Underground Distribution Method
The method of running cable underground between buildings in campus systems by going through buried conduits.

Underwriters Laboratories (UL)
A private testing laboratory concerned with electrical and fire hazards of equipment.

Uniform Service Order Code (USOC)
Bell System term used on Universal System service orders and to denote varying pin configurations on registered jacks (RJs).

Unilay
A conductor with more than one layer of helically laid wires with the direction of lay and length of lay the same for all layers.

Unshielded Twisted Pair (UTP)
Copper cable with no foil or metallic/braid shielding, capable of high-speed voice and data transmission. Techniques exist to address the signal impairments due to the transmission characteristics of copper media and to limit radiated emissions. Standard predominant telecommunications media. See UTP-categories.

UTP-Categories
Category One - Intended for basic communications power limited circuit cable. There is no performance criteria specified in EIA/TIA 568A (UL Level One.)

Category Two - Low performance UTP. Typical applications include voice and low-speed data (IBM Type 3.) Not specified in 568A (UL Level Two.)

Category Three - Applies to UTP cables and associated connecting hardware with transmission characteristics up to 16 MHz. Typical applications include ARCNet 2.5 Mbps, 4 Mbps Token Ring, 10BASE-T and 100BASE-VG

Category Four - Applies to UTP cables and associated connecting hardware with transmission characteristics up to 20 MHz. Typical applications include 16 Mbps Token Ring and low loss 10BASE-T.

Category Five - Applies to UTP cables and associated connecting hardware with transmission characteristics up to 100 MHz (155 Mbps). Typical applications include 16 Mbps Token Ring, TP-PMD and CDDI and, of course, other (future) ATM.

V
volt

VA
Volt-ampere. A designation of power in terms of voltage and current.

Value Added Network (VAN)
A data communications network leased from a common carrier with extra equipment, such as an interface computer with a database storage designed to provide additional services.

Velocity of Propagation, also Nominal Velocity of Propagation (NVP)
The transmission speed of electrical energy in a length of cable compared to the speed of light in free space (vacuum). Usually expressed as a percentage, e.g. .65c, .70c, etc.

VHF
Very high frequency. The spectrum extending from 30 to 300 MHz as designated by the Federal Communications Commission.

Video
Pertaining to visual information in an integrated system.

Video Display Terminal (VDT)
A display screen terminal that permits the viewing of text or graphics for operator manipulation. A cathode ray tube (CRT) terminal.

VLF
Very low frequency. The spectrum extending from 10 to 30 KHz, as designated by the Federal Communications Commission.

Voice Channel
A transmission channel usually limited to the bandwidth of the human voice (300 - 3400 Hz).

Voice Grade Line
A communications channel which can transmit and receive voice frequencies (300 - 3400 Hz).

Volt
The standard unit of electromotive force or electrical pressure. One volt (1 V) is the amount of pressure that will cause one ampere (1 A) of current to flow through one ohm of resistance.

Voltage
Electrical potential or electromotive force (EMF) expressed in volts.

Voltage Drop
The voltage developed across a component or conductor by the current flow through the resistance or impedance of the component or conductor.

VSB
vault splice box

Wangnet
Wang Laboratories' proprietary broadband local area network.

Watt (W)
A unit of electrical power. One watt is equivalent to the power represented by one ampere of current with a pressure of one volt in a dc circuit, or to one joule (1 J) per second.

Wave Form
A graphical representation of a varying electrical quantity. Usually, time is represented on the horizontal axis, and the current or voltage value is represented on the vertical axis.

Wavelength
The physical distance of one electromagnetic wave cycle, or the distance between successive peaks of a wave.

Wavelength Division Multiplexers
Passive fiber optic components which combine or separate optical channels.

Webbed Conductors
The manufacturing process that physically binds the conductor insulation of the wire pairs of an unshielded twisted pair cable.

WCE
wall mount cable enclosure

White Field
The field used in the telecommunications closet or equipment room to terminate cables that connect first level backbone cables.

Wide Area Network (WAN)
Any physical network technology that spans large geographic distances through telco facilities or IECs. WANs usually operate at slower speeds than local area networks (LANs).

Wideband
A communications channel or medium having a bandwidth sufficient to carry multiple voice/video or data signals simultaneously. Also see Broadband.

Winch
A powerful machine for pulling heavy weights. It has a rotating drum around which a pulling cable is coiled.

Wiring Block
A molded plastic block that is designed in various pair configurations to terminate cable pairs and establish pair locations on connector systems. See Terminal Block.

Wiring Closet
See Telecommunications Closet.

Work Area Cable (Line Cord)
A cable assembly used to connect equipment to the telecommunications outlet in the work area. Max. 10 ft. (EIA/TIA).

Work Area Outlet (WAO) aka: Workstation
In general, a workstation is any designated location where constructive activities occurs. In communications, a workstation is an input/output device at which a user can send data to or receive data from a device for the purpose of performing a job. Usually a personal computer or a terminal. WAOs are considered to include voice, data (LAN), video or other applications. See definition for Telecommunications. Reference EIA/TIA 568A.

Wrap
A mechanism in the fiber distributed data interface (FDDI) standard by which the primary and secondary ring are joined by a station wrapping the primary input onto the secondary output (or vice versa). Through automatic failure detection mechanisms, downstream and upstream neighbors can identify faults and wrap away from the fault. The wrap results in an insulation of the fault, and the dual rings are converted into a single continuous ring until the fault is removed and the neighbors unwrap.

X
(a) Symbol for reactance, or (b) on floor plans, denotes a cross-connect location.

X.21
CCITT Standard defining the interface specifications between DTE and DCE for synchronous operation on public data networks, using a 15 pin connector (DB15).

X.25
A communication architecture developed by the International Telegraph and Telephone Consultative Committee (CCITT).

Z
Symbol for impedance.

Zero-Dispersion Wavelength
Wavelength at which the chromatic dispersion of an optical fiber is zero. Occurs when waveguide dispersion cancels out material dispersion.

Zone Method
A ceiling distribution method in which serving areas are divided into sections or zones. Cable is then run to the center of each zone to serve the information outlets nearby. Variations of Star.

Zone of Proection
An area provided by a grounded vertical rod or mast, and creates a surface of a hypothetical sphere of 50 meters (54.68 yds.) that touches the aerial rod and earth tangentially. The surface is concave upward. See NFPA-78. Also referred to as cone of protection.

Standards

Following is a summary of applicable optical fiber cabling standards as well as the most popular optical fiber LAN standards.

Af-phy-0046.000:
622.08 Mbps Physical Layer Specification. Three fiber-based physical medium dependent sublayers are defined for the 622.08 Mbps interface. A singlemode fiber interface, a light emitting diode (LED) based multimode fiber interface and a short wavelength (SW) laser based multimode fiber interface.

ANSI/TIA/EIA 568A (CSA T529):
Commercial Building Telecommunications Cabling Standard. It addresses the telecommunications cabling system requirements for commercial buildings that support various LAN, data, voice and image/video systems.

ANSI/TIA/EIA 569 (CSA T530):
Commercial Building Standard for Telecommunications Pathways and Spaces. The purpose of this standard is to standardize the design and construction practices for raceways and conduit, within and between buildings.

ANSI/TIA/EIA 606 (CSA T528):
Administration Standard for the Telecommunications Infrastructure of Commercial Building. It provides a uniform administration scheme for telecommunications infrastructure.

ANSI/TIA/EIA 607 (CSA T527):
Grounding and Bonding for Telecommunications in Commercial Buildings.

ANSI X 3.230: (Fibre Channel)

IEEE 802.3u (Fast Ethernet):
Media Access Control (MAC) Parameters, Physical Layer, Medium Attachment Units and Repeater for 100 Mbps Operation, Type 100BASE-T. The IEEE 802.3 Physical Layer specification for a 100 Mbps CSMA/CD LAN over optical fiber is 100BASE-FX.

IEEE 802.3z (Gigabit Ethernet):
Media Access Control (MAC) Parameters, Physical Layer, Repeater and Management Parameters for 1000 Mbps Operation, Type 1000BASE-T. The IEEE 802.3 Physical Layer specifications for a 1000 Mbps CSMA/CD LAN over optical fiber are 100BASE-LX (long wavelength laser) and 1000BASE-SX (short wavelength laser).

IEEE 803.12 (100 VG-AnyLAN):
Demand priority Access Method Physical Layer and Repeater Specification for 100 Mbps Operation. The IEEE 803.12 Physical Layer specifications include a 100 Mbps demand priority access LAN over optical fiber cabling.

ISO 9314-3 (FDDI):
Information processing systems - Fiber Distributed Data Interface (FDDI) - Part 3: Physical Layer Medium Dependent. This standard includes a Physical Layer specification for a 100 Mbps FDDI LAN over optical fiber cabling.

ISO/IEC 8802-5 (ANSI/IEEE 802.5):
Token Ring Access Method and Physical Layer Specification. The purpose of this standard is to define media and distance requirements for 4 and 16 Mbps Token Ring LANs.

ISO/IEC 8802-3 (ANSI/IEEE 802.3):
Carrier Sense Multiple Access with Collision Detection (CSMA/CD) Access Method and Physical Layer Specification. The IEEE 802.3 Physical Layer specification for a 10 Mbps CSMA/CD LAN over optical fiber is 10BASE-F which includes 10BASE-FB 10BASE-FP and 10BASE-FL.

ISO/IEC IS 11801:
Generic Cabling for Customer Premises. It specifies generic cabling for use with commercial premises, which may comprise single or multiple buildings on a campus. Cabling defined by this standard supports a wide range of services including voice, data and image/video systems.

UL/NEC
In a building installation, the requirements of the local electrical code must be followed. Most local codes if applicable follow the National Electrical Code closely. The NEC requires that all communication cables be listed by Underwriters Laboratories, which specifies their suitability for general, riser and plenum installation. A "UL" listed mark must appear on the cable you are installing, followed by a designator signifying the cable's rating for a particular type or usage;
- CMP for plenum
- CMR for riser or vertical shafts
- CM for general purpose

General purpose cables are those intrabuilding applications that are neither riser nor plenum. Most wiring limited to a single floor and not installed in an air handling space is considered general purpose.

Riser rated cables are to be installed in an opening or shaft leading from the MC to the TC or from floor to floor in a building.

A plenum-rated cable is suitable for installation in any space used for handling environmental air. The most common example of this type of space is that area between a false ceiling and the floor above it or the roof that is used for HVAC air return. Be sure to check that the space between the base floor and a raised computer floor does not fall into this category.

UL TYPE DESIGNATOR FOR ARTICLE 770: For OPTICAL FIBER CABLE		
CABLE DESCRIPTION	**DESIGNATOR**	**APPLICATION TEST**
Conductive Optical Fiber Cable	OFC	General/Purpose UL-1581 Vertical/Tray Flame
Non-conductive Optical Fiber Cable	OFN	Same as above
Conductive Riser	OFCR	Riser/Ul-1666
Non-conductive Riser	OFNR	Riser/UL-1666
Conductive Plenum	OFCP	Plenum/NFPA 262-1985 (UL910)
Non-conductive Plenum	OFNP	Same as above